Optical Filter Design and Analysis

	Constant	Value	Units
Speed of light in vacuum	c	2.9979×10^8	m/s
Permittivity of free space	ε_0	8.8542×10^{-12}	F/m
Permeability of free space	μ_0	1.2566×10^{-6}	H/m
Electron charge	q	1.6022×10^{-19}	C
Planck's constant	h	6.6262×10^{-34}	$J{\cdot}s$
Boltzmann's constant	k_B	1.3807×10^{-23}	J/K

Optical Filter Design and Analysis
A Signal Processing Approach

CHRISTI K. MADSEN
Bell Laboratories
Lucent Technologies

JIAN H. ZHAO
Rutgers University

A WILEY-INTERSCIENCE PUBLICATION

JOHN WILEY & SONS, INC.

NEW YORK / CHICHESTER / WEINHEIM / BRISBANE / SINGAPORE / TORONTO

Copyright © 1999 by John Wiley & Sons, Inc. All rights reserved.

Published simultaneously in Canada.

Library of Congress Cataloging-in-Publication Data:

Madsen, Christi K.
  Optical filter design and analysis : a signal processing approach
/ Christi K. Madsen, Jian H. Zhao.
     p. cm. — (Wiley series in microwave and optical engineering)
    "A Wiley-Interscience publication."
    Includes index.
    ISBN 0-471-18373-3 (cloth : alk. paper)
    1. Optical communications.   2. Optical wave guides.   3. Digital
filters.   4. Multiplexing.   5. Signal processing—Digital
techniques.   I. Zhao, Jian H.   II. Title.   III. Series.
TK5105.59.M36   1999
621.382′7—dc21   98-34993

Printed in the United States of America

10 9 8 7 6 5 4 3 2

To our families:
Eric
Lucia, Suqing, and Yumao

Contents

Preface

Optical filters whose frequency characteristics can be tailored to a desired response are an enabling technology for exploiting the full bandwidth potential of optical fiber communication systems. Optical filter design is typically approached with electromagnetic models where the fields are solved in the frequency or time domain. These techniques are required for characterizing waveguide properties and individual devices such as directional couplers; however, they can become cumbersome and non-intuitive for filter design. A higher level approach that focuses on the filter characteristics providing insight, fast calculation of the filter response, and easy scaling for larger and more complex filters is addressed in this book. The important filter characteristics are the same as those for electrical and digital filters. For example, passband width, stopband rejection, and the transition width between the passband and stopband are all design parameters for bandpass filters. For high bitrate optical communication systems, a filter's dispersion characteristics must also be understood and controlled. Given the large body of knowledge about analog and digital filter design, it is advantageous to analyze optical filters in a similar manner. In particular, this book is unique in presenting digital signal processing techniques for the design of optical filters, providing both background material and theoretical and experimental research results.

The optical filters described are fundamentally generalized interferometers which split the incoming signal into many paths, in an essentially wavelength independent manner, delayed and recombined. The splitting and recombining ratios, as well as the delays, are varied to change the frequency response. With digital filters, the splitting and recombining are done without concern for loss or the required gain; whereas, filter loss is a major design consideration for optical filters. The delays are typically integer multiples of a smallest common delay. A well-known example is a stack of thin-film dielectric materials where each layer is a quarter-wavelength thick. In this case, the splitters and combiners are the partial reflectances at each interface. Just as capacitors, inductors, and resistors have underlying electromagnetic models but are treated as lumped elements in analog filter designs, each splitting and combining element is modeled from basic electromagnetic theory and then treated as a lumped element in the optical filter design.

Another similarity with analog filters, but a major difference from digital filters, is the level of precision and accuracy that can be achieved in the design parameters for optical filters. For example, analog electrical components and optical components cannot be specified to the tenth decimal place; whereas, such numerical precision is commonplace for digital filters. Thus, a filter's sensitivity to variations in the design parameters must be considered. In addition, measurement and analysis techniques are needed to identify where variations have occurred in the fabrication process and what parameters are causing a filter to deviate from its nominal design. These issues, which are characteristic of optical filters, are addressed in detail.

This book is intended for researchers and students who are interested in optical filters and optical communication systems. Problem sets are given for use in a graduate level course. The main focus is to present the theoretical background for various architectures that can approximate any filter function. Planar waveguide devices realized in silica are used as examples; however, the theory and underlying design considerations are applicable to optical filters realized in other platforms such as fiber, thin-film stacks, and microelectro-mechanical (MEMs) systems. We are at an early point in the evolution of optical filters needed for full capacity optical communication systems and networks. Many filters need experimental investigation, so this book should be valuable to people interested in furthering their theoretical understanding as well as those who are fabricating filters using a wide range of material systems and fabrication techniques.

A detailed introduction to electromagnetic and signal processing theory is given in Chapter 1. In Chapter 2 on electromagnetic theory, a complete discussion is provided on waveguide modes, coupled-mode theory, and dispersion. In Chapter 3 on signal processing theory, Fourier transforms, Z transforms, and digital filter design techniques are discussed. The next three chapters (Chapters 4–6) cover optical filters and include design examples that are relevant to wavelength division multiplexed (WDM) optical communication systems. The examples include bandpass filters, gain equalization filters for compensating the wavelength dependent gain of optical amplifiers, and dispersion compensation filters. A particularly important filter for WDM systems is the waveguide grating router (WGR), which is fundamentally an integrated diffraction grating, because it filters many channels simultaneously. Its operation is examined using Fourier transforms to provide insight into its periodic frequency and spatial behavior. Filters using thin-film dielectric stacks, Bragg gratings, acousto-optic coupling, and long period gratings are also examined. Filters with a large number of periods such as Bragg and long period gratings are typically analyzed using coupled-mode theory. We include the coupled-mode solutions for these filters, thus offering the reader a comparison between signal processing techniques and the coupled-mode approach. Measurement techniques and filter analysis algorithms, which extract the filter's component values from its spectral or time domain response, are addressed in Chapter 7. Finally, areas that are expected to have a dramatic effect on the evolution of optical filters are highlighted.

The authors gratefully acknowledge the review and suggestions provided by G. Lenz, Y. P. Li, W. Lin, D. Muelhner, and A. E. White of Bell Laboratories Lucent Technologies, S. Orfanidis of Rutgers University, B. Nyman of JDS Fitel, and T. Erdogan of the University of Rochester.

CHAPTER ONE

Introduction

The field of signal processing provides a host of mathematical tools, such as linear system theory and Fourier transforms, which are used extensively in optics for the description of diffraction, spatial filtering, and holography [1–2]. Optical filters can also benefit from the research already done in signal processing. In this book, digital signal processing concepts are applied to the design and analysis of optical filters. In particular, digital signal processing provides a readily available mathematical framework, the Z-transform, for the design of complex optical filters. More conventional approaches, such as coupled-mode theory, are also included for comparison. The relationship between digital filters and optical filters is explored in Section 1.1, and a brief historical overview of optical waveguide filters is given. Previously, spectrum analysis was the main application for optical filters. Recently, the demand for optical filters is increasing rapidly because of the deployment of commercial wavelength division multiplexed (WDM) optical communication systems. With low loss optical fibers and broadband optical amplifiers, WDM systems have the potential to harness a huge bandwidth, and optical filters are essential to realizing this goal. In addition to traditional designs such as bandpass filters, new applications have emerged such as the need for filters to perform gain equalization and dispersion compensation. Filter applications for WDM systems are discussed in Section 1.2. These applications are used for filter design examples in later chapters. The scope of this book is outlined in Section 1.3.

1.1 OPTICAL FILTERS

Optical filters are completely described by their frequency response, which specifies how the magnitude and phase of each frequency component of an incoming

1

signal is modified by the filter. While there are many types of optical filters, this book addresses those that allow an arbitrary frequency response to be approximated over a frequency range of interest. For example, thin-film interference filters can approximate arbitrary functions. An illustration of this capability is Dobrowolski's filter approximating the outline of the Taj-Mahal [3]. A complete set of functions is needed to closely approximate an arbitrary function. A well-known set consists of the sinusoidal functions whose weighted sum yields a Fourier series approximation. The Fourier series can be written in terms of exponential functions with complex arguments as follows:

$$H(\nu) = \sum_{n=0}^{N} [c_n e^{-j(2\pi\nu n - \phi_n)}] \tag{1}$$

where $H(\nu)$ is the frequency response of the filter, N is the filter order, and the $c_n e^{j\phi_n}$ terms are the complex weighting coefficients. A weighted sum is common to any basis function decomposition. An incoming signal is split into a number of parts that are individually weighted and then recombined. The optical analog is found in interferometers.

Interferometers come in two general classes, although there are many variations of each. The first class is simply illustrated by the Mach–Zehnder interferometer (MZI). An incoming signal is split equally into two paths. One path is delayed with respect to the other by a time $T = \Delta L n/c$ where n is the refractive index of the paths, c is the velocity of light in vacuum, and ΔL is the path length difference. The effective group index is used for waveguide delay paths. The signals in the two paths are then recombined as shown in Figure 1-1a. A partially reflecting mirror, indicated by the dashed line, acts as a beam splitter and combiner. A waveguide version with directional couplers for the splitter and combiner is shown in Figure 1-1b. Each implementation has two outputs that are power complementary. Coherent interference at the combiner leads to a sinusoidal frequency response whose period is inversely proportional to the path length difference. A representative transmission response is shown in Figure 1-1c. Its Fourier series is given by $H(\nu) = \frac{1}{2}[1 - e^{-j2\pi\nu T}]$. The frequency is referenced to a center frequency ν_0 and normalized to one period, called the free spectral range (FSR). Changing the splitting or combining ratios adjusts the coefficients in the Fourier series. The number of terms in the Fourier series is increased by splitting the incoming signal into more paths. An N term series results when the light is split into N paths, with each path having a relative delay that is T longer than the previous path. The distinguishing feature of this general class of interferometers is that there are a finite number of delays and no recirculating (or feedback) delay paths. The interfering paths are always feeding forward even though the interferometer may be folded such as with a Michelson interferometer. The signal processing term used to designate this filter type is moving average (MA) or finite impulse response (FIR).

The second class of interferometers is illustrated by the Fabry–Perot interferometer (FPI). The FPI consists of a cavity surrounded by two partial reflectors that are parallel to each other as shown in Figure 1-2a. The frequency response results from

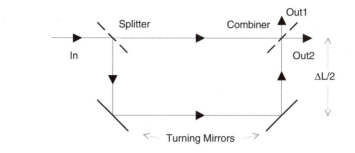

(a)

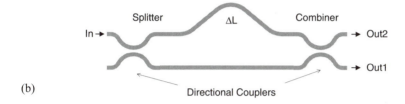

(b)

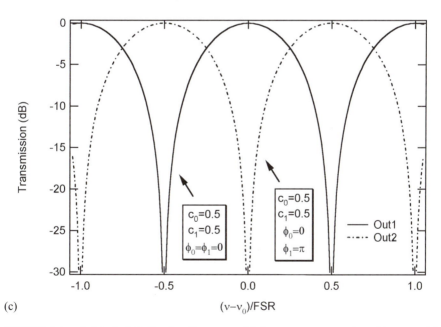

(c)

FIGURE 1-1.   A Mach–Zehnder interferometer: (a) free-space propagation, (b) waveguide device, and (c) transmission response.

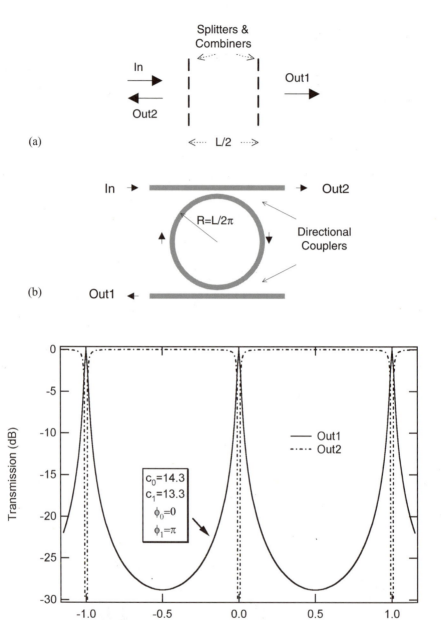

FIGURE 1-2.   A Fabry–Perot interferometer: (a) free-space propagation, (b) waveguide analogue, and (c) transmission response.

the interference of multiple reflections from the mirrors. The output is the infinite sum of delayed versions of the input signal weighted by the roundtrip cavity transmission. A representative transmission response is shown in Figure 1-2c. The waveguide analog is a ring resonator with two directional couplers as shown in Figure 1-2b. There are two outputs: Out1 corresponds to the transmission response of the FPI, and Out2 corresponds to its reflection response. Filters with feedback paths are classified as autoregressive (AR) or infinite impulse response (IIR) filters in signal processing. When several stages are cascaded, the resulting transmission response is described by one over a Fourier series as follows:

$$H(\nu) = \cfrac{1}{\displaystyle\sum_{n=0}^{N} [c_n e^{-j(2\pi\nu n - \phi_n)}]} \tag{2}$$

The reflection response is given by the sum of an AR and MA response and is classified as an autoregressive moving average (ARMA) filter. The IIR classification is ambiguous since the filter may either be AR or ARMA and still have an infinite impulse response; consequently, the terminology MA, AR, and ARMA is preferred over the FIR and IIR designation. Thin-film interference filters and Bragg reflection gratings are coupled cavity FPI's with an order N equal to the number of layers or periods, respectively. They have a transmission response that is AR and a reflection response that is ARMA.

The theory presented in this book applies to generalized interferometers. The concept of a generalized interferometer refers to the ability to tailor the frequency response of the interferometer to approximate any desired function by choosing the number and type (feed-forward or feedback) of paths that the incoming signal is split between and the relative weighting and delay for each path. Propagation in the interferometer may be governed by diffraction or by waveguiding. The former case includes thin-film and micro electro mechanical (MEMs) devices. The latter case encompasses both optical fiber and planar waveguide devices. In either case, the filter response arises from the interference of two or more electromagnetic waves; so, phase changes on the order of a fraction of a wavelength dramatically change the filter function. It is critical, therefore, that the optical path lengths are stable over time, temperature and exposure to mechanical vibrations. Planar waveguide filters are advantageous over optical fiber implementations since many devices can be integrated on a common substrate that can easily be temperature controlled and is mechanically rigid. In addition, several material systems can be used ranging from glass to semiconductors to polymers.

The fundamental relationships between optical waveguide and digital filters were developed by Moslehi et al. [4] in 1984. Both digital and optical filters consist of splitters, delays, and combiners. These parts are identified in Figures 1-1 and 1-2 for the MZI and FPI, respectively. Many stages are formed by concatenating single stages or combining stages in various architectures. The optical path lengths are typically integer multiples of the smallest path length difference. The unit delay is defined as $T = L_U n/c$ where L_U is the smallest path length and is called the unit delay length. Digital signal processing techniques are relevant to optical filters because they are linear, time-invariant systems that have discrete delays. In glass, with a refractive index of 1.5, a unit delay of 100 ps corresponds to a unit length of 2 cm. Because the delays are discrete values of the unit delay, the frequency response is periodic. One period is defined as the FSR, which is related to the unit delay and unit length as follows:

$$\text{FSR} = \frac{1}{T} = \frac{c}{nL_U} \tag{3}$$

The FSR of a filter with a 100 ps unit delay is 10 GHz. At 1550 nm, 10 GHz corresponds to $\Delta\lambda = 0.08$ nm. The shorter the unit delay, the larger the FSR. Because several centimeters of fiber are needed for splicing, optical fiber filters can have very small FSRs (for example, $L_U = 50$ cm and $n = 1.5$ corresponds to a FSR = 400 MHz) compared to several gigahertz (GHz) needed for separating high bit-rate communication channels.

Thin-film and Bragg grating filter design and fabrication are mature, having well established design techniques [5–7]. Their description using a digital filter approach complements these techniques by demonstrating properties that are less obvious with other mathematical approaches. These filters are important stand-alone devices because nearly square bandpass filter responses can be achieved with a short filter length. In addition, they are key elements in more complex filter architectures.

The first optical waveguide filters were realized using optical fibers and discrete components such as tapered fused-fiber couplers for splitters and combiners. They had small FSRs and environmental fluctuations changed the optical path length difference significantly. It was therefore advantageous to use a source with a coherence length shorter than the unit delay so that the combining functions were linear in intensity instead of the field. In this case, referred to as incoherent processing, the filter operates on the modulated signal on an optical carrier. Only positive filter coefficients are achievable, limiting the filter response to low pass designs. Optical MA filters using fiber delay lines were first proposed for high-speed correlators and pulse compression by Wilner et al. [8] in 1976. Other applications included matched filters for pulse encoders and decoders [9–10], frequency filters [11], and matrix-vector multiplication [12–13]. For AR filters, it is interesting to note that Marcatili proposed using an integrated ring resonator for a bandpass filter in 1969 [14].

Advances in planar waveguide fabrication have been critical for waveguide filters. In particular, low loss and process control, whereby fabricated devices closely approximate the design intent, enable the successful integration of multi-stage filters on a chip. The first coherent MA filter was demonstrated in 1984 using optical fibers [15]. The first multi-stage planar waveguide coherent MA filter was demonstrated in 1991 [16], and the first coherent multi-stage AR filter was demonstrated in 1996 [17]. Since then, many new architectures have been proposed and some have been demonstrated. The fundamental architectures for realizing general MA, AR, and ARMA filters that rely on coherent processing are covered in this book.

1.2   FILTER APPLICATIONS IN WDM SYSTEM

Optical filters are an enabling technology for WDM systems. The most obvious application is for demultiplexing very closely spaced channels; however, they also play major roles in gain equalization and dispersion compensation. A simplified WDM system showing one direction of signal transmission is outlined in Figure 1-3. Multiplexing and demultiplexing filters are found in the terminals. Gain equalizers and dispersion compensating filters may be deployed at intermediate points along the transmission line or at the endpoints. Most of the traffic travels from Terminal A to B; however, it may be advantageous to drop a limited bandwidth to one or more intermediate points, represented by Node 1. A filter referred to as an add/drop filter is required to separate the channel to be dropped from those that pass through unaffected. Node 1 receives the dropped channel and may transmit its own information on a new signal at the same wavelength as that dropped or a new wavelength that does not interfere with those already used by the other channels on the through-path.

Optical fibers have an enormous transmission capacity. For example, the low loss wavelength range for the AllWave™ optical fiber spans 73.5 THz ($1200 \leq \lambda \leq 1700$ nm) as shown in Figure 1-4 [18]. Erbium-doped fiber amplifiers (EDFAs) have a gain bandwidth covering approximately 5 terahertz (THz) and centered around 1540 nm. Several system demonstrations with total capacities in the THz range have been reported using EDFAs [19–22]. Amplifiers that can cover wavelength regions not accessible by EDFAs include cascaded Raman amplifiers [23] and semiconductor optical amplifiers [24]. Two ways to increase the capacity are to increase the useable wavelength range and to use the bandwidth already covered more efficiently, for example, by decreasing the channel spacing. Closer channel spacings require sharper filter responses to separate the channels without introducing cross-talk from the other channels. Channel spacings are standardized based on the International Telecommunications Union (ITU) grid defining 100 GHz spaced frequencies, $f = 193.1 \pm m \times 0.1$ THz where m is an integer [25]. The center of the

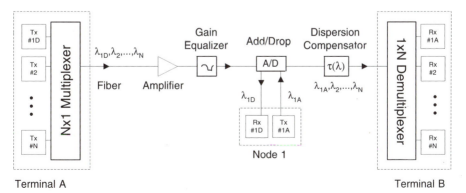

FIGURE 1-3.   Filter applications in a simplified WDM system.

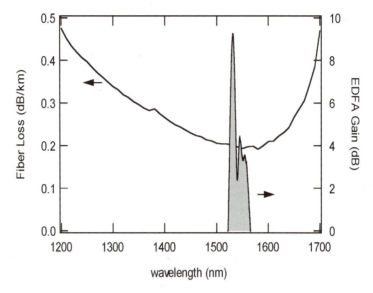

FIGURE 1-4. Bandwidth available with low loss for AllWave™ fiber [18], and the bandwidth covered by the erbium-doped fiber amplifier gain spectrum.

grid is 193.1 THz, which corresponds to a wavelength of 1552.524 nm in vacuum. Channel spacings for commercial systems are currently on the order of 100 GHz with bit rates up to 10 Gb/s per channel. This translates to a bandwidth efficiency of 0.1 bit/s/Hz. For comparison, the theoretical limit is 2 bits/s/Hz.[1]

1.2.1 Bandpass Filters for Multiplexing, Demultiplexing, and Add/Drop

Bandpass optical filters are needed for multiplexing, demultiplexing, and add/drop functions. Current optical filter technologies include Fabry–Perot cavities, diffraction gratings, thin-film interference filters, fiber Bragg gratings, and various planar waveguide filters. A bandpass filter is characterized by its passband width, loss, flatness, and dispersion as well as its stopband rejection. A schematic of a multiplexer and demultiplexer is shown in Figure 1-5. The devices discussed in this book are reciprocal, so the same device can be used for both functions; however, the requirements can be significantly different. For a multiplexer, a low loss method of combining N wavelengths is needed. Since system transmitters are narrowband, the multiplexer's stopband rejection is not critical. For demultiplexers, a large stopband rejection is critical to minimize cross-talk from neighboring channels. Figure 1-6 il-

---

[1]This limit assumes a perfectly bandlimited pulse having a time domain response equal to $T \sin(\pi t/T)/\pi t$, which has an infinite time duration. Additional assumptions include the following: no dispersion or noise in the transmission line, single sideband transmission on the optical carrier, binary transmission where a 1 is represented by a pulse sent and a 0 by no pulse sent, no polarization multiplexing, and ideal bandpass filters to separate the channels.

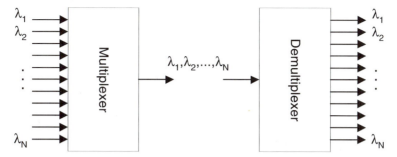

FIGURE 1-5.   Wavelength division multiplexer and demultiplexer.

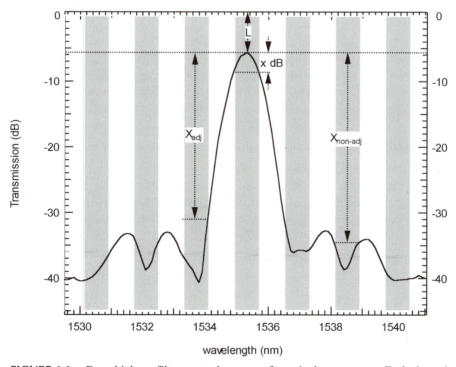

FIGURE 1-6.   Demultiplexer filter spectral response for a single output port. Each channel passband is shaded, L is the minimum passband loss, X_{adj} and $X_{non-adj}$ are the adjacent and non-adjacent channel cross-talk losses, and the filter passband width is defined by the full width at x db down from the peak transmission.

lustrates a demultiplexer response with a 50 GHz channel width ($\Delta\lambda_{fw}$) and 100 GHz channel spacing ($\Delta\lambda_{ch}$). The cross-talk loss is defined for each pair of channels as the minimum difference in loss between the passband of one channel and the stopband of the other. Cross-talk requirements are often quoted for adjacent and non-adjacent channels separately, with stricter requirements for non-adjacent channels. An MA planar waveguide filter known as a waveguide grating router (WGR) is well suited for multiplexing and demultiplexing large numbers of channels. It is discussed in Chapter 4.

An add/drop filter allows one or more channels to be dropped to an intermediate system node. A multi-wavelength add/drop filter is represented schematically in Figure 1-7. It is a four port device. Express channels pass through the device from the input to the through-port. Ideally, these channels experience no loss or dispersion due to the filter. The drop channels exit the drop port. The add channels enter the device through the add port and exit the through-port. The add and drop wavelengths may be different or the same. When the same wavelengths are used, stringent requirements are needed on the attenuation of the dropped signal in the through path. Otherwise, coherent cross-talk between leakage of the dropped signal into the through-port and the add signal will degrade the add channel's performance. An idealized add/drop filter response for the express and add/drop channels is shown in Figure 1-8.

For bandpass filter applications, several factors influence the required passband width beyond the fundamental limit set by the bit rate. For example, the passband width must accommodate fabrication tolerances on the filter and laser center wavelengths as well as their polarization, temperature, and aging characteristics. For high bit-rate long-distance systems, filter dispersion and broadening of the signal due to nonlinearities in the fiber can also become an issue.

1.2.2   Gain Equalization Filters

Besides packing the channels more closely, the bandwidth of WDM systems can be increased by using a larger total wavelength range. In WDM systems, the optical signal-to-noise ratio (SNR_0) of each channel depends on the power out of the amplifier per channel P_{ch} as well as the number of amplifiers N in the system as described by the following formula for a linear system [26]:

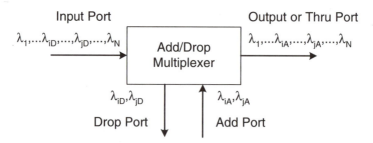

FIGURE 1-7.   Multi-wavelength add/drop filter schematic.

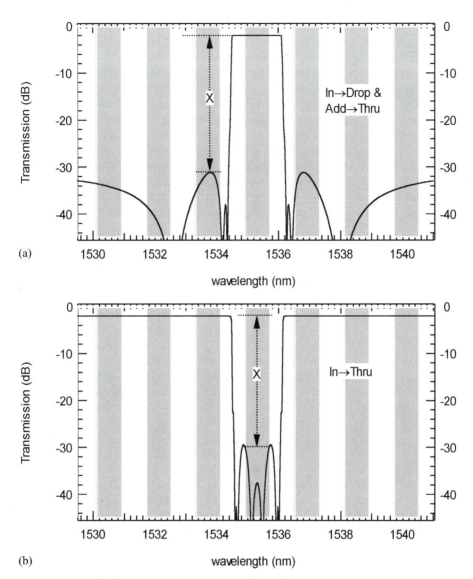

(a)

(b)

FIGURE 1-8. Single channel add/drop filter spectral response.

$$\mathrm{SNR}_0(\mathrm{dB}) \approx 58 + P_{\mathrm{ch}} - L - NF - 10 \log_{10} N \qquad (4)$$

where NF is the amplifier noise figure and L is the span loss between amplifiers. The power for each channel is limited by the total output power of the amplifier, the number of channels, and the wavelength dependence of the amplifier gain. The best use of the total power is to split it evenly among all of the channels. The loss uniformity of multiplexers, add/drop filters, the transmission fiber, and other components also contributes to the power imbalance between channels. Channels at wavelengths that have less power will have a lower SNR_0. If the SNR_0 drops sufficiently, transmission errors will be introduced. Gain equalization filters (GEFs) play an important role in WDM systems by compensating the wavelength dependent amplifier gain and system losses to equalize the power among channels.

Optical amplifiers, EDFAs in particular, have revolutionized optical communications by providing gain at intermediate points along a system that would have required regeneration in prior systems. Regenerators are very expensive by comparison because they require signal detection, electrical clock recovery circuitry, and a new optical signal to be generated. These operations are bit rate and format dependent and are required for each channel. Optical amplifiers are bit rate and format independent, thereby providing flexibility for channel provisioning and capacity upgrades. The EDFAs supply gain over many channels in the 1550 nm region, have low polarization dependence, and can be efficiently spliced to optical fibers. The EDFA gain bandwidth covers the range from 1525 to 1565 nm; however, it has significant wavelength dependence. As an example, a gain response for a single amplifier is shown in Figure 1-9. The gain bandwidth can be increased by changing the amplifier design. For example, a two stage EDFA providing gain over an 80 nm range was demonstrated by splitting the incoming signals into two 40 nm ranges after the first stage and amplifying them separately in the second stage [27].

Several technologies have been used for realizing gain equalization filters. For example, long period fiber gratings use mode coupling in the forward direction between the fundamental and one or more cladding modes to approximate a desired

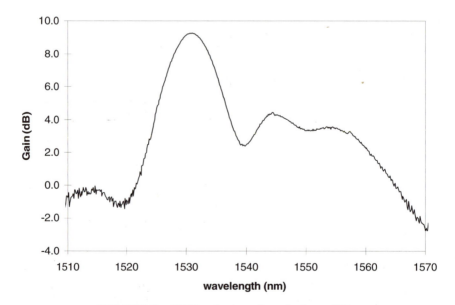

FIGURE 1-9. EDFA gain shape for a single amplifier.

filter response [28]. Planar waveguide MA filters have also been demonstrated [29]. These approaches are discussed in Chapter 4. In addition, GEF designs are explored using AR and ARMA responses in Chapters 5 and 6.

1.2.3   Dispersion Compensation Filters

The previous two applications focused on the filter's amplitude response; however, the time domain aspects of pulse propagation are of equal importance, particularly as the bit rates per channel continue to increase. Dispersion causes the pulse width to broaden. If the dispersion is large enough, adjacent pulses may overlap in time resulting in intersymbol interference. For a system penalty of 1 dB, the bit rate, distance and dispersion are related as follows [30]:

$$B^2DL \approx 10^5 \text{ ps/nm (Gb/s)}^2 \tag{5}$$

where B is the bit rate, D is the fiber dispersion, and L is the length of fiber. For every factor of two increase in the bit rate, the allowable cumulative dispersion over the same distance reduces by a factor of four. For example, a 10 Gb/s signal can tolerate a cumulative dispersion of 1000 ps/nm; whereas, a 40 Gb/s has an allowable cumulative dispersion of only 63 ps/nm!

Standard singlemode optical fibers have a dispersion zero around 1300 nm. In the 1550 nm wavelength range, the typical dispersion is +17 ps/nm-km with a slope of +0.08 ps/nm²-km. Dispersion shifted fibers are made by increasing the waveguide dispersion to move the dispersion minimum to the 1550 nm band. Nonlinear four-wave-mixing in the fiber prohibits system operation at zero dispersion. In practical systems, dispersion management schemes are used that combine fibers with opposite signs of dispersion in the appropriate ratio of lengths. For example, sections of positive dispersion fiber are interspersed with negative dispersion fiber to bring the total dispersion to zero periodically and at the receiver. The dispersion slope becomes important as the wavelength range of the channels increases. An example of the cumulative dispersion for the extreme and middle wavelengths of a system with 8 channels separated by 0.8 nm is shown in Figure 1-10. The cumulative dispersion is returned to zero at the center wavelength every 150 km in this example. The dispersion slopes of the two fiber types are not compensated, resulting in large cumulative dispersions for the remaining wavelengths even at lengths where the cumulative dispersion is perfectly canceled for the center wavelength.

There are several methods for performing dispersion compensation. One approach is to use dispersion compensating fiber (DCF) which can be made with a dispersion on the order of –200 ps/nm-km, but at the expense of larger loss compared to standard transmission fiber [31]. A figure of merit (FOM) for dispersion compensating fiber is the ratio of the dispersion to the loss. Typical FOMs around 200 ps/nm-km-dB have been demonstrated using DCF. Several kilometers of DCF may be necessary and different lengths may be required for different wavelengths. There are also several filter approaches. One is to use chirped Bragg gratings in optical fibers [32]. To achieve negative dispersion across the channel passband, the

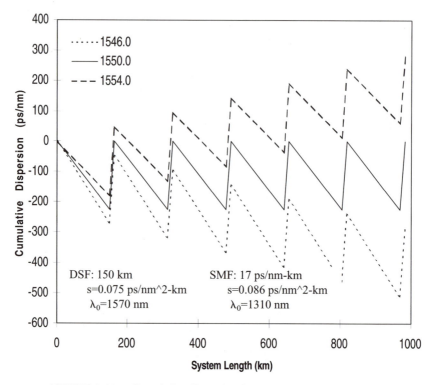

FIGURE 1-10. Cumulative dispersion for an 8 channel WDM system.

longest wavelength is reflected first and the shortest wavelength last with a linear variation in reflected wavelength along the length of the grating. Such filters offer dispersion compensation for a single channel and are much shorter compared to DCF. Planar waveguide filters of the MA type have also been demonstrated for dispersion compensation [33]. Optical all-pass filters, which are analogous to digital all-pass filters, are also discussed for dispersion compensation applications [34]. The MA dispersion compensator is addressed in Chapter 4, the chirped Bragg grating approach in Chapter 5, and the all-pass filter in Chapter 6.

1.3   SCOPE OF THE BOOK

Fundamentals of dielectric waveguides are presented in Chapter 2. A brief review of plane waves, the wave equation, and modal solutions for slab and rectangular waveguide modes is given. Common devices for splitting and combining signals are covered including directional couplers, star couplers, and multi-mode interference couplers. In addition, materials and fabrication related issues are discussed as they relate to realizing practical waveguide filters.

Chapter 3 provides an overview of digital signal processing concepts for optical filter design. Basic linear system and frequency analysis techniques are presented including Fourier and Z transforms. The correspondence between digital and optical waveguide architectures is introduced along with the necessary approximations to analyze optical filters using digital filter techniques. Digital filter design methods are discussed, and several multi-stage digital architectures are reviewed, which will find their optical analog in later chapters.

Chapters 4–6 focus on multi-stage filter architectures. Filter design examples are given for bandpass filters, gain equalizers, and dispersion compensators. While perfect filters can be designed on the computer, the actual implementation must consider expected tolerances resulting from fabrication variations; consequently, we discuss the sensitivity of various filter architectures to fabrication variations. Chapter 4 covers MA optical filters, which include cascade, transversal, multi-port (diffraction gratings and the WGR), lattice, and Fourier filters. The single stage MZI is treated in detail first, since it illuminates many practical issues involving loss, wavelength dependence of the filter coefficients, and fabrication tolerances. In Chapter 5, the design of optical AR filters is addressed. General insights into thin-film interference filters and Bragg gratings are drawn, including their amplitude and dispersion properties. Chapter 6 addresses the design of general ARMA responses. A general lattice architecture is discussed along with architectures which are based on optical all-pass filters.

Waveguide measurements and filter characterization are necessary for the successful realization of complex optical filters. Chapter 7 introduces filter measurement techniques. Given the complexity of the filter designs discussed in the previous chapters and the uncertainty introduced by fabrication variations, methods to determine the filter parameters after fabrication are desirable. Algorithms to determine the filter parameters from the frequency response are presented.

There are two dominant drivers that push the state-of-the-art in optical filter technology: (1) applications such as dense WDM systems and (2) progress in materials and fabrication processes. In Chapter 8, we briefly review recent work and trends in these areas that are likely to have an impact on optical filters in the future.

REFERENCES

1. J. Goodman, *Introduction to Fourier Optics*. San Francisco: McGraw-Hill, 1968.

2. A. Papoulis, *Systems and Transforms with Applications in Optics*. New York: McGraw-Hill, 1968.

3. J. Dobrowolski, "Numerical methods for optical thin films," *Opt. & Photon. News,* pp. 25–33, 1997.

4. B. Moslehi, J. Goodman, M. Tur, and H. Shaw, "Fiber-Optic Lattice Signal Processing," *Proc. IEEE,* vol. 72, no. 7, pp. 909–930, 1984.

5. H. Kogelnik, "Filter Response of Nonuniform Almost-Periodic Structures," *Bell System Techn. J.,* vol. 55, no. 1, pp. 109–126, 1976.

6. H. Macleod, *Thin-film Optical Filters*. New York: McGraw-Hill, 1989.

7. A. Thelen, *Design of Optical Interference Coatings.* New York: McGraw-Hill, 1989.

8. K. Wilner and A. Van den Heuvel, "Fiber-optic delay lines for microwave signal processing," *Proc. IEEE,* vol. 64, pp. 805–807, 1976.

9. E. Marom, "Optical delay line matched filtering," *IEEE Trans. Circuits Systems,* vol. 25, pp. 360–364, 1978.

10. R. Ohlhaber and K. Wilner, "Fiber optic delay lines for pulse coding," *Electro-optical System Design,* pp. 33–35, 1977.

11. C. Chang, J. Cassaboom, and H. Taylor, "Fibre-optic delay-line devices for R.F. signal processing," *Electron. Lett.,* vol. 13, pp. 678, 1977.

12. K.P. Jackson, S.A. Newton, B. Moslehi, M. Tur, C.C. Cutler, J.W. Goodman, and H.J. Shaw, "Optical Fiber Delay-Line Signal Processing," *IEEE Trans. Microwave Theory Techniques,* vol. 33, no. 3, pp. 193–209, 1985.

13. K. Jackson and H. Shaw, "Fiber-Optic Delay-Line Signal Processors," in *Optical Signal Processing,* J. Horner, Ed. San Diego: Academic Press, 1987, pp. 431–475.

14. E. Marcatili, "Bends in Optical Dielectric Guides," *Bell System Techn. J.,* vol. 48, no. 7, pp. 2103–2132, 1969.

15. D. Davies and G. James, "Fibre-optic tapped delay line filter employing coherent optical processing," *Electron. Lett.,* vol. 20, pp. 96–97, 1984.

16. K. Sasayama, M. Okuno, and K. Habara, "Coherent Optical Transversal Filter Using Silica-Based Waveguides for High-Speed Signal Processing," *J. Lightw. Technol.,* vol. 9, no. 10, pp. 1225–1230, 1991.

17. C. Madsen and J. Zhao, "A General Planar Waveguide Autoregressive Optical Filter," *J. Lightw. Technol.,* vol. 14, no. 3, pp. 437–447, 1996.

18. D. Kalish, M. Pearsall, K. Chang, and T. Miller, Lucent Technologies. Norcross, GA, 1998, private communication.

19. H. Onaka, H. Miyata, Ishikawa, K. Otsuka, H. Ooi, Y. Kai, S. Kinoshita, M. Seino, H. Nishimoto, and T. Chikama, "1.1 Tb/s WDM transmission over a 150 km 1.3 micron zero-dispersion single-mode fiber," PD–19, Vol. 2 of 1996 Technical Digest Series (Optical Society of America, Washington, DC), Optical Fiber Communication Conference1996.

20. A. Gnauck, A. Chraplyvy, R. Tkach, J. Zyskind, J. Sulhoff, A. Lucero, Y. Sun, R. Jopson, F. Forghieri, R. Derosier, C. Wolf, and A. McCormick, "One Terabit/s transmission experiment," PD–20, Vol. 2 of 1996 Technical Digest Series (Optical Society of America, Washington, DC), Optical Fiber Communication Conference1996.

21. T. Morioka, H. Takara, S. Kawanishi, O. Kamatani, K. Takiguchi, K. Uchiyama, M. Saruwatari, H. Takahashi, M. Yamada, T. Kanamori, and H. Ono, "100 Gbit/s x 10 channel OTDM/WDM transmission using a single supercontinuum WDM source," PD–21, Vol. 2 of 1996 Technical Digest Series (Optical Society of America, Washington, DC), Optical Fiber Communication Conference1996.

22. Y. Yano, T. Ono, K. Fukuchi, T. Ito, H. Yamazaki, M. Yamaguchi, and K. Emura, "2.6 Terabit/s WDM transmission experiment using optical duobinary coding," 22nd European Conference on Optical Communication. Oslo, Norway, 1996, p. ThB.3.1.

23. A. White and S. Grubb, "Optical Fiber Components and Devices," in *Optical Fiber Telecommunications IIIB,* I. Kaminow and T. Koch, Eds. San Diego: Academic Press, 1997, pp. 267–318.

24. D. Fishman and B. Jackson, "Transmitter and Receiver Design for Amplified Lightwave

Systems," in *Optical Fiber Telecommunications IIIB,* I. Kaminow and T. Koch, Eds. San Diego: Academic Press, 1997, pp. 69–114.

25. T. Koch, "Laser sources for amplified and WDM lightwave systems," in *Optical Fiber Telecommunications IIIB,* I. Kaminow and T. Koch, Eds. San Diego: Academic Press, 1997, pp. 115–162.

26. J. Zyskind, J. Nagel, and H. Kidorf, "Erbium-Doped Fiber Amplifiers for Optical Communications," in *Optical Fiber Telecommunications IIIB,* I. Kaminow and T. Koch, Eds. San Diego: Academic Press, 1997, pp. 13–68.

27. Y. Sun, A. Saleh, J. Zyskind, D. Wilson, A. Srivastava, and J. Sulhoff, "Modeling of Small-Signal Cross-Modulation in WDM Optical Systems with Erbium-Doped Fiber Amplifiers," in Vol. 6 of 1997 OSA Technical Digest Series (Optical Society of America, Washington, D.C., 1997). Optical Fiber Communication Conference, 1997, pp. 106–107.

28. A. Vengsarkar, J. Pedrazzani, J. Judkins, P. Lemaire, N. Bergano, and C. Davidson, "Long Period Fiber Grating Based Gain Equalizers," *Opt. Lett.,* vol. 21, no. 5, pp. 336–338, 1996.

29. Y. Li, C. Henry, E. Laskowski, C. Mak, and H. Yaffe, "Waveguide EDFA Gain Equalisation Filter," *Electron. Lett.,* vol. 31, no. 23, pp. 2005–2006, 1995.

30. T. Li, "The Impact of Amplifiers on Long-Distance Lightwave Telecommunications," *Proc. IEEE,* pp. 1568–1579, 1993.

31. A. Vengsarkar, A. Miller, M. Haner, A. Gnauck, W. Reed, and K. Walker, "Fundamental-mode dispersion-compensating fibers: design considerations and experiments," Optical Fiber Conference, San Jose, CA, paper ThK2, 1994.

32. F. Ouellette, "Dispersion cancellation using linearly chirped Bragg grating filters in optical waveguides," *Opt. Lett.,* vol. 12, pp. 847–849, 1987.

33. K. Takiguchi, K. Okamoto, and K. Moriwaki, "Dispersion Compensation Using a Planar Lightwave Circuit Optical Equalizer," *IEEE Photon. Technol. Lett.,* vol. 6, no. 4, pp. 561–564, 1994.

34. C. Madsen and G. Lenz, "Optical All-pass Filters for Phase Response Design with Applications for Dispersion Compensation," *IEEE Photon. Technol. Lett.,* vol. 10, no. 7, pp. 994–996, 1998.

CHAPTER TWO

Fundamentals of Electromagnetic Waves and Waveguides

The plane wave is the most fundamental electromagnetic wave and is a particular solution of Maxwell's equations. It will be used as an example to develop the fundamentals of electromagnetic waves in dielectric media including solutions to Maxwell's equations, phase velocity, group velocity, polarization, boundary conditions, and total internal reflection, which makes it possible for optical signals to propagate and be guided in waveguides. Based on the concepts developed for plane waves, planar one-dimensional waveguides will be presented, followed by the study of rectangular two-dimensional waveguides. Fundamental concepts for wave propagation in waveguides such as modes, orthogonality and completeness of modes, dispersion, and losses are discussed. A review of waveguide-based directional couplers, the fundamental building block for planar waveguide filters, is followed by a brief presentation of star couplers and multi-mode interference couplers, which are important for multi-port filters. The chapter concludes with a discussion of material properties and fabrication processes that are important to filter design, with a focus on planar waveguide implementations.

2.1 THE PLANE WAVE

2.1.1 Maxwell's Equations

The governing equations for electromagnetic wave propagation are Maxwell's equations. In fact, Maxwell's equations describe not only wave propagation at high frequencies, but also zero Hz static electromagnetic problems. The four equations in their differential forms are

$$\nabla \times \vec{E} = -\frac{\partial \vec{B}}{\partial t} \tag{1}$$

$$\nabla \times \vec{H} = \vec{J} + \frac{\partial \vec{D}}{\partial t} \tag{2}$$

$$\nabla \cdot \vec{B} = 0 \tag{3}$$

$$\nabla \cdot \vec{D} = \rho \tag{4}$$

where $\vec{E}$ is the electric field vector, $\vec{B}$ the magnetic flux density vector, $\vec{D}$ the electric flux density or electric displacement vector, and $\vec{H}$ the magnetic field vector. The parameter ρ is the volume density of free charge, and $\vec{J}$ is the density vector of free currents. The flux densities $\vec{D}$ and $\vec{B}$ are related to the fields $\vec{E}$ and $\vec{H}$ by the constitutive relations. For linear, isotropic media, the relations are given by $\vec{D} = \varepsilon\vec{E}$ and $\vec{B} = \mu\vec{H}$, where ε is the dielectric permittivity of the medium and μ is the magnetic permeability of the medium. A dielectric medium is linear if the permittivity ε and permeability μ are independent of field strengths. Generally speaking, most dielectrics become nonlinear when the electric field intensity E is comparable to the Coulomb fields, typically in the range of 10^{10}V/cm, that bind electrons to the central nucleus.

The boundary conditions that can be derived directly from Maxwell's equations are

$$\vec{n} \times (\vec{E}_2 - \vec{E}_1) = 0 \tag{5}$$

$$\vec{n} \times (\vec{H}_2 - \vec{H}_1) = \vec{\alpha} \tag{6}$$

$$\vec{n} \cdot (\vec{D}_2 - \vec{D}_1) = \sigma \tag{7}$$

$$\vec{n} \cdot (\vec{B}_2 - \vec{B}_1) = 0 \tag{8}$$

where $\vec{\alpha}$ and σ are the linear free current density vector and free surface charge density, respectively, and $\vec{n}$ is a unit normal vector pointing from medium 1 to medium 2.

2.1.2   The Wave Equation in a Dielectric Medium

Let us consider a linear (ε and μ independent of $\vec{E}$ and $\vec{H}$), isotropic, non-magnetic ($\mu = \mu_0$) dielectric medium free of current and charge sources ($J = 0$ and $\rho = 0$). Maxwell's equations become

$$\nabla \times \vec{E} = -\frac{\partial \vec{B}}{\partial t} \tag{9}$$

$$\nabla \times \vec{H} = \frac{\partial \vec{D}}{\partial t} \tag{10}$$

$$\nabla \cdot \vec{B} = 0 \tag{11}$$

$$\nabla \cdot \vec{D} = 0 \tag{12}$$

These equations are first-order differential equations in $\vec{E}$ and $\vec{H}$. A second-order equation in $\vec{E}$ or $\vec{H}$ can be obtained by combining Maxwell's equations. To do this, first take the curl of both sides of Eq. (9) and assume position independent (or weakly position-dependent) as well as time invariant ε and μ.

$$\nabla \times \nabla \times \vec{E} = -\nabla \times \frac{\partial \vec{B}}{\partial t} = -\nabla \times \frac{\partial \mu \vec{H}}{\partial t} \tag{13}$$

Since $\vec{H}$ is continuous, the order of the curl and time derivative operators can be reversed. Equation (13) becomes

$$\nabla \times \nabla \times \vec{E} = -\mu \frac{\partial}{\partial t} \nabla \times \vec{H} = -\mu \frac{\partial}{\partial t}\left(\frac{\partial \vec{D}}{\partial t}\right) = -\mu\varepsilon \frac{\partial^2 \vec{E}}{\partial t^2} \tag{14}$$

Using the vector identity

$$\nabla \times \nabla \times \vec{E} = \nabla(\nabla \cdot \vec{E}) - \nabla^2 \vec{E} \tag{15}$$

and the fact that

$$\nabla \cdot \vec{D} = \nabla \cdot (\varepsilon \vec{E}) = \nabla \varepsilon \cdot \vec{E} + \varepsilon \nabla \cdot \vec{E} = 0 \tag{16}$$

or

$$\nabla \cdot \vec{E} = -\frac{\vec{E}}{\varepsilon} \cdot \nabla \varepsilon \approx 0 \tag{17}$$

we have the following wave equation for $\vec{E}$ by combining Eqs. (14), (15) and (17):

$$\nabla^2 \vec{E} - \mu\varepsilon \frac{\partial^2 \vec{E}}{\partial t^2} = 0 \tag{18}$$

Similarly, we can derive the wave equation for $\vec{H}$ as

$$\nabla^2 \vec{H} - \mu\varepsilon \frac{\partial^2 \vec{H}}{\partial t^2} = 0 \tag{19}$$

It should be stressed that the above wave equations are valid for $\nabla \varepsilon \approx 0$ (for homogeneous or near homogeneous dielectric materials). In general, μ and ε are both functions of angular frequency, that is, $\mu = \mu(\omega)$ and $\varepsilon = \varepsilon(\omega)$. This frequency dependence of μ and ε is called the material dispersion. The following equations

$$\vec{D}(\omega) = \varepsilon(\omega)\vec{E}(\omega) \tag{20}$$

and

$$\vec{B}(\omega) = \mu(\omega)\vec{H}(\omega) \tag{21}$$

hold true for linear dielectric materials only.

2.1.3   Solutions of the Wave Equation

In many practical cases, the excitation source has a single oscillation frequency ω. The electromagnetic wave radiated as a result of such an excitation has the same, single frequency ω. In general, even if the electromagnetic wave is not of a single frequency, it can be treated as consisting of many single-frequency waves by using Fourier transforms. Therefore, the following discussion concentrates on solutions to the wave equations with a given frequency in the form of

$$\vec{E}(\vec{r}, t) = \vec{E}(\vec{r})\exp(j\omega t) \tag{22}$$

and

$$\vec{B}(\vec{r}, t) = \vec{B}(\vec{r})\exp(j\omega t) \tag{23}$$

The use of the complex notation $\exp(j\omega t)$ for the time variation is standard in electromagnetic theory, but it should be noted that only the real parts of the above equations are significant.

By using Eqs. (22) and (23), the wave equation for $\vec{E}$ reduces to

$$\nabla^2\vec{E}(\vec{r}) + \omega^2\mu\varepsilon\vec{E}(\vec{r}) = 0 \tag{24}$$

or

$$\nabla^2\vec{E}(\vec{r}) + k^2\vec{E}(\vec{r}) = 0 \tag{25}$$

Note that

$$\nabla \cdot \vec{E}(\vec{r}) = 0 \tag{26}$$

and

$$\vec{B} = \frac{j}{\omega}\nabla \times \vec{E}(\vec{r}) \tag{27}$$

Similarly, the wave equation for $\vec{B}$ reduces to

$$\nabla^2 \vec{B}(\vec{r}) + k^2 \vec{B}(\vec{r}) = 0 \tag{28}$$

where

$$\nabla \cdot \vec{B}(\vec{r}) = 0 \tag{29}$$

and

$$\vec{E} = -\frac{j}{\omega\mu\varepsilon} \nabla \times \vec{B}(\vec{r}) \tag{30}$$

Equations (25) and (28) are referred to as the Helmholtz equations. Expanding Eq. (25) yields

$$\left(\frac{\partial^2}{\partial x^2} + \frac{\partial^2}{\partial y^2} + \frac{\partial^2}{\partial z^2} + k^2 \right) \vec{E}(\vec{r}) = 0 \tag{31}$$

For a plane wave, there is no variation in the transverse directions, so the derivatives with respect to x and y are zero, leaving

$$\frac{d^2\vec{E}}{dz^2} + k^2 \vec{E} = 0 \tag{32}$$

The solution of this ordinary differential equation is

$$\vec{E}(z) = \vec{E}_{0+}e^{-jkz} + \vec{E}_{0-}e^{+jkz} \tag{33}$$

where $\vec{E}_{0+}$ and $\vec{E}_{0-}$ are two constants to be determined by the boundary conditions for the second-order differential equation. By applying the condition $\nabla \cdot \vec{E}(r) = 0$ to Eq. (33), it is found that $\vec{k} \cdot \vec{E} = 0$, where vector $\vec{k}$ is defined as $\vec{k} = k\hat{z}$. The parameter $\vec{k}$ is called the wave vector, and its direction is along the wave propagation direction $\hat{z}$. Obviously, $\vec{E}$ is a transverse field and $E_z = 0$.

The full solution, including the time variation, is

$$\vec{E} = \vec{E}_{0+}e^{j(\omega t - kz)} + \vec{E}_{0-}e^{j(\omega t + kz)} \tag{34}$$

By plotting the first and second terms as a function of z at different times, it becomes obvious that the first term describes a plane wave traveling along the positive z direction while the second term depicts a plane wave traveling along the negative z direction. The physically meaningful electric field is therefore the real part of Eq. (34):

$$\vec{E} = \vec{E}_{0+} \cos(\omega t - kz) + \vec{E}_{0-} \cos(\omega t + kz) \tag{35}$$

Notice that a special orientation with z along the propagation direction has been chosen to lead to the solution of Eq. (33). For a general situation, the solution for a forward traveling wave will be of the following form

$$\vec{E}(z) = \vec{E}_{0+} e^{j(\omega t - \vec{k} \cdot \vec{r})} \tag{36}$$

where $\vec{r}$ is the position vector. Again

$$\nabla \cdot \vec{E} = j\vec{k} \cdot \vec{E} = 0 \tag{37}$$

Equation (37) requires that $\vec{E}$ be a transverse field. The direction of $\vec{E}$ is called the direction of polarization.

The magnitude of wave vector $\vec{k}$ is equal to $2\pi/\lambda$ because Eq. (35) states that the phase difference is 2π for two positions separated by $2\pi/k$ along the propagation direction.

The plane wave solution for $\vec{B}$ can be derived by using Eq. (27)

$$\vec{B} = \frac{j}{\omega} \nabla \times \vec{E} = \frac{j}{\omega} j\vec{k} \times \vec{E} \tag{38}$$

From Eq. (38), it is obvious that $\vec{k} \cdot \vec{B} = 0$. So, the magnetic field is also a transverse field. A few apparent conclusions are

1. $\vec{E}$, $\vec{B}$ and $\vec{k}$ are orthogonal to each other;
2. $\vec{E}$ and $\vec{B}$ are in phase and their amplitude ratio is equal to $1/\sqrt{\mu\varepsilon}$;
3. $\vec{E} \times \vec{B}$ is along the propagation direction.

An important concept for characterizing electromagnetic waves is the measure of power flowing through a surface, called the Poynting vector, defined as

$$\vec{S} = \vec{E} \times \vec{H} \tag{39}$$

which represents the instantaneous power intensity of the wave. For plane waves,

$$\vec{S} = \vec{E} \times \vec{H} = \sqrt{\frac{\varepsilon}{\mu}} \vec{E} \times (\vec{n} \times \vec{E}) = \sqrt{\frac{\varepsilon}{\mu}} E^2 \vec{n} \tag{40}$$

Besides, the energy density stored in an electromagnetic wave can be shown to be

$$w = \frac{1}{2} (\vec{E} \cdot \vec{D} + \vec{H} \cdot \vec{B}) = \frac{1}{2} \left(\varepsilon E^2 + \frac{1}{\mu} B^2 \right) \tag{41}$$

For plane waves it is straightforward to show, using Eq. (38), that the energy stored in the electric field, $\frac{1}{2} \varepsilon E^2$, is equal to the energy stored in the magnetic field, $\frac{1}{2} \mu B^2$. Therefore,

$$\vec{S} = \sqrt{\frac{1}{\varepsilon\mu}}\, w\vec{n} = v_p w\vec{n} \tag{42}$$

where v_p is the phase velocity to be discussed next.

2.1.4 Phase Velocity and Group Velocity

Consider the phase term $(\omega t - kz)$ in Eq. (35). At time $t = 0$, the wave peak is at $z = 0$; At a later time t, the wave peak travels to $(\omega t - kz) = 0$ or $z = \omega t / k$. The plane of constant phase, $\omega t - kz = $ constant, is therefore moving at a phase velocity of

$$v_p = \frac{z}{t} = \frac{\omega}{k} = \frac{1}{\sqrt{\mu\varepsilon}} \tag{43}$$

In vacuum, this leads to the electromagnetic wave phase velocity

$$c = \frac{1}{\sqrt{\mu_0\varepsilon_0}} = 2.99792458 \times 10^{10} \text{ cm/s} \tag{44}$$

In non-magnetic dielectric media of relative dielectric constant ε_r, the phase velocity is

$$v_p = \frac{c}{\sqrt{\varepsilon_r}} = \frac{c}{n} \tag{45}$$

where n is the refractive index. Because ε_r is a function of frequency, the phase velocity is frequency dependent, which is the source of material dispersion.

Generally speaking, an information-bearing signal contains a small spread of frequencies around a carrier frequency. Because of interference among such a group of waves of slightly different frequencies, a wave packet is formed. Group velocity is the velocity of propagation of the envelope of the wave packet.

Let us consider the simplest case of a signal consisting of two traveling waves of equal amplitude E_0 with slightly different angular frequencies of $\omega_1 = \omega + \Delta\omega$ and $\omega_2 = \omega - \Delta\omega$, where $\Delta\omega \ll \omega$. The two associate wave vectors, being functions of frequency, will also be slightly different. Let the corresponding wave vectors be $k_1 = k + \Delta k$ and $k_2 = k - \Delta k$. The signal is

$$E(t, z) = E_0 \cos[(\omega + \Delta\omega)t - (k + \Delta k)z] + E_0 \cos[(\omega - \Delta\omega)t - (k - \Delta k)z]$$

$$= 2E_0 \cos(\Delta\omega t - \Delta k z) \cos(\omega t - kz) \tag{46}$$

Since $\Delta\omega$ is much less than ω, the superposition of these two waves leads to a temporal beat of frequency $\Delta\omega$ and a spatial beat of period Δk. Figure 2-1 depicts such a signal consisting of a rapidly oscillating wave of $\cos(\omega t - kz)$ with amplitude $2E_0 \cos(\Delta\omega t - \Delta kz)$ that varies slowly with an angular frequency $\Delta\omega$.

The group velocity is the velocity at which the envelope, or the pulse, travels. Because the envelope is described by $2E_0 \cos(\Delta\omega t - \Delta kz)$ and its constant phase condition is $(\Delta\omega t - \Delta kz) =$ constant, we obtain the position, $z(t)$, of constant phase as a function of time

$$z(t) = \frac{\Delta\omega t}{\Delta k} + \text{constant} \tag{47}$$

The group velocity is therefore

$$v_g = \frac{dz}{dt} = \frac{\Delta\omega}{\Delta k} \tag{48}$$

or in the limit when $\Delta\omega$ approaches zero

$$v_g = \frac{d\omega}{dk} \tag{49}$$

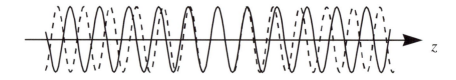

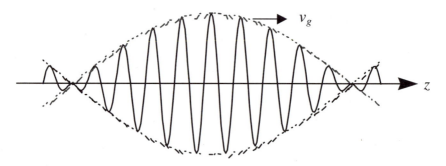

FIGURE 2-1   A signal consisting of two traveling waves of equal amplitude with slightly different angular frequencies of $\omega_1 = \omega + \Delta\omega$ and $\omega_2 = \omega - \Delta\omega$, leading to a temporal beat of frequency $\Delta\omega$.

Thus, v_g is the velocity of the envelope of the wave packet, as shown in Figure 2-1. In free space where $\omega = k_0 c$, the group and phase velocities are therefore the same. For waves in medium of dielectric constant ε, which usually is a function of frequency, the group velocity is

$$v_g = \frac{1}{dk/d\omega} = \frac{1}{d(\omega n/c)/d\omega} = \frac{1}{n/c + (\omega/c)dn/d\omega} = \frac{c}{n - \lambda dn/d\lambda} \tag{50}$$

The quantity $n - \lambda dn/d\lambda$ is called the group index. From this expression, we see three possible cases:

1. No dispersion: $dn/d\lambda = 0$, $v_g = v_p$;
2. Normal dispersion: $dn/d\lambda < 0$, $v_g < v_p$;
3. Anomalous dispersion: $dn/d\lambda > 0$, $v_g > v_p$.

Optical materials display normal dispersion throughout their transparent regions, so the envelope (the pulse or the energy) travels slower than the phase. The dispersion of dielectric waveguides will be discussed in more detail in Section 2.2.5.

2.1.5 Reflection and Refraction at Dielectric Interfaces

2.1.5.1 Laws of Reflection and Refraction   Consider an infinite planar interface shown in Figure 2-2 between two lossless dielectric materials of dielectric constants ε_1 and ε_2. A plane wave of frequency ω and wave vector k propagates from medium 1 onto medium 2. Depending on the properties of the two dielectric media and the plane wave incident angle, part of the wave is reflected back to the original medium and part of the wave gets transmitted into the second medium. The rules connecting the reflected and the transmitted fields to the incident field are the boundary conditions of Eqs. (5)–(8). For interfaces free of surface charge and current, the boundary conditions become

$$\vec{n} \times (\vec{E_2} - \vec{E_1}) = 0 \tag{51}$$

$$\vec{n} \times (\vec{H_2} - \vec{H_1}) = 0 \tag{52}$$

$$\vec{n} \cdot (\vec{D_2} - \vec{D_1}) = 0 \tag{53}$$

$$\vec{n} \cdot (\vec{B_2} - \vec{B_1}) = 0 \tag{54}$$

As Maxwell's equations are not independent for waves of a given frequency in space free of charge and current, the four boundary conditions are also not independent. Boundary conditions described by Eqs. (53) and (54) can be derived from Eqs.

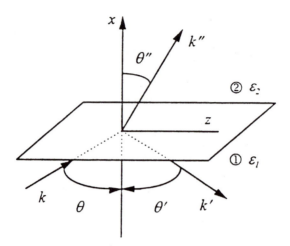

FIGURE 2-2 A plane wave of wave vector $\vec{k}$ incident onto an infinite planar interface between two lossless dielectric materials, resulting in a reflected wave of wave vector $\vec{k}'$ and a refracted wave of wave vector $\vec{k}''$.

(51) and (52). Therefore, in discussing waves of a given frequency at boundaries free of surface charge and current, we only need to consider the tangential continuity of $\vec{E}$ and $\vec{H}$.

Let us assume that the reflected, $\vec{E}'$, and refracted, $\vec{E}''$, waves are also plane waves, and their wave vectors are, respectively, $\vec{k}'$ and $\vec{k}''$. (These assumptions will be proved to be true soon.) Expressions for the incident, reflected and refracted waves are therefore

$$\vec{E} = \vec{E}_0 e^{j(\omega t - \vec{k}\cdot\vec{r})} \tag{55}$$

$$\vec{E}' = \vec{E}'_0 e^{j(\omega t - \vec{k}'\cdot\vec{r})} \tag{56}$$

$$\vec{E}'' = \vec{E}''_0 e^{j(\omega t - \vec{k}''\cdot\vec{r})} \tag{57}$$

To obtain the relationship among the wave vectors, we apply the boundary condition of Eq. (51) to the interface to get

$$\vec{n} \times (\vec{E} + \vec{E}') = \vec{n} \times \vec{E}'' \tag{58}$$

Note that the total field intensity in medium 1 is the sum of the incident and reflected fields. By inserting Eqs. (55)–(57) into Eq. (58), we find

$$\vec{n} \times (\vec{E}_0 e^{-j\vec{k}\cdot\vec{r}} + \vec{E}'_0 e^{-j\vec{k}'\cdot\vec{r}})\vec{n} \times \vec{E}''_0 e^{-j\vec{k}''\cdot\vec{r}} \tag{59}$$

The three exponents must be equal because Eq. (59) must hold true for the entire interface of $x = 0$. Therefore,

$$\vec{k}\cdot\vec{r} = \vec{k}'\cdot\vec{r} = \vec{k}''\cdot\vec{r} \quad \text{for } x = 0 \tag{60}$$

Because Eq. (60) is valid for arbitrary y and z, the coefficients have to be identical, i.e.,

$$k_y = k_y' = k_y'' \tag{61}$$

and

$$k_z = k_z' = k_z'' \tag{62}$$

As shown in Figure 2-2, let the incident wave vector be on the xz plane or $k_y = 0$, it follows from Eq. (61) that k_y' and k_y'' both must be zero. In other words, the incident, reflected and refracted wave vectors should lie on the same plane. Denoting the incident, reflected and refracted angles by θ, θ', and θ'', we have

$$k_z = k\sin\theta, \ k_z' = k'\sin\theta', \ k_z'' = k''\sin\theta'' \tag{63}$$

Relating wave vectors to their group velocities, we have

$$k = k' = \frac{\omega}{v_{g1}} \quad \text{and} \quad k'' = \frac{\omega}{v_{g2}} \tag{64}$$

Inserting Eqs. (63) and (64) into Eq.(62), we find

$$\theta = \theta' \quad \text{and} \quad \frac{\sin\theta}{\sin\theta''} = \frac{v_{g1}}{v_{g2}} \tag{65}$$

These are the familiar laws of reflection and refraction. For electromagnetic waves, $v_g = 1/\sqrt{\varepsilon\mu}$, so

$$\frac{\sin\theta}{\sin\theta''} = \frac{\sqrt{\varepsilon_2\mu_2}}{\sqrt{\varepsilon_1\mu_1}} = \frac{n_2}{n_1} \equiv n_{21} \tag{66}$$

where n_1 and n_2 are the indices of refraction of media 1 and 2 and n_{21} is defined as the index of refraction of medium 2 with respect to medium 1. Non-magnetic materials have been assumed such that $\mu_{1,2} \approx \mu_0$. Equation (66) is called Snell's law.

2.1.5.2   *Fresnel Formulas*

Let us now use the boundary conditions to establish the amplitude relationship among the incident, reflected and refracted fields.

Although all the wave vectors lie on the same incident plane, each wave vector has associated with it two independent directions of polarization. It is therefore necessary to discuss the two cases separately: $\vec{E}$ perpendicular and $\vec{E}$ parallel to the incident plane. Alternate designations are transverse electric (TE) and transverse magnetic (TM) for the perpendicular and parallel $\vec{E}$ field orientations, respectively.

Case (1): $\vec{E} \perp$ to the incident plane as shown in Figure 2-3*a*. By applying Eqs. (51)–(52) to the interface, we have

$$E + E' = E'' \tag{67}$$

and

$$H \cos\theta - H' \cos\theta' = H'' \cos\theta'' \tag{68}$$

Because $H = \sqrt{\varepsilon/\mu}\, E$ and $\mu = \mu_0$, Eq. (68) becomes

$$\sqrt{\varepsilon_1}(E - E') \cos\theta = \sqrt{\varepsilon_2}E'' \cos\theta'' \tag{69}$$

Combining Eqs. (66), (67), and (69), we have

$$\frac{E'}{E} = -\frac{\sin(\theta - \theta'')}{\sin(\theta + \theta'')} \tag{70}$$

and

$$\frac{E''}{E} = \frac{2 \cos\theta \sin\theta''}{\sin(\theta + \theta'')} \tag{71}$$

Case (2): $\vec{E}\ /\!/$ to the incident plane as shown in Figure 2-3*b*. By applying Eqs. (51)–(52) to the interface, we have

$$E \cos\theta - E' \cos\theta' = E'' \cos\theta'' \tag{72}$$

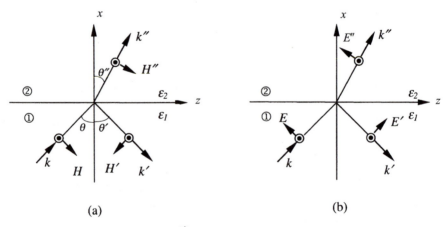

(a)   (b)

FIGURE 2-3   Incident plane wave $\vec{E}$ perpendicular to the incident plane (a) and incident plane wave $\vec{E}$ parallel to the incident plane (b).

and

$$H + H' = H''$$ (73)

Equation (73) can be expressed by the electric field components as

$$\sqrt{\varepsilon_1}(E + E') = \sqrt{\varepsilon_2}E''$$ (74)

By combining Eqs. (66), (72) and (74), we have

$$\frac{E'}{E} = \frac{\tan(\theta - \theta'')}{\tan(\theta + \theta'')}$$ (75)

and

$$\frac{E''}{E} = \frac{2\cos\theta\sin\theta''}{\sin(\theta + \theta'')\cos(\theta - \theta'')}$$ (76)

Equations (70), (71), (75), and (76) are called Fresnel formulas. They describe the ratio of reflected and refracted field amplitudes to the incident field amplitude. Equations (70) and (75) are plotted in Figure 2-4 for an air to glass interface (a) and vice versa in (b). It is obvious that the perpendicular and parallel incident fields yield very different reflection and refraction results. If the incident wave is unpolarized, the reflected and refracted waves become partially polarized because the amount of reflected and refracted waves are different for the two directions of polarization as described by the Fresnel formulas. In the special case of

$$\theta + \theta'' = \pi/2$$ (77)

there is no reflection component ($E' = 0$) for the parallel polarization, according to Eq. (75). The reflected wave of a natural light is therefore completely polarized to be perpendicular to the incident plane. Equation (77) is the familiar Brewster's law in optics. This particular incident angle is called the Brewster angle and is denoted by θ_B in Figure 2-4.

Fresnel formulas provide not only the amplitude relationship but also the relative phase among the incident, reflected, and refracted waves. In the case of $\vec{E} \perp$ incident plane, Eq. (70) indicates that the reflected and incident waves are out of phase if $\theta > \theta''$, or, according to the Snell's law, if $\varepsilon_2 > \varepsilon_1$.

2.1.5.3   *Total Internal Reflection*   According to Snell's law, the refraction angle θ'' becomes larger than the incident angle θ for $\varepsilon_1 > \varepsilon_2$. If the incident angle θ is so chosen that $\sin\theta = n_{21}$, the refraction angle θ'' becomes equal to 90°. If the incident angle is further increased such that $\sin\theta > n_{21}$, no real refraction angle can be defined. This phenomena is called total internal reflection for reasons discussed as follows.

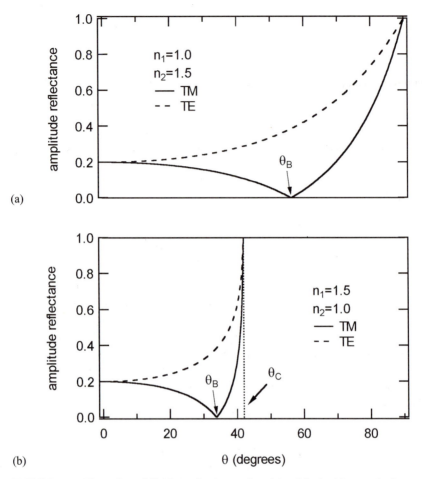

FIGURE 2-4   The reflected field amplitude as a function of the incident angle for an air to glass interface (a) and a glass to air interface (b).

If the same expressions [Eqs. (55)–(57)] are used to describe the case of $\sin\theta > n_{21}$, the boundary conditions still lead to $k_z'' = k_z = k\sin\theta$ and $k'' = kv_{g1}/v_{g2} = kn_{21}$. Because $\sin\theta > n_{21}$, we have $k_z'' > k''$, so

$$k_x'' = \sqrt{k''^2 - k_z''^2} = -jk\sqrt{\sin^2\theta - n_{21}^2} \tag{78}$$

which is imaginary. The refracted wave expression becomes

$$E'' = E_0'' e^{-k_a x} e^{j(\omega t - k_z'' z)} \tag{79}$$

where k_a is defined as

$$k_a = k\sqrt{\sin^2\theta - n_{21}^2} \tag{80}$$

Equation (79) describes a wave propagating along the z-direction with an amplitude decaying along the x-direction. In other words, this refracted wave exists only at regions very close to the interface. The depth of this region is proportional to $1/k_a$, which is generally on the order of the wavelength in medium 1.

Consider the case of $\vec{E} \perp$ to the incident plane ($E'' = E_y''$), the refracted field intensity can also be calculated, using Eq. (38), to be

$$H_x'' = -n_2 \frac{k_z''}{k''} E_y'' = -n_1 \sin\theta \, E'' \tag{81}$$

and

$$H_z'' = n_2 \frac{k_x''}{k''} E_y'' = -jn_1 \sqrt{\sin^2\theta - n_{21}^2} \, E'' \tag{82}$$

It is seen that H_x'' and E'' are in phase while H_z'' and E'' have a phase difference of $\pi/2$. The average refracted power flow along the x-direction can be calculated as

$$S_x'' = -\tfrac{1}{2}\text{Re}(E_y''^* H_z'') = 0 \tag{83}$$

where Re( ) represents the real part of $E_y''^* H_z''$ and $E_y''^*$ is the complex conjugate of E_y''. So, on average, zero power flows into the second medium, which is the reason it is called total internal reflection. Optical waveguide devices are possible because of the phenomena of total internal reflection. It is through the total internal reflection that waves bounce to and fro within the guiding media and travel down a waveguide.

The Fresnel formulas [Eqs. (70), (71), (75), and (76)] derived earlier, are still valid in the case of $\sin\theta > n_{21}$, if we make the following replacements in the formulas:

$$\cos\theta'' \to -j\sqrt{\sin^2\theta/n_{21}^2 - 1} \tag{84}$$

and

$$\sin\theta'' \to \frac{\sin\theta}{n_{21}} \tag{85}$$

For example, with $\vec{E} \perp$ to the incident plane, Eq. (70) becomes

$$\frac{E'}{E} = \frac{\cos\theta + j\sqrt{\sin^2\theta - n_{21}^2}}{\cos\theta - j\sqrt{\sin^2\theta - n_{21}^2}} = e^{2j\phi} \tag{86}$$

where

$$\tan\phi = \frac{\sqrt{\sin^2\theta - n_{21}^2}}{\cos\theta} \tag{87}$$

These expressions show that the reflected and incident wave amplitudes are the same, but there is a phase difference of 2ϕ. Similarly, it can be shown that in the case of $\vec{E} \, //$ to the incident plane

$$\frac{E'}{E} = e^{2j\delta} \tag{88}$$

where

$$\tan\delta = \frac{\tan\phi}{n_{21}^2} \tag{89}$$

2.2. SLAB WAVEGUIDES

2.2.1 The Guided Wave Condition

The simplest optical waveguide is the step-index infinite planar slab waveguide shown in Figure 2-5. The waveguide consists of three layers: a higher-index (n_1) guiding layer (or core layer) sandwiched by the lower-index substrate (n_2) and cover or cladding (n_3) layers. The cover layer could be air ($n_3 = 1$). The structure is usually called an asymmetric slab waveguide with $n_1 > n_2 > n_3$.

As discussed earlier, in general both $\vec{E}$ and $\vec{H}$ are orthogonal to each other as well as to the propagation vector $\vec{k}$, and there are two possible electric field polarizations. The two possible polarizations in a slab waveguide, one with $\vec{E}$ polarized along the y axis (out of the page in Fig. 2-5) and the other with $\vec{E}$ lying in the x-z plane, establish the so-called transverse electric (TE) and transverse magnetic (TM) fields. A wave is called a TE wave when its electric field $\vec{E}$ is transverse to the direction of propagation, i.e., there is no longitudinal electric field component. Similarly, a wave is called a TM wave when there is no longitudinal magnetic field component. If a wave does not have logitudinal electric or logitudinal magnetic components, it is

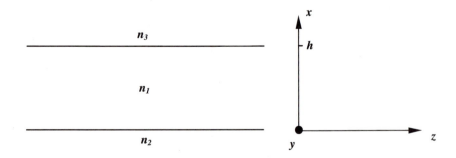

FIGURE 2-5   Step-index infinite planar slab waveguide.

called a TEM wave. Generally speaking, waves in a slab waveguide can be classified as either TE or TM waves. For simplicity, we will consider one of the two waves, the TE wave, in the following derivation. The case for the TM wave can be solved similarly by using the corresponding boundary conditions.

Using the coordinate axes shown in Fig. 2-5 and assuming a sinusoidal wave of angular frequency ω_0 with the field $\vec{E}$ polarized along the y axis, the scalar wave equation for the TE wave electric field, E_{TE} becomes

$$\nabla^2 E_{\mathrm{TE}}(x, z) + k_0^2 n_i^2 E_{\mathrm{TE}}(x, z) = 0 \qquad (i = 1, 2, 3) \qquad (90)$$

$E_{\mathrm{TE}}(x, z)$ is not a function of y because the slab extends to infinity in the y direction. The solution to Eq. (90) should have the following form for the TE wave traveling along the z direction:

$$E_{\mathrm{TE}}(x, z) = E_{\mathrm{TE}}(x)e^{-j\beta z} \qquad (91)$$

The amplitude is independent of z due to the translational invariance of the structure along the z direction, while the phase varies along the propagation direction. The new parameter introduced here, β, is a propagation coefficient along the z direction, namely, the wave vector along the z direction. It is often called the longitudinal wave vector. Inserting this solution into Eq. (90), we have

$$\frac{d^2 E_{\mathrm{TE}}(x)}{dx^2} + (k_0^2 n_i^2 - \beta^2)E_{\mathrm{TE}}(x) = 0 \quad (i = 1, 2, 3) \qquad (92)$$

The solution to Eq. (92) depends on the relative magnitude of $k_0 n_i$ with respect to β. The possible solutions are

$$E_{\mathrm{TE}}(x) = E_0 e^{\pm j\sqrt{k_0^2 n_i^2 - \beta^2}\,x} \qquad \text{if } \beta < k_0 n_i \qquad (93)$$

or

$$E_{\mathrm{TE}}(x) = E_0 e^{\pm\sqrt{\beta^2 - k_0^2 n_i^2}\,x} \quad \text{if } \beta > k_0 n_i \qquad (94)$$

For $\beta < k_0 n_i$, the solution is oscillatory with a transverse wave vector κ defined by

$$\kappa = \sqrt{k_0^2 n_i^2 - \beta^2} \qquad (95)$$

For $\beta > k_0 n_i$, the solution is exponentially decaying, and the attenuation coefficient is

$$\gamma = \sqrt{\beta^2 - k_0^2 n_i^2} \qquad (96)$$

The geometrical relation among the longitudinal propagation coefficient β, the total wave vector k, and the transverse wave vector κ are shown in Figure 2-6.

Because β is the longitudinal component of wave vector k, as shown in Figure

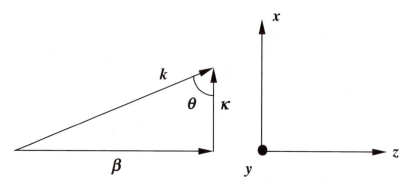

FIGURE 2-6 Geometrical relationship among the longitudinal, transverse, and total wave vectors.

2-6, it may take values ranging from 0 to k, depending on the incident angle of k to the guiding interface. Each possible solution of β is called a mode. For an asymmetric waveguide, Figure 2-7 depicts the physical picture of different possible modes. In the range of $\beta = 0$ to $\beta = k_0 n_3$, oscillatory waves may exist in all three regions of the waveguide [Eq. (93)] because the incident angle is less than the critical angle for total internal reflection. The wave is therefore not guided. In the extreme case when $\beta = 0$, the wave is incident perpendicular to the interface with part of the wave going through the interface and the rest reflected back to the original medium. In this region, the value of β is obviously continuous and the number of modes is therefore infinite. Modes that are not guided are called radiation modes.

In the range of $k_0 n_3 < \beta < k_0 n_2$, solutions to the wave equation in the cover layer

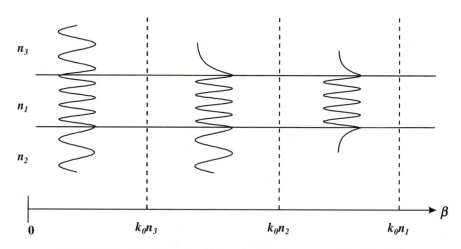

FIGURE 2-7 Condition of guided and radiated waves as a function of β.

are exponentially decaying [Eq. (94)] and the average power flow into the cover along the x-direction is equal to zero [Eq. (83)]. However, oscillatory waves form in the substrate layer because $\beta < k_0 n_2$ [Eq. (93)]. In the case of $\beta > k_0 n_2$, the waves in the guiding layer can be totally reflected from both interfaces because the incident angle can be larger than the critical angles for total internal reflection for both interfaces. The wave can therefore be guided. The possible value of β is described by Eq. (92), and there is a limited number of solutions. In other words, β is discrete and there is a limited number of modes that can be guided by the waveguide. Each mode is described by a mode number m to be discussed in Section 2.2.3. The maximum possible β value is $k_0 n_1$ beyond which the picture would be unphysical. At $\beta = k_0 n_1$, k and β are both along the z direction and the ray in this case is called the central ray.

From the above discussion, it is obvious that the condition for the existence of a guided wave in a planar slab dielectric waveguide with $n_1 > n_2 > n_3$ is

$$k_0 n_2 < \beta < k_0 n_1 \tag{97}$$

For bounded waves in waveguides, the group velocity now takes the form of

$$v_g = \frac{d\omega}{d\beta} \tag{98}$$

2.2.2   Characteristic Equations for the Slab Waveguide

To determine the possible values of the longitudinal wave vector β, which characterizes the possible guided waves, we have to apply the proper boundary conditions to the general solutions of Eqs. (93) and (94). Since we are interested in standing waves inside and evanescent fields outside the guiding layer, the transverse electric field amplitudes in the three regions of the slab waveguide can be written as

$$E_{TE}(x) = A \cos(\kappa x - \phi) \quad \text{for} \quad 0 < x < h \tag{99}$$

$$E_{TE}(x) = B e^{\gamma_1 x} \quad \text{for} \quad x < 0 \tag{100}$$

$$E_{TE}(x) = C e^{-\gamma_2 (x-h)} \quad \text{for} \quad x > h \tag{101}$$

where A, B, and C are amplitude coefficients to be determined and γ_1, γ_2 and κ are defined as

$$\gamma_1 = \sqrt{\beta^2 - k_0^2 n_2^2} \tag{102}$$

$$\gamma_2 = \sqrt{\beta^2 - k_0^2 n_3^2} \tag{103}$$

$$\kappa_1 = \sqrt{k_0^2 n_1^2 - \beta^2} \tag{104}$$

By applying boundary conditions for the continuous tangential E and H (equivalent to continuous dE_{TE}/dx) across each interface, we have, at $x = 0$

$$B = A \cos(\phi) \tag{105}$$

$$\gamma_1 B = -\kappa A \sin(-\phi) \tag{106}$$

which lead to

$$\tan(\phi) = \frac{\gamma_1}{\kappa} \tag{107}$$

Similarly, by applying the boundary conditions to the $x = h$ interface, we get

$$C = A \cos(\kappa h - \phi) \tag{108}$$

and

$$\gamma_2 C = \kappa A \sin(\kappa h - \phi) \tag{109}$$

which lead to

$$\tan(\kappa h - \phi) = \frac{\gamma_2}{\kappa} \tag{110}$$

By using the following trigonometrical indentity

$$\tan(A_1 - A_2) = (\tan A_1 - \tan A_2)/(1 + \tan A_1 \tan A_2) \tag{111}$$

we can combine Eqs. (107) and (110) to get

$$\tan(\kappa h) = \frac{\kappa(\gamma_1 + \gamma_2)}{\kappa^2 - \gamma_1 \gamma_2} \tag{112}$$

Equation (112) is called the characteristic equation for the TE modes of a slab dielectric waveguide, or, the eigenvalue equation for β because all the parameters in the equation depend on β. Since this is a transcendental equation, the eigenvalues for β have to be calculated numerically. As there is a set of discrete solutions to β, the waveguide can only support a set of discrete modes.

Similarly, the characteristic equation for the TM mode can be derived by setting up the initial problem for transverse magnetic fields. Without repeating very much of the same derivation, the TM-mode eigenvalue equation for β can be written as

$$\tan(\kappa h) = \frac{\kappa \left[\left(\dfrac{n_1}{n_2} \right)^2 \gamma_1 + \left(\dfrac{n_1}{n_3} \right)^2 \gamma_2 \right]}{\kappa^2 - \left(\dfrac{n_1^2}{n_2 n_3} \right)^2 \gamma_1 \gamma_2} \tag{113}$$

In the case of a symmetrical waveguide where a guiding or core layer of index n_{co} is surrounded on both sides by a cladding layer index n_{cl} which is lower than n_{co}, the TE mode electric fields can be shown to take the following forms:

$$E_{TE}(x) = A\cos(\kappa x)/\cos(\kappa h/2) \text{ or } A\sin(\kappa x)/\sin(\kappa h/2) \quad \text{for} \quad -\frac{h}{2} < x < \frac{h}{2} \quad (114)$$

$$E_{TE}(x) = Ae^{-\gamma(x-h/2)} \quad \text{for} \quad x \geq \frac{h}{2} \tag{115}$$

$$E_{TE}(x) = \pm Ae^{\gamma(x+h/2)} \quad \text{for} \quad x \leq -\frac{h}{2} \tag{116}$$

where $x = 0$ is chosen at the center of the guiding layer for reasons of symmetry and $\gamma = \sqrt{\beta^2 - k_0^2 n_{cl}^2}$. Notice that there are two possibilities for the field in the guiding layer, a symmetric even mode described by a cosine function or an antisymmetric odd mode represented by the sine function.

The characteristic or eigenvalue equation for the TE modes in such a symmetric waveguide becomes

$$\tan\left(\frac{\kappa h}{2}\right) = \frac{\gamma}{\kappa} \quad \text{for even modes} \tag{117}$$

or

$$\tan\left(\frac{\kappa h}{2}\right) = -\frac{\kappa}{\gamma} \quad \text{for odd modes} \tag{118}$$

Similarly, the characteristic or eigenvalue equation for TM modes becomes

$$\tan\left(\frac{\kappa h}{2}\right) = \left(\frac{n_{co}}{n_{cl}}\right)^2 \frac{\gamma}{\kappa} \quad \text{for even modes} \tag{119}$$

or

$$\tan\left(\frac{\kappa h}{2}\right) = -\left(\frac{n_{cl}}{n_{co}}\right)^2 \frac{\kappa}{\gamma} \quad \text{for odd modes} \tag{120}$$

2.2.3   Waveguide Modes

A straightforward numerical calculation can be done to show that at least one mode can be supported by a symmetric waveguide. The same conclusion can also be reached by examining the cutoff condition as a function of mode frequency to be shown later. The field distribution as a function of frequency can be found from numerical calculation. At low frequencies with large wavelengths and small wave vec-

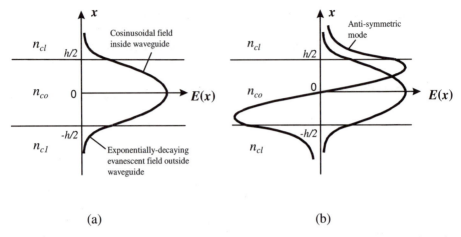

(a)   (b)

FIGURE 2-8   Fundamental symmetrical mode (a) and the development of the second mode (antisymmetric) when the frequency is increased.

tors, the guide is singlemoded and supports only the symmetric even pattern shown in Figure 2-8a. The lower the frequency the larger the evanescent field outside the guide. As the frequency increases, the field distributes more towards the center of the guide, and then the second antisymmetric mode develops at higher frequencies as shown in Figure 2-8b. At even higher frequencies, more modes can be supported by the guide. Without resorting to numerical calculation, we can determine the number of modes that can propagate at a given frequency.

Recall that a mode ceases to be guided when the incident angle θ to the guiding interface approaches the critical angle θ_c for total internal reflection ($\theta'' = 90°$). Because

$$\beta = k_0 n_{co} \sin \theta_c \qquad (121)$$

and

$$n_{co} \sin \theta_c = n_{cl} \sin \theta'' = n_{cl} \qquad (122)$$

The cutoff occurs when β approaches $k_0 n_{cl}$, which leads to $\gamma \to 0$. For TE even modes, we have, from Eq. (117)

$$\tan(\kappa h/2) = 0 \quad \text{at cutoff} \qquad (123)$$

Therefore

$$\kappa h/2 = m\pi \quad \text{for } m = 0, 1, 2, \ldots \quad \text{at cutoff} \qquad (124)$$

Similar arguments can be made for the antisymmetric TE modes to get, from Eq. (118),

$$\kappa h/2 = (2m-1)\,\pi/2 \quad \text{for } m = 1, 2, 3, \dots \quad \text{at cutoff} \tag{125}$$

Combining Eqs. (124) and (125), we have, in general,

$$\kappa h/2 = m\pi/2 \quad \text{for } m = 0, 1, 2, \dots \quad \text{at cutoff} \tag{126}$$

where m is called the mode number. Note that the larger the incident angle, the smaller the mode number m as illustrated in Figure 2-9. Note that at cutoff

$$\kappa = \sqrt{k_0^2 n_{co}^2 - \beta^2} = k_0 \sqrt{n_{co}^2 - n_{cl}^2} \tag{127}$$

A dimensionless parameter V, called the normalized frequency, can be defined as follows:

$$V = \frac{k_0 h}{2}\sqrt{n_{co}^2 - n_{cl}^2} = \frac{k_0 h n_{co}}{2}\sqrt{2\Delta} \tag{128}$$

where the relative refractive index is given by $\Delta = (n_{co}^2 - n_{cl}^2)/2n_{co}^2 \approx (n_{co} - n_{cl})/n_{co}$. By inserting Eqs. (126)–(127) into Eq. (128), we obtain the normalized cutoff frequency for each mode

$$V_c = \frac{k_0 h}{2}\sqrt{n_{co}^2 - n_{cl}^2} = \frac{m\pi}{2} \tag{129}$$

This expression describes the cutoff conditions for all the possible guided modes for symmetric waveguides. It is straightforward to show that at least one mode can be supported by a symmetric waveguide, although the mode can be very poorly guided. For the lowest mode with $m = 0$, we have

$$V_c = \frac{k_0 h}{2}\sqrt{n_{co}^2 - n_{cl}^2} = 0 \tag{130}$$

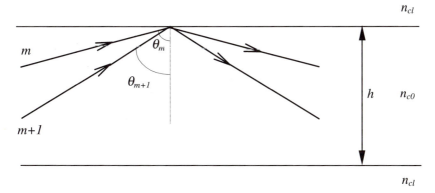

FIGURE 2-9   Ray picture showing the larger the incident angle θ, the larger the mode number m.

which is possible only if $k_0 = 0$. In other words, there is no cutoff and at least one mode, the lowest-order mode, can always be supported. To design a singlemode waveguide, we set $m = 1$ and rearrange Eq. (129) to get

$$h = \frac{\lambda_0}{2}\sqrt{n_{co}^2 - n_{cl}^2} \tag{131}$$

which is the design criterion for the guiding layer thickness for given n_{co} and n_{cl} values. As long as the film thickness h is less than the value defined by Eq. (131), only the $m = 0$ mode can be supported.

To numerically solve the eigenvalue equations, two other normalized parameters are often used. They are the asymmetry parameter a and the normalized effective index or normalized propagation vector b_n defined as

$$a = \frac{n_2^2 - n_3^2}{n_1^2 - n_2^2} \tag{132}$$

and

$$b_n = \frac{n_{eff}^2 - n_2^2}{n_1^2 - n_2^2} \tag{133}$$

where n_{eff} is the effective index of the waveguide defined as $n_{eff} = \beta/k_0$.

2.2.4 Orthogonality and Completeness of Modes

It should be obvious that any physically possible wave can be expressed in terms of the whole set of modes for a waveguide because all the modes together form a complete set of solutions to Maxwell's equations. Furthermore, it can be shown that all modes are orthogonal to each other. In other words, the integral over the entire waveguide cross section for the product of the transverse fields of two modes is equal to zero, i.e.,

$$\iint_{area} (\vec{E}_m \times \vec{H}_l^*) \cdot d\vec{A} = 0 \qquad m \neq l \tag{134}$$

To prove the orthogonal property, we start with the expression for power propagating along the z direction, which can be obtained by integrating the z component of the time-averaged Poynting vector over the waveguide cross section

$$P = \frac{1}{2}\text{Re}\left[\iint_{area} (\vec{E} \times \vec{H}^*) \cdot d\vec{A}\right] \tag{135}$$

where $\vec{E}$ and $\vec{H}$ can be expressed in terms of the eigenfunctions with

$$\vec{E} = \sum_m c_m \vec{E}_m e^{-j\beta_m z} \qquad (136)$$

and

$$\vec{H} = \sum_m c_m \vec{H}_m e^{-j\beta_m z} \qquad (137)$$

The expansion coefficients c_m describe the magnitude of each mode. For a lossless waveguide, P should be independent of z, or, $dP/dz = 0$. This leads to

$$\frac{dP}{dz} = \frac{d}{dz}\left(1/4 \iint_{\text{area}} \sum_{m,l} (c_m c_l^* e^{-j(\beta_m - \beta_l)z} \vec{E}_m \times \vec{H}_l^* + c_m c_l^* e^{-j(\beta_m - \beta_l)z} \vec{E}_l^* \times \vec{H}_m) \cdot d\vec{A}\right) = 0$$

$$(138)$$

or

$$\sum_{m,l} c_m c_l^* j(\beta_l - \beta_m) e^{-j(\beta_m - \beta_l)z} \iint_{\text{area}} (\vec{E}_m \times \vec{H}_l^* + \vec{E}_l^* \times \vec{H}_m) \cdot d\vec{A} = 0 \qquad (139)$$

where the identity of $\text{Re}(z) = 1/2(z + z^*)$ has been used. For arbitrary expansion coefficients and unequal mode constants, the above equation is satisfied for all z if and only if

$$\iint_{\text{area}} (\vec{E}_m \times \vec{H}_l^* + \vec{E}_l^* \times \vec{H}_m) \cdot d\vec{A} = 0 \qquad m \neq l \qquad (140)$$

This orthogonality condition is derived using only the forward-traveling modes. If the backward-traveling modes are also included in the derivation, it can be shown that

$$\iint_{\text{area}} (\vec{E}_m \times \vec{H}_l^*) \cdot d\vec{A} = 0 \qquad m \neq l \qquad (141)$$

Because the guided and radiation modes form a complete set of solutions to Maxwell's equations for the entire waveguide space, and any electromagnetic wave should satisfy Maxwell's equations, it follows that any continuous distribution of the electric field can be expressed as a superposition of weighted modes of the waveguide. In other words,

$$\vec{E}(\vec{r}) = \sum_m c_m \vec{E}_m(\vec{r}) + \int a(\beta) \vec{E}(\vec{r}, \beta) d\beta \qquad (142)$$

The first term is the sum over all the discrete modes of the waveguide, and the second term integrates over the whole range of β for the radiation modes of the waveguide. The expansion coefficients for the radiation modes are given by the continuous function $a(\beta)$.

The orthogonality and completeness of the waveguide modes are very useful for the description of complicated waveguide systems and will be used in the perturbation analysis discussed in Section 2.3.3. For example, the coupling of power between two waveguides can be described by using a superposition based on the completeness of the waveguide modes. As another example, the transmitted power in a waveguide can be described by a weighted sum over the modal powers by using the orthogonality property, shown as follows.

From Eqs. (135)–(137), we know the total power flow along the z direction is

$$P = \frac{1}{4} \iint_{\text{area}} \sum_{m,l} (c_m c_l^* e^{-j(\beta_m - \beta_l)z} \vec{E}_m \times \vec{H}_l^* + c_m c_l^* e^{-j(\beta_m - \beta_l)z} \vec{E}_l \times \vec{H}_m) \cdot d\vec{A} \quad (143)$$

Due to the orthogonality property of the modes, all the terms where $m \neq l$ vanish. The expression is therefore simplified to

$$P = \frac{1}{4} \sum_m c_m c_m^* \iint_{\text{area}} (\vec{E}_m \times \vec{H}_m^* + \vec{E}_m^* \times \vec{H}_m) \cdot d\vec{A} \quad (144)$$

Note that the modal power is

$$P_m = \frac{1}{4} \iint_{\text{area}} (\vec{E}_m \times \vec{H}_m^* + \vec{E}_m^* \times \vec{H}_m) \cdot d\vec{A} \quad (145)$$

So, we have

$$P = \sum_m c_m c_m^* P_m \quad (146)$$

It is seen that due to the orthogonality property, the total power carried by all the forward-traveling modes can be expressed simply as the weighted sum of the individual mode powers. The weighting coefficients are the modulus squared of the individual mode amplitudes.

2.2.5 Dispersion

The most important fundamental bandwidth limitation for optical waveguides is dispersion, the spreading of a signal in the time domain. Dispersion causes broadening of signal pulses and leads to overlapping of adjacent pulses, resulting in erroneous signal transmission.

There are basically three types of dispersion in waveguides: material dispersion, modal dispersion (or inter-modal dispersion), and waveguide dispersion (or intra-modal dispersion). Among the three dispersions, waveguide dispersion is normally the smallest, which does not mean it can always be neglected in practical waveguides. For example, near the vicinity of the zero material dispersion point in a singlemode waveguide, waveguide dispersion plays the dominant role. If the dispersions have different signs, a careful design of the waveguide dispersion can be used to cancel the material dispersion near the zero-material dispersion point.

Material dispersion is due to the dependence of the refractive index on wavelength. For a pulse traveling in a dielectric material, each wavelength component of the pulse will "see" a different refractive index and therefore travel at a different velocity.

Inter-modal dispersion is a result of multi-mode propagation where each mode travels at a different group velocity. Recall that each mode has a different incident angle that corresponds to a different velocity component along the z direction. Because a lower order mode has a larger incident angle θ, its velocity component along the z direction is therefore larger. So, lower order modes travel faster than higher order modes in an optical waveguide. Inter-modal dispersion can be eliminated by using singlemode waveguides.

Intra-mode waveguide dispersion is a direct consequence of the dependence of the propagation constant β on wavelength. This dependence exists even in a single-mode waveguide. For a pulse of finite width, each wavelength component of the pulse will have a different propagation constant β and travel at a different velocity.

If more than one polarization mode exists in a birefringent waveguide, polarization mode dispersion should also be considered. Polarization mode dispersion is due to the refractive index difference between different polarizations. The refractive index difference or the birefringence is often due to the inevitable stress between the cladding and guiding core materials in dielectric waveguides.

2.2.5.1 *Material Dispersion*

The dependence of the dielectric constant on wavelength (or frequency) is a well-known phenomena. Basically, a displacement of electrons from their equilibrium positions occurs when an external electric field is applied to the dielectric material. Each displaced charge results in a dipole moment $q\vec{r}$ where $\vec{r}$ is the displacement of charge. The sum of the dipoles gives rise to a macroscopic polarization $\vec{P}$, which in turn leads to a frequency dependent dielectric constant, shown as follows.

If the applied electric field is

$$\vec{E}(t) = \vec{E}_0 e^{j\omega t} \tag{147}$$

the displacement of charge can be found to be [1]

$$\vec{r}(t) = \frac{-q/m}{\omega_0^2 - \omega^2 + j\omega\gamma} \vec{E}_0 e^{j\omega t} \tag{148}$$

where γ relates to energy radiation of the forced moving charge, m is the mass of charge and q is the elemental charge. According to fundamental electromagnetic theory, the displacement vector is

$$\vec{D} = \varepsilon\vec{E} = \varepsilon_0\vec{E} + \vec{P} = \varepsilon_0\vec{E} - Nq\vec{r} \tag{149}$$

where N is the density of dipoles per unit volume. By inserting Eq. (148) into Eq. (149), we have

$$\vec{D} = \left(\varepsilon_0 + \frac{Nq^2/m}{\omega_0^2 - \omega^2 + j\omega\gamma} \right) \vec{E}_0 e^{j\omega t} \tag{150}$$

The relative dielectric constant, defined as $\varepsilon_r = \varepsilon/\varepsilon_0$, is therefore

$$\varepsilon_r = \frac{\vec{D}}{\varepsilon_0 \vec{E}} = 1 + \frac{Nq^2/m}{\varepsilon_0(\omega_0^2 - \omega^2 + j\omega\gamma)} \equiv \varepsilon_r' - j\varepsilon_r'' \tag{151}$$

where the real and imaginary components are

$$\varepsilon_r' = 1 + \frac{Nq^2/m(\omega_0^2 - \omega^2)}{\varepsilon_0[(\omega_0^2 - \omega^2)^2 + \omega^2\gamma^2]} \tag{152}$$

and

$$\varepsilon_r'' = \frac{\gamma\omega Nq^2/m}{\varepsilon_0[(\omega_0^2 - \omega^2)^2 + \omega^2\gamma^2]} \tag{153}$$

If the radiation term γ can be neglected, Eqs. (152) and (153) can be simplified to

$$\varepsilon_r' \approx 1 + \frac{Nq^2/m}{\varepsilon_0(\omega_0^2 - \omega^2)} \tag{154}$$

and

$$\varepsilon_r'' \approx \frac{\gamma\omega Nq^2/m}{\varepsilon_0(\omega_0^2 - \omega^2)^2} \tag{155}$$

Because $Nq^2/m\varepsilon_0$ is generally much smaller than the denominator in Eq. (154), except for $\omega_0 \to \omega$, ε_r' will normally be approximately unity. The region where ε_r' increases with ω is called the normal dispersion region while the region where ε_r' decreases with ω is referred to as the anomalous dispersion region. Because the index of refraction n is equal to $\sqrt{\varepsilon_r}$, we have

$$n = \sqrt{\varepsilon_r' - j\varepsilon_r''} \equiv n' - jn'' \tag{156}$$

Assuming that the radiation term is small, we have, after some manipulation,

$$n' \approx \sqrt{1 + (Nq^2/m\varepsilon_0)/(\omega_0^2 - \omega^2)} \tag{157}$$

and

$$n'' \approx \frac{1}{2}(\omega\gamma)/[(\omega_0^2 - \omega^2)\sqrt{1 + (\omega_0^2 - \omega^2)/(Nq^2/m\varepsilon_0)}] \tag{158}$$

Equation (157) describes the change in the real part of the refractive index near a resonance while Eq. (158) represents the absorption, which is discussed further in Section 2.2.6. Their functional dependence is shown in Figure 2-10. The largest change in the real part of the index occurs at frequencies where the absorption is changing rapidly.

So far, we have considered only one resonant condition case. For real dielectric materials, generally speaking, more than one type of charge resonant frequency exists. The dielectric constant in this case can be simply determined by adding up all the resonance components. The result is

$$\varepsilon \approx \varepsilon_0 + \sum_i^l \frac{f_i N q^2 / m}{\omega_i^2 - \omega^2 + j\gamma\omega} \tag{159}$$

where f_i describes the ith oscillation strength of the system. Neglecting the radiation term γ and replacing frequency by wavelength, we can rearrange Eq. (159) in terms of the index of refraction

$$n^2 - A_0 = \sum_{i=1}^l \frac{A_i \lambda^2}{\lambda^2 - \lambda_i^2} \tag{160}$$

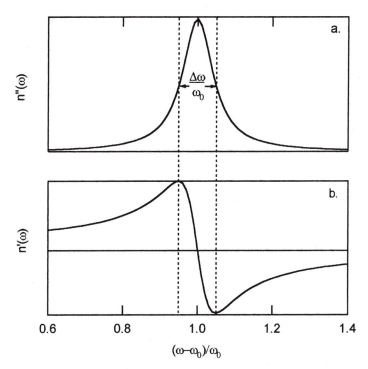

FIGURE 2-10   The absorption (a) and refractive index change (b) as a function of frequency around a resonance.

Equation (160) is referred to as a Sellmeir equation. Generally speaking, A_0 is very close to unity. The other coefficients are normally determined by fitting dispersion data to Eq. (160) over a wide spectral range. The dependence of the index of refraction on frequency or wavelength ($\omega = 2\pi c/\lambda$) explains the material dispersion. The coefficients for fused silica are $A_0 = 1$, $A_1 = 0.6961663$, $A_2 = 0.4079426$, $A_3 = 0.8974794$, $\lambda_1 = 0.0684043$, $\lambda_2 = 0.1162414$, and $\lambda_3 = 9.896161$ where λ is in microns [2]. Using these coefficients, the index as a function of wavelength is plotted in Figure 2-11a. The group index, defined in Eq. (50), is also shown.

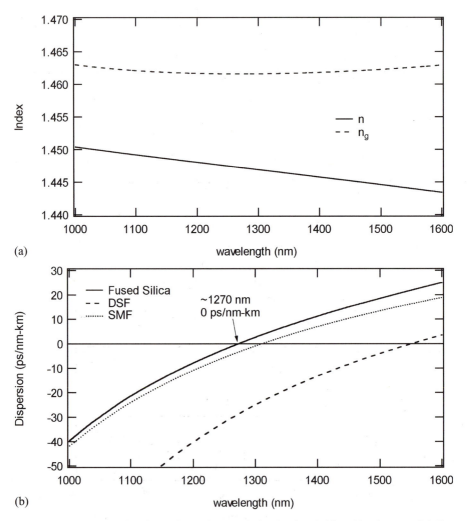

(a)

(b)

FIGURE 2-11   The refractive index and group index for fused silica (a). The material dispersion is compared to the total dispersion for a standard singlemode fiber (SMF) and a dispersion shifted fiber (DSF) in (b).

To derive the expression for material dispersion, let us consider a pulse of finite bandwidth traveling at the group velocity of $v_g = d\omega/dk$. Because

$$\omega = 2\pi v \text{ and } k = \frac{2\pi}{\lambda} \tag{161}$$

we have

$$v_g = \frac{dv}{d(1/\lambda)} = -\lambda^2 \frac{dv}{d\lambda} \tag{162}$$

By substituting

$$v = \frac{c}{n\lambda} \tag{163}$$

Equation (162) becomes

$$v_g = \frac{c}{n}\left(1 + \frac{\lambda}{n}\frac{dn}{d\lambda}\right) \tag{164}$$

Equation (164) describes the dependence of group velocity on the derivative of the refractive index, n, with respect to wavelength. The group velocity dispersion is

$$\frac{dv_g}{d\lambda} = \frac{c\lambda}{n^2}\left[\frac{d^2n}{d\lambda^2} - \frac{2}{n}\left(\frac{dn}{d\lambda}\right)^2\right] \tag{165}$$

or

$$\Delta v_g \approx \frac{dv_g}{d\lambda}\Delta\lambda = \frac{c\lambda}{n^2}\left[\frac{d^2n}{d\lambda^2} - \frac{2}{n}\left(\frac{dn}{d\lambda}\right)^2\right]\Delta\lambda \tag{166}$$

The dispersion in time, $\Delta\tau$, of an initially narrow pulse after traveling a distance L is given by

$$\Delta\tau = \frac{L}{v_{g2}} - \frac{L}{v_{g1}} = -L\frac{v_{g2} - v_{g1}}{v_{g2}v_{g1}} \approx -L\frac{\Delta v_g}{v_g^2} \tag{167}$$

By combining Eqs. (164), (166) and (167), we have

$$\Delta\tau = -L\left(\frac{c\lambda}{n^2}\left[\frac{d^2n}{d\lambda^2} - \frac{2}{n}\left(\frac{dn}{d\lambda}\right)^2\right]\Delta\lambda\right)\Big/\left(\frac{c}{n}\left(1 + \frac{\lambda}{n}\frac{dn}{d\lambda}\right)\right)^2 \tag{168}$$

Since most waveguide materials satisfy

$$\frac{\lambda}{n}\frac{dn}{d\lambda} \ll 1 \tag{169}$$

and

$$\frac{d^2 n}{d\lambda^2} \gg \frac{2}{n}\left(\frac{dn}{d\lambda}\right)^2 \tag{170}$$

Eq. (168) can be simplified to

$$\Delta\tau = -L\frac{\lambda}{c}\frac{d^2 n}{d\lambda^2}\Delta\lambda \tag{171}$$

By rearranging Eq. (171), we have

$$D = \frac{\Delta\tau}{L\Delta\lambda} = -\frac{\lambda}{c}\frac{d^2 n}{d\lambda^2} \tag{172}$$

which is the commonly used parameter of material dispersion measured in the units of ps/nm/km. It describes the pulse broadening in ps for each kilometer traveled by a pulse of spectral width equal to 1 nm. The material dispersion for fused silica, calculated from its Sellmeir equation, is shown in Figure 2-11b. It has zero dispersion at 1270 nm.

Before continuing the discussion of other types of dispersion, we present the Kramers-Kronig relations because they allow the determination of the real part of the response of a linear passive system if the imaginary part of the system response at all frequencies is known, and vice versa. It will be shown that the Kramers-Kronig relations are very important because they permit the determination of the immeasurable $\varepsilon_r(\omega) = \varepsilon_r'(\omega) - j\varepsilon_r''(\omega)$ from the reflectance $R(\omega)$ that can be measured directly as a function of frequency (or wavelength) in an optical experiment.

We start by defining the reflectance at a single interface as the ratio of the reflected to incident light intensity.

$$R(\omega) = r(\omega)r^*(\omega) = \left(\frac{E'}{E}\right)\left(\frac{E'}{E}\right)^* = [A(\omega)e^{j\theta(\omega)}][A(\omega)e^{-j\theta(\omega)}] = A(\omega)^2 \tag{173}$$

where the reflectivity $r(\omega)$ is the ratio of the reflected electric field to the incident electric field.

For normal incidence, it is straightforward to show that, by using the boundary conditions given by Eqs. (51) and (52), the reflectivity $r(\omega)$ can be described by

$$r(\omega) = \frac{\sqrt{\varepsilon_2} - \sqrt{\varepsilon_1}}{\sqrt{\varepsilon_2} + \sqrt{\varepsilon_1}} = \frac{n_2 - n_1}{n_2 + n_1} = \frac{n_{21} - 1}{n_{21} + 1} \tag{174}$$

By using the expression $n_{21} = n_{21}' - jn_{21}''$, we have

$$r(\omega) = \frac{n_{21}'(\omega) - 1 - jn_{21}''(\omega)}{n_{21}'(\omega) + 1 - jn_{21}''(\omega)} \tag{175}$$

In order to find the real and imaginary parts of $\varepsilon_r(\omega)$, it is necessary to find the real and imaginary parts of $n_{21}(\omega)$ because $\sqrt{\varepsilon_r(\omega)} = \sqrt{\varepsilon_r' - j\varepsilon_r''} = n_{21}'(\omega) - jn_{21}''(\omega)$. This can be done by using the Kramers-Kronig relations presented as follows.

Let us consider a Cauchy integral of the function $\varepsilon_r - 1$ in the form of

$$\varepsilon_r(\omega) - 1 = \frac{j}{\pi}P\int_{-\infty}^{\infty}\frac{\varepsilon_r(\omega') - 1}{\omega' - \omega}d\omega' \tag{176}$$

where P means integration over the principal part. This integration comes from Cauchy's theorem and is true for any function $F(\omega)$ that satisfies the following three conditions:

(1) The poles of $F(\omega)$ are all below the real axis;
(2) The real part of $F(\omega)$ is even and the imaginary part of $F(\omega)$ is odd with respect to the real axis ω;
(3) $F(\omega)$ approaches zero uniformly when $\omega \to \pm\infty$.

By equating the real parts of Eq. (176), we have

$$\begin{aligned}\varepsilon_r'(\omega) - 1 &= \frac{1}{\pi}P\int_{-\infty}^{\infty}\frac{\varepsilon_r''(\omega')}{\omega' - \omega}d\omega'\\ &= \frac{1}{\pi}P\left[\int_{-\infty}^{0}\frac{\varepsilon_r''(\omega')}{\omega' - \omega}d\omega' + \int_{0}^{\infty}\frac{\varepsilon_r''(\omega')}{\omega' - \omega}d\omega'\right]\end{aligned} \tag{177}$$

using the fact that $\varepsilon_r''(\omega')$ is odd with respect to the real axis ω, as seen in Eq. (153), we get

$$\begin{aligned}\varepsilon_r'(\omega) - 1 &= \frac{1}{\pi}P\left[\int_{\infty}^{0}\frac{-\varepsilon_r''(\omega')}{\omega' + \omega}d\omega' + \int_{0}^{\infty}\frac{\varepsilon_r''(\omega')}{\omega' - \omega}d\omega'\right]\\ &= \frac{2}{\pi}P\int_{0}^{\infty}\frac{\omega'\varepsilon_r''(\omega')}{\omega'^2 - \omega^2}d\omega'\end{aligned} \tag{178}$$

Similarly, by equating the imaginary parts of Eq. (176) and using the fact that $\varepsilon_r'(\omega') - 1$ is an even function, as seen in Eq. (152), we get

$$\varepsilon_r''(\omega) = -\frac{2\omega}{\pi}P\int_{0}^{\infty}\frac{\varepsilon_r'(\omega') - 1}{\omega'^2 - \omega^2}d\omega' \tag{179}$$

Equations (178) and (179) are the well known Kramers-Kronig relations.

Because Eqs. (178) and (179) can be applied to any function that satisfies the aforementioned three conditions, let us apply them to

$$r(\omega) = A(\omega)e^{j\theta(\omega)} = \sqrt{R(\omega)}e^{j\theta(\omega)} \tag{180}$$

or

$$\ln r(\omega) = \ln\sqrt{R(\omega)} + j\theta(\omega) \tag{181}$$

where $R(\omega)$ can be directly measured as a function of frequency. By using Eq. (179), the imaginary part of Eq. (181) can be determined from the real part by:

$$\theta(\omega) = -\frac{2\omega}{\pi}\mathrm{P}\int_0^\infty \frac{\ln\sqrt{R(\omega')}}{\omega'^2 - \omega^2}d\omega' \tag{182}$$

Once the real and imaginary parts of $r(\omega)$ are known, we can use Eq. (174) to determine $n_{21}(\omega) = n'_{21}(\omega) - jn''_{21}(\omega)$, and $\varepsilon_r(\omega) = n^2_{21}(\omega)$ by equating the real parts and imaginary parts on both sides of Eq. (174).

2.2.5.2 *Inter-Modal Dispersion*   Inter-modal dispersion results from the difference in group velocities of different modes. It is normally the dominant dispersion in multimode waveguides.

Because the higher order mode corresponds to a smaller incident angle θ as illustrated in Figure 2-9, it travels zigzaging through the guiding layer with an effective group velocity $c\sin\theta/n_{co}$ along the z-direction. The time needed to travel a distance L is

$$\tau = \frac{L}{c\sin\theta/n_{co}} \tag{183}$$

The lowest order mode, however, travels parallel to the z-direction with a group velocity of c/n_{co}. The dispersion between the higher order mode and the lowest order mode can be written as

$$\Delta\tau = \frac{L}{c\sin\theta/n_{co}} - \frac{L}{c/n_{co}} \tag{184}$$

For the cutoff mode or the highest order mode, the angle θ is equal to the critical angle for total internal reflection. Therefore

$$\sin\theta = \frac{n_{cl}}{n_{co}} \tag{185}$$

By inserting Eq. (185) into Eq. (184) we have

$$\Delta\tau = \frac{L}{cn_{cl}/n^2_{co}} - \frac{L}{c/n_{co}} = \frac{L}{c}\frac{n_{co}(n_{co} - n_{cl})}{n_{cl}} \tag{186}$$

For low index contrast waveguides, $n_{co} \approx n_{cl}$, we have

$$\Delta\tau = \frac{L}{c}(n_{co} - n_{cl}) \tag{187}$$

which is the maximum inter-modal dispersion for multi-mode waveguides.

Note that in reality the measured inter-modal dispersion is always less than the one predicted by Eq. (187). The reason is that there are many modes between the two extreme modes used for calculation and each mode carries only a small percent of the total energy for a signal pulse. The weighted contribution to the broadening of a signal pulse of the two extreme modes is therefore very small. Furthermore, a phenomena called mode coupling, which describes the energy transfer from one mode to another, leads also to smaller inter-modal dispersion because the fastest and the slowest extreme modes have no other choice but to couple their energy to the slower and faster modes, respectively. This mode coupling, in effect, reduces the inter-modal dispersion.

Mode coupling occurs when small dielectric perturbations such as dielectric non-uniformity and microbending of the waveguide lead to wave scattering and thus cause some rays to strike the cladding and guiding layer interface at angles different than the original ones, leading to energy transfer to other guided modes.

Inter-modal dispersion in a multimode waveguide is inevitable because of the aforementioned mode coupling. In other words, one can not eliminate inter-modal dispersion by transmitting a single mode signal in a multimode waveguide. Because even if a singlemode is excited in a multimode waveguide, multiple modes will propagate after the wave travels a finite distance due to the inevitable mode coupling.

2.2.5.3 *Waveguide Dispersion*

Waveguide dispersion is important because a pulse of finite temporal profile will always have a finite spectral width. Although it is generally the smallest dispersion among the three sources, it plays a critical role in singlemode waveguides, especially near the zero material dispersion region.

Consider a pulse propagating in a singlemode waveguide with a finite spectral bandwidth of Δk (or $\Delta\lambda$), each value of k within Δk will have a specific value of β associated with it, as determined by the eigenvalue equation. Each β will have a different group velocity. The dispersion of this singlemode, after traveling a distance L, is therefore

$$\Delta\tau = \frac{L}{v_{g,\text{slowest}}} - \frac{L}{v_{g,\text{fastest}}} = \frac{L}{c}\left[\left(\frac{d\beta}{dk}\right)_{\text{max}} - \left(\frac{d\beta}{dk}\right)_{\text{min}}\right] \approx \frac{L}{c}\frac{d^2\beta}{dk^2}\Delta k \tag{188}$$

where the dependence of β on k is described by the eigenvalue equation. Obviously, waveguide dispersion can only be obtained through numerical solution. For a standard singlemode fiber (SMF), the presence of germanium to raise the core index and waveguide dispersion moves the zero dispersion wavelength to ~1310 nm as shown in Figure 2-11b. Waveguides can be designed specifically to shift the dispersion zero, as indicated by the dispersion shifted fiber (DSF) response, or to flatten the dispersion response over a wavelength range. When all three dispersions exist simultaneously in the waveguide, the total pulse broadening requires a detailed calculation that involves the input pulse spectrum. A rough estimate of the total pulse broadening, however, can be made assuming the spectral shape can be approximated as a Gaussian pulse. The total effect becomes

$$\Delta\tau_{\text{sum}} = \sqrt{\Delta\tau_{\text{inter-modal}}^2 + (\Delta\tau_{\text{waveguide}} + \Delta\tau_{\text{material}})^2} \qquad (189)$$

Note that the effects of material and waveguide dispersion are added directly because they both depend on wavelength while the inter-modal dispersion depends largely on the characteristics of the modes.

2.2.5.4 Polarization Mode Dispersion

Polarization mode dispersion (PMD) is caused by birefringence. The birefringence of a material is defined as the maximum difference in the refractive index between any two polarization states. We have previously discussed linear polarizations; however, a general polarization state is defined as a combination of an x- and a y-polarized electric field with a relative phase delay φ between the components as follows: $\vec{E} = E_x\hat{x} + e^{j\varphi}E_y\hat{y}$. For example, a circularly polarized state is given by $E_x = E_y$ and $\varphi = \pm\frac{\pi}{2}$. The polarization states which correspond to the maximum and minimum index are the so-called principal states. For planar waveguides, the principal states are the TE and TM modes, and the birefringence is given by the difference in their effective indices $B = n_{TM} - n_{TE}$. In the absence of mode coupling, the PMD is defined as the differential delay resulting from propagation over a distance L and is found by applying Eq. (167) to the birefringence as follows:

$$PMD = \frac{\Delta\tau}{L} = \frac{1}{c}\left|B - \omega\frac{dB}{d\omega}\right| \qquad (190)$$

In isotropic materials, any asymmetrical stress will induce a birefringence through the photoelastic effect. The birefringence in optical fibers for long-haul optical transmission systems can be very small, on the order of $B \approx 10^{-6}$. The PMD in optical fibers is further reduced by mode coupling between polarizations, which arises from microbends and scattering just as for the inter-modal dispersion case. The differential delay must then be described in a statistical manner due to the random nature of the mode coupling. In this case, the delay is proportional to $\sqrt{L}$ instead of L [3]. Values for PMD in optical fibers are on the order of $0.1 \; ps/\sqrt{km}$.

2.2.5.5 The Dispersion Diagram

Because the phase velocity of each mode in a waveguide is

$$v_{p,m} = \frac{\omega}{\beta_m} \qquad (191)$$

and the information-carrying group velocity is

$$v_{g,m} = \frac{d\omega}{d\beta_m} \qquad (192)$$

where the subscript m stands for the mth mode in the waveguide, all the possible modes have a dispersion diagram confined between the maximum phase velocity c/n_{cl} and the minimum phase velocity c/n_{co} as shown in Figure 2-12 for TE modes in a symmetric slab waveguide. Phase velocities between the two boundaries are often described by the effective index of the waveguide such that

$$v_{p,m} = \frac{c}{n_{eff,m}} \qquad (193)$$

Clearly, the effective index for the mth mode is related to the corresponding propagation coefficient β by

$$n_{eff,m} = \frac{\beta_m}{k_0} \qquad (194)$$

where $k_0 = 2\pi/\lambda_0$ is the wave vector in vacuum. Note that for the TE_0 mode shown in Figure 2-12, there is no cutoff region, as discussed earlier. However, for TE_m modes with $m \neq 0$, a cutoff region exists for each mode. When the critical angle for total internal reflection is approached, the wave starts to travel largely outside the guide. Beyond the critical angle, the wave becomes unconfined, or cutoff. The other boundary, c/n_{co}, applies because the maximum velocity occurs when the wave is traveling well within the guiding layer.

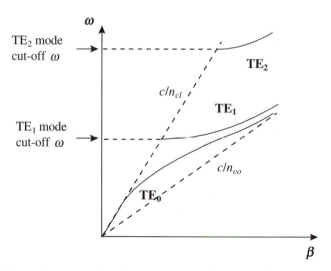

FIGURE 2-12 Schematic relationship between ω and β, i.e., the dispersion diagram, for the first few TE modes of a symmetric slab waveguide.

2.2.6 Loss and Signal Attenuation

In earlier sections we discussed how a mode becomes guided and propagates in a waveguide and how a signal is degraded by dispersion. The next obvious question is—what else may affect the quality of guided wave transmission.

Clearly, signal attenuation is the next issue to be addressed. In fact, while dispersion may sometimes be eliminated or compensated through careful waveguide design, attenuation is inherent to the waveguide and results ultimately in the loss of the signal. Attenuation is therefore a very important issue in the design and analysis of waveguide based devices.

We have been using the classical electromagnetic wave property or the associated ray picture in our earlier discussions. Loss mechanisms, however, can be best described by the particle nature of the electromagnetic fields. In this quantum mechanical picture, electromagnetic energy is carried by photons, a quantized unit of electromagnetic energy. When interacting with materials such as dielectric waveguides, photons can either be absorbed, losing their identity and energy to the absorbing particles, scattered, or radiated. Scattered and radiated photons represent loss of electromagnetic energy in terms of the total energy transmitted through the waveguide.

Basically, there are three fundamental types of loss: material absorption loss, scattering loss, and radiation loss. Other mechanical losses include coupling loss. In this section, we will briefly review the fundamental and mechanical losses and discuss their effects on waveguide devices.

2.2.6.1 Fundamental Absorption Loss

Fundamental absorption loss occurs via atomic or molecular absorption and band-to-band electronic absorption of electromagnetic energy.

In the earlier discussion of material dispersion, we derived an expression for the index of refraction based on the resonant transitions between atomic or molecular states and free carriers due to the external excitation source. We recall that $n = n' - jn''$ and rewrite Eq. (158) here for ease of reference.

$$n'' \approx \frac{1}{2}(\omega\gamma)/[(\omega_0^2 - \omega^2)\sqrt{1 + (\omega_0^2 - \omega^2)/(Nq^2/m\varepsilon_0)}] \qquad (195)$$

We see that n'' leads to an attenuation of the electric field propagating along the z direction, described by

$$E(z, t) = E_0 e^{j(\omega t - k_0 n z)} = (E_0 e^{-k_0 n'' z})e^{j(\omega t - k_0 n' z)} \qquad (196)$$

In other words, the amplitude of the wave will attenuate and the power intensity (power per cross sectional area) can be described by

$$I(z) = I_0 e^{-\alpha z} \qquad (197)$$

where α is related to the imaginary index of n'' and is called the attenuation coefficient. Obviously, the attenuation coefficient is related to n'' by

$$\alpha = 2k_0 n'' \tag{198}$$

Many dielectric materials and semiconductors are characterized by their forbidden energy bandgaps which separate bound electrons in the valence band from free electrons in the conduction band. When the photon energy is high enough, a valence electron can be excited to the conduction band by absorbing the photon energy. The empirical absorption coefficient α for this band-to-band absorption depends on the photon energy E and has the form of

$$\alpha = \alpha_0 e^{E/E_0} \tag{199}$$

where E_0 and α_0 are material dependent constants. Equation (199) is often called Urbach's law of band-to-band absorption.

2.2.6.2 Fundamental Scattering Loss
Scattering loss occurs when electromagnetic waves interact with scattering centers of a size smaller than the wavelength. Because of the small size of the scattering centers, which can be due to impurity clusters and localized dielectric fluctuations, they can be viewed as uniformly excited by the field. This field excitation of scattering centers leads to the formation of dipoles that can "scatter" or radiate electromagnetic energy out of the waveguide.

Two dominant scattering mechanisms exist in dielectric waveguides: Rayleigh (or volume) scattering largely due to fluctuation of index of refraction and surface scattering that results from surface roughness.

Most dielectric waveguide materials do not have single-crystal structures. They are "disordered" amorphous materials. The disorder can be structural, where some basic molecular units are connected in a random way, or compositional in which chemical composition varies from place to place. In both cases, the net effect is a fluctuation of the refractive index over dimensions which can be on the scale of 10% of the wavelength or even less. Such small "particles" act as point scattering centers, and scattering of this type is commonly referred to as Rayleigh scattering.

In Rayleigh scattering, each "particle" is uniformly excited by the field and becomes polarized with a dipole moment. Because the power radiated by a dipole is inversely proportional to λ^4_0, the absorption coefficient also varies with λ_0^{-4} and is given by [4]:

$$\alpha_R = \frac{8\pi^3}{3\lambda_0^4} n^8 C^2 L k_B T_F \tag{200}$$

where n is the index of refraction, C the photoelastic coefficient, L the isothermal compressibility at T_F, the temperature at which the disorder "particle" freezes in the dielectric material, and k_B is Boltzmann's constant.

Rayleigh scattering decreases rapidly as the wavelength increases. For silica fibers, this fundamental Rayleigh scattering limit has been reached for $\lambda \leq 1.5$ μm, except in limited regions where impurity absorption such as the hydroxyl OH⁻ ion absorptions at 0.95, 1.24, and 1.39 μm dominate. For planar waveguides, however, there is an added source of scattering loss resulting from the rough interface between the guiding core and the cladding layer because the core is often formed by etching. Etching processes, as discussed in section 2.5.2, contribute significantly more roughness compared to optical fiber cores which are drawn from a preform. The planar waveguide interface scattering loss increases approximately as the square of the refractive index difference [5,6].

2.2.6.3 Waveguide Bending Loss   Waveguide bending is essential for integrated optical circuits. Each curved path involves two types of losses: pure bending loss and transition loss. Pure bending loss occurs in regions where the waveguide has a constant radius of curvature. Transition loss occurs when the waveguide has a discontinuity in its bending curvature.

Figure 2-13 shows a schematic view of the TE_0 mode propagating in a circular bend of radius R for a symmetrical waveguide. The constant phase front of the mode is represented by the dotted line. The waveguide thickness is h.

It is clear that if the phase front is to remain planar, the tangential speed of the wavefront must increase as the radial distance is increased. At a certain point, the tangential speed approaches the speed of light, beyond which the tail of the mode has to break away from the guided mode and radiate outside the guiding layer. The locus of the breaking points is called the radiation caustic.

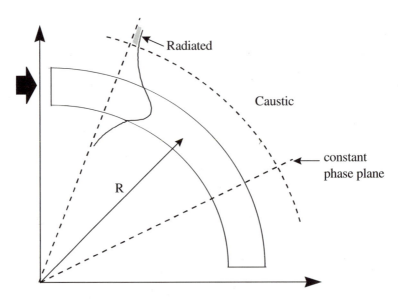

FIGURE 2-13   Schematic view of TE_0 mode propagating in a circular bend. To maintain the constant phase plane, the shadowed area radiates out of the waveguide.

This simple physical picture of pure bending loss can be described by the concept of an equivalent index profile. Because the wavefront of the mode has to travel at a faster and faster speed away from the center of curvature in order for the wave as a whole to be guided, the wavefront behaves as if the index of refraction of the waveguide is varying along the radial direction r. It has been shown that a bent waveguide with radius R behaves like a straight waveguide, to a first order approximation, with an equivalent refractive index profile [7]

$$n_{\text{equ}}(x) = n(x) \sqrt{1 + \frac{2x}{R}} \qquad \text{for } x \ll R \qquad (201)$$

where $n(x)$ is the refractive index profile for the waveguide. Figure 2-14 shows the schematic equivalent index profile $n_{\text{equ}}(x)$ as compared to the waveguide index profile. As discussed earlier, a mode is guided when $n_{\text{cl}} < \beta/k_0 < n_{\text{co}}$. Due to the increase in the equivalent index, we see that n_{equ} becomes larger than β/k_0 at the radiation caustic. Obviously, modes near cutoff will suffer larger radiation loss because of their smaller β value.

A quantitative expression for the pure bend loss coefficient α has been derived by Marcuse [8]. Assuming the power intensity attenuates in the form of Eq. (197) after propagating a length z in a bend of radius R, the attenuation coefficient α is given by

$$\alpha = \frac{\gamma \kappa^2 e^{h\gamma} e^{-U}}{(n_{\text{co}}^2 - n_{\text{cl}}^2) k_0^2 \beta (h/2 + \gamma^{-1})} \qquad (202)$$

where

$$U = \left[\frac{\beta}{\gamma} \ln\left(\frac{1 + \gamma\beta^{-1}}{1 - \gamma\beta^{-1}} \right) - 2 \right] \gamma R \approx \frac{2\gamma^3 R}{3\beta^2} \qquad (203)$$

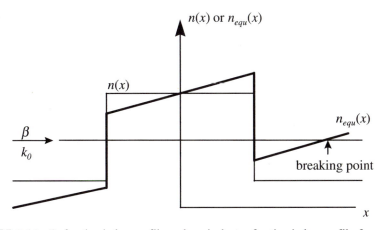

FIGURE 2-14 Refractive index profile and equivalent refractive index profile for a symmetrical slab bending waveguide of radius R.

The approximation is based on the assumption that $\gamma\,\beta^{-1} \ll 1$. This formula assumes a weak-guidance simplification and is valid for

$$n_{cl}k_0\left(1 + \frac{h}{2R}\right) < \beta < n_{co}k_0\left(1 - \frac{h}{2R}\right) \qquad (204)$$

Clearly, bending loss increases exponentially as the bending radius R decreases. Once the bend loss starts to increase, it rises so sharply that a small change in the bend radius can have a dramatic effect on loss. A bend loss model is typically used to estimate the critical radius, below which the bend loss increases dramatically; however, it is not expected that the absolute loss at a particular radius smaller than the critical radius can be closely estimated, since it depends strongly on the fabricated waveguide parameters and wavelength. Measured data on bend loss versus radius is shown in Figure 2-15 for Ge-doped silica planar waveguides with two different core-to-cladding index of refraction differences [6]. To provide a margin of safety from the loss edge, a minimum bend radius of 2 mm for the $\Delta = 1.5\%$ and 4 mm for the $\Delta = 0.75\%$ may be chosen for a design. The bend loss formula for the slab can be applied to a two-dimensional waveguide by employing the effective index method, discussed in Section 2.3.2, to obtain an equivalent slab structure which approximates the two-dimensional waveguide.

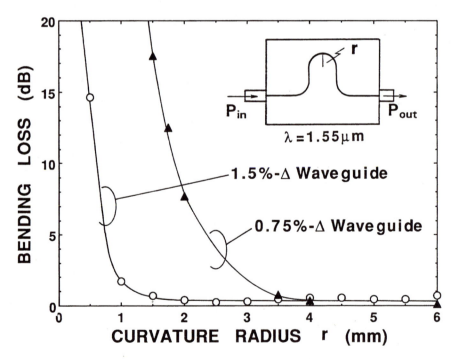

FIGURE 2-15   Bend loss dependence on the radius of curvature for $\Delta = 1.5\%$ guides with $4.5 \times 4.5\ \mu m^2$ cores and $\Delta = 0.75\%$ guides with 6.5×6.5 cores [6]. Reprinted by permission. Copyright © 1994 IEEE.

Transition losses occur at locations where the waveguide's curvature changes abruptly. Because the mode field varies with curvature, there is loss associated with the coupling of modes between waveguide sections with different curvatures. The mode field becomes asymmetric in shape as R decreases and is offset to the outside of the bend by a distance x_0 that can be estimated as follows [9]:

$$x_0 = \frac{8\pi^2 n_{cl}^2 W_w^4}{\lambda^2 R} \tag{205}$$

where w_w is the waveguide mode spot size (defined in Section 2.2.6.4). The transition loss increases with core size [10], which is a result of the increase in spot size. For a transition with a change in the sign of the curvature, the loss is quadrupled since the offsets will be in opposite directions and the loss is proportional to the square of the offset. Transition losses can be minimized by introducing lateral offsets between the waveguides which compensate for the mode offsets between the two sections.

2.2.6.4 Mode Coupling Losses

Mode coupling losses between different waveguides result from mismatches in the spatial dependence of the modes and from any translational or angular offsets between the guides. The loss between two modes is given by the overlap integral

$$\text{Loss} = -10 \log_{10} \left[\frac{\left(\int_{\text{area}} \phi_1 \phi_2 dA \right)^2}{\int_{\text{area}} \phi_1^2 dA \int_{\text{area}} \phi_2^2 dA} \right]$$

where the mode fields are denoted by $\phi_{1,2}$ and the integrals are over the waveguide cross-section. A convenient method for estimating mode mismatch loss between fundamental modes is to approximate each mode by a Gaussian function $\hat{\phi}(r) \approx e^{-(r/w)^2/2}$ where w represents the mode spot size and r is the radius or transverse coordinate. Note that there is no azimuthal dependence. The coupling losses due to mode field mismatch and losses between identical waveguides due to a transverse offset of x_0 microns or an angular offset of θ_0 radians are given in Table 2–1 [11]. For the offsets, $w = \sqrt{(w_w^2 + w_f^2)/2}$.

Simple formulas are available for estimating the spot size for the fundamental modes of step index planar waveguides and fibers. The following spot size approximation is used for step index fibers [12]:

$$w_f = \frac{a}{\sqrt{2}} [0.65 + 1.619 V^{-3/2} + 2.879 V^{-6}] \tag{206}$$

where a is the core radius. This approximation is good to within 1% error over a range of $1.20 \leq V \leq 3.06$ [12]. The spot size for a standard singlemode fiber with $\Delta = 0.27\%$ and $a = 4.45$ µm is estimated to be 3.59 µm at $\lambda = 1300$ nm and 4.15 µm

TABLE 2-1   Coupling Loss Mechanisms and Formulas

Loss Mechanism	Loss in dB
Mode Mismatch	$-20\log\left(\dfrac{2w_f w_w}{w_f^2 + w_w^2}\right)$
Angular Offset	$8.69\left(\dfrac{\pi n_{cl} w \theta_0}{\lambda}\right)^2$
Transversal Offset	$2.17\left(\dfrac{x_0}{w}\right)^2$

at $\lambda = 1550$ nm. For dispersion shifted fiber with equivalent step parameters of $\Delta = 0.75\%$ and $a = 1.85\,\mu m$, the spot size is estimated to be $2.15\ \mu m$ at $\lambda = 1300$ nm and $2.92\,\mu m$ at $\lambda = 1550$ nm. The spot size for a square waveguide can be estimated using the following transcendental equation [13]:

$$\exp\left(\frac{a^2}{w_w^2}\right) = \frac{2w_w V^2}{a\sqrt{\pi}} \tag{207}$$

where $a = h/2$. For a $6.0 \times 6.0\ \mu m$ waveguide with $\Delta = 0.7\%$, the spot size is estimated to be $2.32\ \mu m$ at $\lambda = 1300$ nm and $2.52\ \mu m$ at $\lambda = 1550$ nm.

2.2.6.5 *Whispering Gallery Mode Loss*   In this section, we discuss bend loss as a function of the core width. As the core width increases, the field is guided only by the outer core-to-cladding interface and is called a whispering gallery mode (WGM). The name was first used by Lord Rayleigh for sound waves guided along a cathedral ceiling [14]. The approximations for Eqs. (202) and (203) are not valid in the wide waveguide regime as $h \rightarrow R$ as seen in Eq. (204), so an alternate approach is needed. The asymptotic limit for bend loss as $h \rightarrow R$ from Marcatili's analysis [10] is

$$\alpha_c R = \left(\frac{n_{cl}}{n_{co}}\right)^2\left[1 - \left(\frac{n_{cl}}{n_{co}}\right)^2\right]^{-\frac{1}{2}}\exp\left\{\frac{-\Re}{3}\left[1 - \left(\frac{9\pi}{2\Re}\right)^{\frac{2}{3}} + \frac{4}{\Re}\left(\frac{n_{cl}}{n_{co}}\right)^2\right]^{\frac{3}{2}}\right\} \tag{208}$$

where

$$\Re \cong 2k\,n_{co}\left[1 - \left(\frac{n_{cl}}{n_{co}}\right)^2\right]^{\frac{3}{2}} R \tag{209}$$

and α_c is the field attenuation in nepers/meter, R is measured from the center of curvature to the outside waveguide wall, and the loss per loop (dB/loop) is $8.686\alpha_c L$ where L is the loop perimeter in meters.

Figure 2-16 plots the predicted bend radius for a WGM having a loss of

0.1dB/loop versus index of refraction difference. A plot of bend radius for a quasi-singlemode, square waveguide having 0.1 dB/loop and a core size varying with Δ to maintain a constant $V = 2.3$ is also included, which is based on the effective index method (discussed in Section 2.3.2) and Eqs. (202) and (203). At $\Delta = 0.75\%$, the improvement between the singlemode square waveguide and the WGM limit is a factor of 1.65. Design methods have been reported for transitioning between single-mode and WGM (highly multimoded) waveguide sections so that smaller bend radii can be used while minimizing transition losses and the potential for exciting higher order modes [15].

Marcatili [10] also provides an estimated width, given in Eq. (210), beyond which the reduction in loss for a wider guide is negligible.

$$w_{\mathrm{WGM}} \cong \left(\frac{R\lambda^2}{\pi n_{\mathrm{co}}^2} \right)^{\frac{1}{3}} \qquad (210)$$

For a bend radius of 2.75 mm and $\lambda = 1550$ nm, $w_{\mathrm{WGM}} \cong 10$ μm. The improvement in bend loss performance for WGM operation versus quasi-singlemode opera-

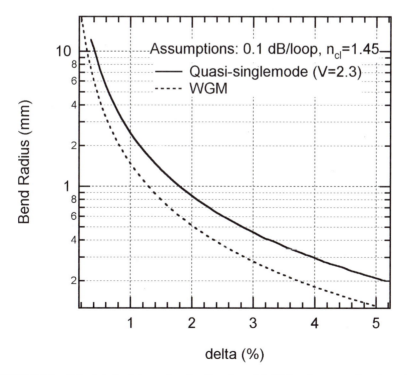

FIGURE 2-16   Calculated minimum bend radius for a 0.1 dB loss/loop as a function of the relative core-to-cladding index difference assuming $n_{\mathrm{cl}} = 1.45$ for quasi-singlemode operation versus the WGM regime.

tion has been explored experimentally [16]. The measurement approach and results are discussed in Chapter 7.

2.3   RECTANGULAR WAVEGUIDES

Planar one-dimensional waveguides confine waves only in one dimension and are therefore not practical for planar integration. By introducing the second dimensional confinement, rectangular or two-dimensional waveguides with other cross-sections are formed which lay the foundation for planar integrated optics.

There are many types of rectangular waveguides. Three of the commonly used structures are shown in Figure 2-17. The raised strip waveguide shown in Figure 2-17a is formed by an etching process, removing the high index thin films on both sides of the strip. The cladding material can be formed by air or another dielectric material with a dielectric constant n_{cl} lower than that of the strip. The buried strip or channel waveguide shown in Figure 2-17b can be formed by ion implantation, diffusion, or overcladding. The guiding layer can be completely surrounded by a cladding material if formed by deep ion implantation. Figure 2-17c depicts a ridge waveguide where the film on both sides of the ridge is not completely removed.

Whatever the structure may be, all rectangular waveguides confine waves in both lateral dimensions and waves are essentially bounced back and forth between four boundaries that may or may not have the same dielectric constants.

In Sections 2.3.1–2.3.3, several methods will be presented for analyzing rectangular waveguides. The first method is an approximate analytical solution to the wave equation developed by Marcatili [17], which ignores the corner regions of the rectangular waveguide based on the assumption that the modes are far from cutoff or the index difference between the cladding and guiding materials is small so that the fields are essentially transverse. The second method is known as the effective index method, developed by Hocker and Burns [18], which converts a two-dimensional waveguide to two orthogonally oriented one-dimensional waveguides with direct interaction. The final method to be discussed is the perturbation method, which includes the effect of corner fields neglected in the analytical method.

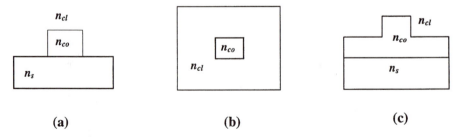

(a)   **(b)**   **(c)**

FIGURE 2-17   Three commonly used rectangular waveguide structures: raised strip waveguide (a), buried strip or channel waveguide (b), and rib or ridge waveguide (c).

2.3.1   Wave Equation Analysis

Consider a general rectangular waveguide structure shown in Figure 2-18 where four different dielectric materials surround the waveguide core. If the shadowed corner regions are neglected, there will be no coupling between the three components of the wave and the vector wave equation can be decomposed into a scalar equation for each of the three components. Furthermore, if the index difference between the cladding and core materials is small or if modes far from cutoff are considered, the waves are essentially transverse waves. Consequently, the scalar wave equation for each of the wave components becomes, assuming an $\exp(-j\beta z)$ dependence for the longitudinal component,

$$\frac{\partial^2 E(x, y)}{\partial x^2} + \frac{\partial^2 E(x, y)}{\partial y^2} + [k_0^2 n_i^2 - \beta^2]E(x, y) = 0 \qquad i = 1,2,3,4,5 \quad (211)$$

where n_i is the index of refraction for the region under consideration, and β is the eigenvalue for each possible mode. By employing the separation of variables method, we can solve Eq. (211) in a straightforward manner for each region of the waveguide. Let $E_i(x, y) = A_i X_i(x) Y_i(y)$ with $X_i(x)$ and $Y_i(y)$ describing the field amplitudes along the x- and y-direction, respectively. The solution for each of the five regions ($i = 1,2,3,4,5$) can be obtained as follows.

$$X_1(x) = \cos(\kappa_x x + \phi_x) \quad \text{and} \quad Y_1(y) = \cos(\kappa_y y + \phi_y) \quad \text{for } n_1 \text{ region} \quad (212)$$

$$X_2(x) = \exp(\gamma_2(x + h)) \quad \text{and} \quad Y_2(y) = \cos(\kappa_y y + \phi_y) \quad \text{for } n_2 \text{ region} \quad (213)$$

$$X_3(x) = \exp(-\gamma_3 x) \quad \text{and} \quad Y_3(y) = \cos(\kappa_y y + \phi_y) \quad \text{for } n_3 \text{ region} \quad (214)$$

$$X_4(x) = \cos(\kappa_x x + \phi_x) \quad \text{and} \quad Y_4(x) = \exp(-\gamma_4(y - w)) \quad \text{for } n_4 \text{ region} \quad (215)$$

$$X_5(x) = \cos(\kappa_x x + \phi_x) \quad \text{and} \quad Y_5(y) = \exp(\gamma_5 y) \quad \text{for } n_5 \text{ region} \quad (216)$$

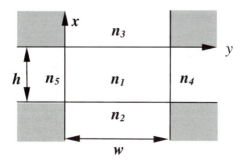

FIGURE 2-18   A general rectangular waveguide of cross sectional area $h \times w$.

where the separation constants κ_x and κ_y satisfy

$$\kappa_x^2 + \kappa_y^2 = k_0^2 n_1^2 - \beta^2 \tag{217}$$

and γ_2, γ_3, γ_4, γ_5 are defined as

$$\gamma_2 = \sqrt{k_0^2(n_1^2 - n_2^2) - \kappa_x^2} \tag{218}$$

$$\gamma_3 = \sqrt{k_0^2(n_1^2 - n_3^2) - \kappa_x^2} \tag{219}$$

$$\gamma_4 = \sqrt{k_0^2(n_1^2 - n_4^2) - \kappa_y^2} \tag{220}$$

$$\gamma_5 = \sqrt{k_0^2(n_1^2 - n_5^2) - \kappa_y^2} \tag{221}$$

Notice that the solutions in the guiding core are oscillatory while at least one of the wave amplitude components, $X(x)$ or $Y(y)$, must be exponentially decaying in the other four regions outside the guiding core to satisfy the boundary conditions. In the shadowed four corners, the solutions for both x and y components should be exponentially decaying.

By utilizing the proper boundary conditions for the transverse fields with the electric field polarized along the x direction and the magnetic field polarized along the y direction, we can obtain the characteristic equations for both κ_x and κ_y as

$$\tan \kappa_x h = \frac{n_1^2(n_2^2 \gamma_3 + n_3^2 \gamma_2)}{n_2^2 n_3^2 \kappa_x^2 - n_1^4 \gamma_2 \gamma_3} \kappa_x \tag{222}$$

and

$$\tan \kappa_y w = \frac{\gamma_4 + \gamma_5}{\kappa_y^2 - \gamma_4 \gamma_5} \kappa_y \tag{223}$$

By combining these characteristic equations, we can solve for κ_x and κ_y and therefore the propagation constant β using Eq. (217). The phase constants ϕ_x and ϕ_y are determined by [19]

$$\tan \phi_x = -\frac{n_3^2 \kappa_x}{n_1^2 \gamma_3} \tag{224}$$

and

$$\tan \phi_y = -\frac{\gamma_5}{\kappa_y} \tag{225}$$

For transverse fields with the electric field polarized along the y direction and the magnetic field polarized along the x direction, the characteristic equations can similarly be derived as

$$\tan \kappa_x h = \frac{\gamma_2 + \gamma_3}{\kappa_x^2 - \gamma_2 \gamma_3} \kappa_x \tag{226}$$

and

$$\tan \kappa_y w = \frac{n_1^2(n_5^2 \gamma_4 + n_4^2 \gamma_5)}{n_4^2 n_5^2 \kappa_y^2 - n_1^4 \gamma_4 \gamma_5} \kappa_y \tag{227}$$

The phase constants of ϕ_x and ϕ_y in this case are determined by

$$\tan \phi_x = \frac{\gamma_3}{\kappa_x} \tag{228}$$

and

$$\tan \phi_y = \frac{n_5^2 \kappa_y}{n_1^2 \gamma_5} \tag{229}$$

For a square, channel waveguide of dimension $h \times h$, the normalized frequency V is defined by

$$V = k_0 \frac{h}{2} \sqrt{n_{co}^2 - n_{cl}^2} \tag{230}$$

Caution should be taken when applying the results of the wave equation solutions near cutoff where the assumptions are not valid. In such a case, either the effective index method or the perturbation approach should be taken to more closely estimate the propagation constant β. Numerical techniques, such as the modified Fourier decomposition method and finite element method, yield a second-mode cutoff of $V = 2.136$ for singlemode operation of a square channel waveguide [13].

2.3.2   The Effective Index Method

When the modes are near cutoff where the direct wave equation analysis can not be used to obtain accurate solutions, the effective index method becomes attractive. In this approach, the rectangular waveguide is viewed as two orthogonally oriented slab waveguides, and the problem becomes one-dimensional, instead of two-dimensional. The effective index method allows the determination of universal dispersion curves for rectangular waveguides in terms of the normalized slab waveguide parameters (V, b_n, and a).

Figure 2-19 depicts the basic idea of the effective index method: a general rectangular waveguide is decomposed into two orthogonally-oriented slab waveguides. The thinner horizontal slab waveguide maintains the same core index of refraction while the thicker vertical slab waveguide is solved by replacing the core index of refraction by the effective index of the horizontal slab waveguide, defined by $n_{eff} = \beta_s/k_0$. The procedure for the effective index method is therefore to solve the charac-

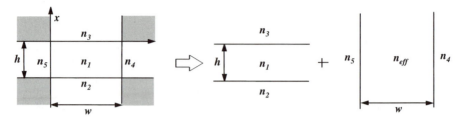

FIGURE 2-19   Basic idea of the effective index method: decomposing a rectangular waveguide into two slab waveguides.

teristic equation for the TE or TM mode of the thinner slab waveguide as discussed in Section 2.2. Once β_s is found, the effective index n_{eff} is calculated as β_s/k_0 and used as the core index of refraction for the thicker slab waveguide. The corresponding characteristic equation for the mode under consideration in the vertical slab thicker waveguide is then solved to find the propagation constant β, which is the approximate propagation constant for the rectangular waveguide. By using the effective index of refraction, the interaction between the two slab waveguides is taken into consideration. In other words, the effect of the corner regions neglected in the former analysis is now considered.

2.3.3   Perturbation Corrections

Another way to obtain more accurate solutions to the rectangular waveguide problem is the perturbation approach. The basic idea of the perturbation approach is to replace the index of refraction profile $n(x,y)$ by an approximate index of refraction $n_{\text{app}}(x,y)$ for which an analytical solution to the wave equation can be obtained. If $n_{\text{app}}(x,y)$ is very close to $n(x,y)$, the analytical result will be very close to the true wave equation solution. Based on the completeness of modes, the true wave mode can be expressed approximately in terms of the superposition of the complete set of the analytical solutions. If $E_t(x,y)$ is the true orthonormal solution to the wave equation

$$\nabla^2 E_t(x,y) + [k_0^2 n^2(x,y) - \beta^2]E_t(x,y) = 0 \tag{231}$$

and $E_p(x,y)$ is one of the analytical orthonormal solutions to the wave equation with $n(x,y)$ replaced by $n_{\text{app}}(x,y)$

$$\nabla^2 E_j(x,y) + [k_0^2 n_{\text{app}}^2(x,y) - \beta_j^2]E_j(x,y) = 0 \quad j = 1, 2, 3, \ldots \tag{232}$$

$E_t(x,y)$ can be expanded in terms of the set of analytical solutions because of their completeness. Therefore,

$$E_f(x,y) = E_p(x,y) + \sum_{j}^{p \neq j} c_j E_j(x,y) \qquad (233)$$

If $E_p(x,y)$ is very close to $E_f(x,y)$, the second term in Eq. (233) can be regarded as perturbation correction to $E_p(x,y)$ due to the small difference between $n(x,y)$ and $n_{app}(x,y)$. It can be shown that the perturbation correction to the value of the analytical solution β_p is [20]

$$\Delta\beta^2 = \beta^2 - \beta_p^2 = k_0^2 \frac{\int_A (n^2(x,y) - n_{app}^2(x,y))E_p^2(x,y)dA}{\int_A E_p^2(x, y)dA} \qquad (234)$$

and the expression for the expansion coefficients can be written as

$$c_j = \frac{k_0^2}{\beta_p^2 - \beta_j^2} \int_A [n^2(x,y) - n_{app}^2(x,y)]E_p(x,y)E_j(x,y)dA \qquad (235)$$

where $j \neq p$. The above perturbation expressions can be applied to waveguides of arbitrary cross section, and the integration only needs to be carried out over the regions where the approximated index profile is different from the actual one.

2.4 SPLITTERS AND COMBINERS

Splitters and combiners are essential components for optical filters. A directional coupler is one example of a splitter and combiner for guided waves. Another example is a partial reflector, which divides the incident light between a reflected and transmitted output. Partial reflectors are discussed in Chapter 5. This section presents the basic analysis of the directional coupler, starting with the coupled-mode equations. Both the directional coupler and the partial reflector have two inputs and two outputs. Star couplers and multi-mode interference couplers are also discussed in this section. They are used to realize splitters and combiners with a larger number of input and output ports.

2.4.1 Directional Couplers

A directional coupler is formed by placing two waveguides in close proximity. Coupling occurs as a result of the overlap between the evanescent field of one waveguide and the core of the other waveguide. Figure 2-20 shows the field distributions and the refractive index profiles for two symmetrical slab waveguides along the z-direction with guiding layer refractive indices of n_{c1} and n_{c2} and a refractive index n_s everywhere outside the guiding layers. We assume that the waveguides are single-

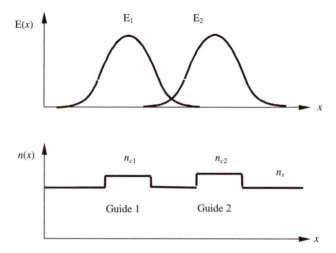

FIGURE 2-20   Schematic illustration of field distributions (top) and refractive index profile (bottom) for two closely positioned symmetric slab waveguides.

moded and the electric field is polarized along the y-direction. The wave equation for each of the singlemode isolated waveguide is

$$\frac{\partial^2 E_i(x,y)}{\partial x^2} + \frac{\partial^2 E_i(x,y)}{\partial y^2} + [k_0^2 n_{ci}^2(x,y) - \beta_i^2]E_i(x,y) = 0 \quad (i = 1, 2) \tag{236}$$

and the isolated modes can be written as,

$$E_i(x,y,z) = E_i(x,y)e^{-j\beta_i z} \quad (i = 1, 2) \tag{237}$$

E_i and β_i have been solved in Section 2.3.1. We use them to determine the coupling when the two waveguides are not isolated from each other.

The coupler wave equation for the y-polarized electric field can be written as

$$\nabla^2 E(x,y,z) + k_0^2 n^2(x,y)E(x,y,z) = 0 \tag{238}$$

where $n(x,y)$ is the refractive index profile for the whole coupler. Since the coupling from one waveguide to the other occurs slowly due to the overlap of the evanescent waves, the solution for Eq. (238) can be described as a superposition of the two isolated modes:

$$E(x,y,z) = C_1(z)E_1(x,y)e^{-j\beta_1 z} + C_2(z)E_2(x,y)e^{-j\beta_2 z} \tag{239}$$

where C_1 and C_2 describe the modification to the mode amplitude as a result of coupling. Notice that they are assumed to be a function of z, as we expect the power coupling between the two waveguides to be z-dependent.

By inserting Eq. (239) for $E(x,y,z)$ into the coupler wave equation and making use of the solutions for the isolated waveguides, we have

$$
\left[\frac{d^2C_1}{dz^2} - 2j\beta_1 \frac{dC_1}{dz} + k_0^2(n^2 - n_{c1}^2)C_1 \right] E_1 e^{-j\beta_1 z}
$$

$$
+ \left[\frac{d^2C_2}{dz^2} - 2j\beta_2 \frac{dC_2}{dz} + k_0^2(n^2 - n_{c2}^2)C_2 \right] E_2 e^{-j\beta_2 z} = 0 \qquad (240)
$$

The second-order derivative terms can be neglected because the coupling mode envelopes vary slowly with distance. Equation (240) simplifies to

$$
\left[-2j\beta_1 \frac{dC_1}{dz} + k_0^2(n^2 - n_{c1}^2)C_1 \right] E_1 e^{-j\beta_1 z}
$$

$$
+ \left[-2j\beta_2 \frac{dC_2}{dz} + k_0^2(n^2 - n_{c2}^2)C_2 \right] E_2 e^{-j\beta_2 z} = 0 \qquad (241)
$$

Equation (241) can be further simplified by multiplying on both sides by the complex conjugate of the transverse field in waveguide 1, E_1^*(or in waveguide 2, E_2^*) and integrating the equation over the whole coupler cross section. The result is

$$
\left[-2j\beta_1 \frac{dC_1}{dz} \int_A E_1 \cdot E_1^* dA + C_1 k_0^2 \int_A (n^2 - n_{c1}^2)E_1 \cdot E_1^* dA \right] e^{-j\beta_1 z}
$$

$$
+ \left[-2j\beta_2 \frac{dC_2}{dz} \int_A E_2 \cdot E_1^* dA + C_2 k_0^2 \int_A (n^2 - n_{c2}^2)E_2 \cdot E_1^* dA \right] e^{-j\beta_2 z} = 0 \quad (242)
$$

Note that $(n^2 - n_{c1}^2)$ describes the perturbation to the dielectric constant of waveguide 1 caused by the presence of nearby waveguide 2. This perturbation needs to be considered only in the neighborhood of waveguide 2 where E_1 exists in the form of an evanescent wave and is very small. The second integral in Eq. (242) can, therefore, be neglected. The third integral can also be neglected if we assume the spatial overlap of the two waveguide modes is small, which is true if the waveguides are not too close to each other. Consequently, we have

$$
-2j\beta_1 \frac{dC_1}{dz} \int_A E_1 \cdot E_1^* dA + C_2 \left[k_0^2 \int_A (n^2 - n_{c2}^2)E_2 \cdot E_1^* dA \right] e^{-j\Delta\beta z} = 0 \qquad (243)
$$

where $\Delta\beta = \beta_2 - \beta_1$ describes the difference in the propagation constants between the two waveguides. Defining κ_{12}, known as the coupling coefficient from waveguide 2 to 1, by

$$
\kappa_{12} = \frac{k_0^2}{2\beta_1} \int_A (n^2 - n_{c2}^2)E_2 \cdot E_1^* dA \bigg/ \int_A E_1 \cdot E_1^* dA \qquad (244)
$$

we get the first coupled-mode equation

$$\frac{dC_1}{dz} + j\kappa_{12}C_2e^{-j\Delta\beta z} = 0 \tag{245}$$

When $\Delta\beta$ is small, we can assume that $\beta_1 \cong \beta_2 \cong \beta$ in the definition of κ_{12} in Eq. (244). It is straightforward to see that, as a result of the symmetry of the coupler, the second coupled-mode equation can be derived as

$$\frac{dC_2}{dz} + j\kappa_{21}C_1e^{+j\Delta\beta z} = 0 \tag{246}$$

and the coupling coefficient, κ_{21}, can also be calculated by

$$\kappa_{21} = \frac{k_0^2}{2\beta}\int_A (n^2 - n_{c1}^2)E_1 \cdot E_2^* dA \bigg/ \int_A E_2 \cdot E_2^* dA \tag{247}$$

For identical waveguides, $\kappa_{12} = \kappa_{21} = \kappa$. The mode amplitudes, C_1 and C_2, are coupled through the coupling coefficient κ. The critical factor that affects the magnitude of κ is the integral in the numerator. It is obvious that, in order for κ to be large, the evanescent tail of field E_2 has to reach a significant region of waveguide 1 and vice versa. Basically, it suggests that the spacing between the two waveguides has the largest effect on the coupling coefficient, as expected physically. It should also be obvious that κ depends exponentially on the waveguide spacing because the evanescent tails of E_1 and E_2 decrease exponentially. In order for the assumptions made earlier to be true and for κ not to be too small, the spacing between the waveguides of a coupler should be in the order of the waveguide width.

The coupled-mode Eqs. (245) and (246) provide the basis for the design of directional couplers because power coupling between the two waveguides in the coupler can be obtained by solving these two equations with the proper boundary conditions. By assuming that light is input into waveguide 1 only with unit modal amplitude, we can show that, when $\Delta\beta = 0$ (identical waveguides), the solutions to the coupled-mode equations are

$$C_1(z) = \cos \kappa z \tag{248}$$

and

$$C_2(z) = -j \sin \kappa z \tag{249}$$

It is obvious that the power in waveguide 1 is

$$P_1(z) = |C_1(z)|^2 = \cos^2 \kappa z \tag{250}$$

and the power in waveguide 2 is

$$P_2(z) = |C_2(z)|^2 = \sin^2 \kappa z \tag{251}$$

Note that the total power in the coupler is conserved,

$$P = P_1(z) + P_2(z) = 1 \tag{252}$$

because a lossless case is assumed in the derivation.

The power input into waveguide 1 is coupled completely into waveguide 2 when

$$\kappa z = (2n - 1)\frac{\pi}{2} \quad n = 1, 2, 3, \ldots \tag{253}$$

In other words, complete coupling or switching of power from waveguide 1 to waveguide 2 occurs if the distance of coupling, the distance over which the two parallel waveguides are closely positioned, is designed to be $L_c = \pi/2\kappa$. If $\Delta\beta \neq 0$, i.e., the two parallel waveguides are not identical, the coupled-mode equations can be solved to yield the following expression for power coupled into waveguide 2, assuming once again unit modal power input into waveguide 1:

$$P_2(z) = \sin^2[\sqrt{\chi^2 + \xi^2}]/(1 + \xi^2/\chi^2) \tag{254}$$

where $\chi = \kappa z$ and $\xi = \frac{1}{2}\Delta\beta z$. The first zero for $P_2(z)$ occurs at

$$\chi^2 + \xi^2 = \pi^2 \tag{255}$$

or, for a normalized coupling length of $\chi = \pi/2$, it occurs at

$$\xi = \frac{\sqrt{3}}{2}\pi \tag{256}$$

Figure 2-21 plots the power distributions as a function of ξ for waveguides whose $\Delta\beta$ varies linearly as a function of frequency. Note that if $\xi = 0$, i.e., $\Delta\beta = 0$, all the power is transferred to waveguide 2 for the normalized coupling length of $\chi = \pi/2$, which is consistent with the earlier discussion.

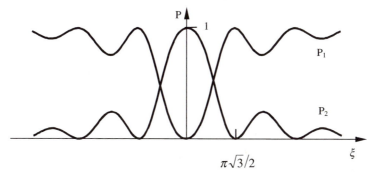

FIGURE 2-21   Directional coupler power switch characteristics.

The sidelobes in Figure 2-21 can be reduced to improve the coupler performance. We note that by varying κ as a function of z, the sidelobes can effectively be suppressed. Consider a coupler whose waveguide spacing is a parabolic function of z, the coupling coefficient $\kappa(z)$, which depends exponentially on the waveguide spacing, should have a Gaussian variation over z. At the center of the Gaussian function, the spacing between the waveguides is minimum, which corresponds to a maximum coupling coefficient. From the second mode coupled equation, it is straightforward to show that

$$P_2(L) = \left[\int_0^L \kappa(z)C_1(z)e^{j\Delta\beta z}dz \right]^2 \tag{257}$$

It can be seen qualitatively from Eq. (257) that the sidelobes in P_2 are reduced if $\kappa(z)$ has a Gaussian variation instead of being a constant.

The two waveguides are positioned closely (and parallel to each other) over a length L called the coupling (or interaction) length. The power is coupled from one waveguide to the other, and the power coupling ratio η, defined as

$$\eta = \frac{P_2}{P_1 + P_2} \tag{258}$$

is equal to

$$\eta = \sin^2(\kappa L) \tag{259}$$

for symmetrical $\Delta\beta = 0$ couplers. Of course, there must be some transition regions at the ends of the coupling region, as shown in Figure 2-22, over which the two waveguides separate. These input and output transition regions lead to some power coupling and therefore contribute to an increase in the total coupling length by L_{end}. In this case, the power coupling ratio becomes

$$\eta = \sin^2[\kappa(L + L_{end})] = \sin^2\left[\frac{\pi}{2} \frac{L + L_{end}}{L_c} \right] \tag{260}$$

where

$$L_c = \frac{2\pi}{\kappa} \tag{261}$$

is the coupling length needed to completely couple light from one waveguide to the other.

As discussed previously, the coupling coefficient κ should depend exponentially on the waveguide spacing because the evanescent field tails decrease exponentially away from the waveguide cores. Therefore, in general, L_c can be assumed to have the following dependence [11]

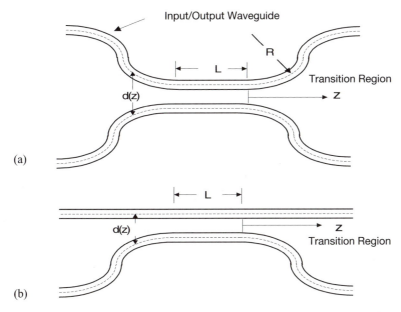

FIGURE 2.22 Coupler configurations: curved input/output waveguide (a) and straight input/output waveguide (b).

$$L_c = L_{c0} \exp\left(\frac{d(z)}{d_0}\right) \tag{262}$$

where L_{c0} and d_0 are parameters that can be obtained from modeling the coupler for a given waveguide design and wavelength. For a constant curvature transition as shown in Figure 2-22a, the spacing between the two waveguides can be approximated by

$$d(z) \cong d_{min} + \frac{z^2}{R} \tag{263}$$

where R is the bending radius and d_{min} is the minimum separation between the two waveguides in the parallel region. By combining the expressions for κ_c and L_c and setting $L = 0$, we have

$$\kappa_c L_{end} = \frac{\pi}{2L_{c0}} \int_{-\infty}^{+\infty} \exp\left[-\frac{d(z)}{d_0}\right] dz \tag{264}$$

or

$$L_{end} = \frac{\pi}{2\kappa_c L_{c0}} \int_{-\infty}^{+\infty} \exp\left[-\frac{d(z)}{d_0}\right] dz \tag{265}$$

where the integration is over the transition regions only, because the parallel regions are included separately by the parameter L in Eq. (260). By carrying out the integration for the configuration shown in Figure 2-22a, we have

$$L_{end} = \sqrt{\pi R d_0} \qquad (266)$$

For the coupler depicted in Figure 2-22b, it can be shown that

$$L_{end} = \sqrt{2\pi R d_0} \qquad (267)$$

The dependence of L_c and L_{end} on the waveguide separation d_{min} at different wavelengths has been investigated and compared with theoretical predictions [21]. Figure 2-23 shows the experimental dependence of L_c on d_{min} as well as predictions by Marcatili [17], Kumar et al. [22], and point matching methods [23] for couplers shown in Figure 2-22a based on silica waveguides of cross section 10×8 μm and Δ = 0.24%. The dependence of L_{end} on the radius of curvature for the same coupler with $d_{min} = 2.2$ μm is shown in Figure 2-24 for $\lambda = 1.29$ μm.

2.4.2   Star Couplers

Although splitters with a large number of input and output ports can be realized by cascading 2×2 couplers, the star coupler offers a simpler, wavelength independent

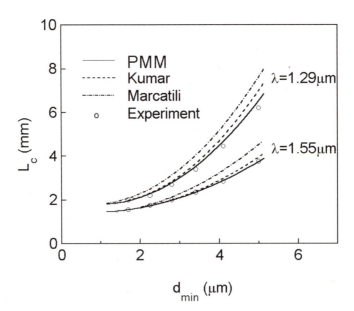

FIGURE 2-23   Dependence of the coupling length L_c on the waveguide separation d_{min} [21]. Reprinted by permission. Copyright © 1988 IEEE.

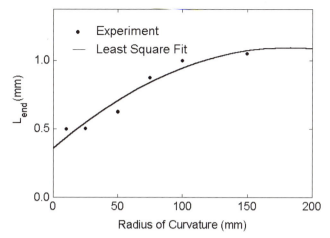

FIGURE 2-24 Dependence of L_{end} on the radius of waveguide curvature [21]. Reprinted by permission. Copyright © 1988 IEEE.

approach. A star coupler consists of an input and output array of waveguides separated by a slab diffraction region as shown in Figure 2-25a [24]. The goal is to split the power from each one of the input ports evenly among the N output ports. By conservation of power, the minimum loss per port is $10 \log_{10}(N)$, known as the splitting loss. The input and output waveguide arrays are arranged on a radius of curvature R pointing radially to a focal point. As the waveguides approach the slab region, they become strongly coupled. At the entrance and exit of the slab, the waveguide separation in each array is denoted by a. As discussed in Chapter 3, the field distribution at the output array is the Fourier transform of the field at the input array. Since the waveguide arrays are periodic, the output field exhibits an angular periodicity of $\theta_{BZ} \approx \lambda/n_s a$ where n_s is the slab effective index. The region subtended by θ_{BZ} is known as the first Brillouin zone. The output array must cover θ_{BZ} to assure an efficient design with low excess loss, so $Na/R \approx \theta_{BZ}$ where N is the number of outputs. When the center guide of the input array is excited, the center guide of the output array receives the most power. The waveguides furthest away from the center receive the least power. The experimental results for a 19×19 star coupler are shown in Figure 2-25b for the central input port and for the ninth input port from the center [25]. An excess loss of 1.5dB for the center guide and 3.5dB for the marginal guides was achieved.

Large arrays, up to 144×144, have been fabricated in silica waveguides with excess losses of 2.0 dB and standard deviations between output waveguides of 1.5dB [26]. To make the outputs more uniform, the input waveguides are up-tapered near the slab region. The coupling between input waveguides is designed to give a uniform far field pattern. Since the coupling is wavelength dependent, the uniformity and loss are also wavelength dependent. Optimum tapers have been designed to minimize the wavelength dependence. Devices with average losses of 1.4 dB at $\lambda =$

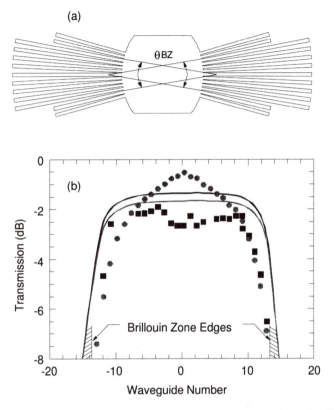

FIGURE 2-25 A star coupler. (a) Schematic showing the slab coupler and Brillouin zone angle. (b) The power transmission for a 19×19 device for the central input port (circles) and the ninth input waveguide from the center (squares) [25]. Reprinted by permission. Copyright © 1989 IEEE.

1.3 μm and 1.55 dB at 1.55 μm have been demonstrated with nearly identical standard deviations for the loss between the output ports of 0.4 dB [27].

2.4.3 Multi-Mode Interference Couplers

Multi-mode interference (MMI) couplers are based on the self-imaging property of multi-mode propagation in a slab [28,29]. Self-imaging describes the formation of images, single or multiple, of the input field at periodic intervals along the direction of propagation without the aid of a lens. Also known as the Talbot effect [30], this behavior requires that the medium of propagation have an associated periodicity. In free space applications, self-imaging occurs when light transmits through or reflects off of a periodic aperture such as a grating [31]. In multimode waveguides, the input field excites a number of modes. If the propagation constants of the modes are integer multiples of a constant $2\pi/L_0$, then self-imaging occurs at distances that are integer multiples of L_0. Slab waveguide self-imaging was first proposed by Bryngdahl [32] and explained further by Ulrich [33]. A 4 × 4 fiber coupler based on self-imaging was reported by Niemeier et al. [34].

An MMI coupler is shown schematically in Figure 2-26a with a width W and length L. A theoretical description was given by Soldano et al. [35]. Self-imaging in slab waveguides requires the excitation of multiple modes in the slab by the input field. The effective index method may be applied to reduce the description of the modes from two to one dimension. The input field, $\Phi(x, 0)$, is written as a sum of the slab modes, $\phi(x)$, as follows:

$$\Phi(x,0) \approx \sum_{m=0}^{M-1} C_m \phi_m(x) \tag{268}$$

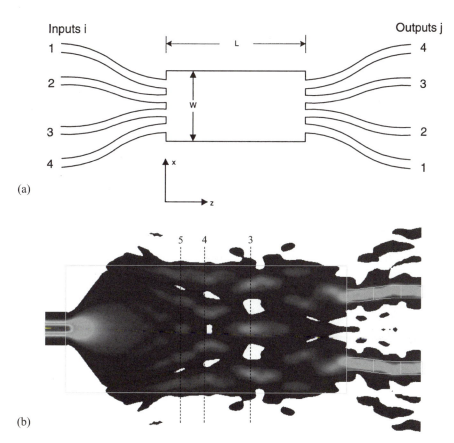

(a)

(b)

FIGURE 2-26   A multi-mode interference coupler schematic (a) and a beam propagation simulation of a 3 dB MMI coupler showing the waveguide outline and the field intensity in the multimode region [obtained using Prometheus by BBV software, The Netherlands (bbv@bbv.nl)] (b).

where C_m represents the amplitude of the mth waveguide mode, which is defined in terms of the overlap integral:

$$C_m = \frac{\int \Phi(x)\phi_m(x)dx}{\sqrt{\int \phi_m(x)^2 dx}} \tag{269}$$

This process is known as eigenmode decomposition, where the $\phi_m(x)$'s are the eigenmodes of the slab waveguide. To more accurately reproduce the input field, a large number of modes is desired. The field as it propagates along z is then described as follows:

$$\Phi(x, z) = \sum_{m=0}^{M-1} C_m \phi_m(x) e^{-j\beta_m z} \tag{270}$$

where the β_m's are the propagation constants of each mode and M modes supported by the slab. The transverse propagation constants for a strongly guided slab are given by $\kappa_{xm} = (m + 1)\pi/W_e$, where m is the mode number and W_e is an effective width which accounts for the lateral penetration of the evanescent mode into the cladding region and the polarization dependence. For high contrast waveguides, $W_e \approx W$. The longitudinal propagation constant is $\beta_m^2 = n_s^2 k_0^2 - \kappa_{xm}^2$ where $k_0 = 2\pi/\lambda$ and n_s is the slab index. The paraxial approximation is applied to obtain $\beta_m \cong n_s k_0 - \kappa_{xm}^2/2n_s k_0$, which can be written as follows:

$$\beta_m \approx n_s k_0 - \frac{(m + 1)^2 \lambda \pi}{4 W_e^2 n_s} \tag{271}$$

Note that $\beta_0 - \beta_1 = 3\lambda\pi/4W_e^2 n_s$. Let the beat length between the fundamental and first order modes be given by

$$L_c \equiv \frac{\pi}{\beta_0 - \beta_1} \cong \frac{4n_s W_e^2}{3\lambda} \tag{272}$$

Now, the propagation constants are written as follows:

$$\beta_0 \approx n_s k_0 - \frac{\pi}{3L_c} \tag{273}$$

$$\beta_m \approx \beta_0 - \frac{m(m + 2)\pi}{3L_c} \tag{274}$$

At a distance of $3L_c$, the modes have a relative phase that differs by an even or odd multiple of π depending on whether m is even or odd. Let the origin of the x-axis be located at the center of the multimode region. Due to the structural symmetry about $x = 0$, the even and odd eigenmodes satisfy the following symmetry conditions:

$$\phi_m(x) = \phi_m(-x) \quad \text{for even } m$$
$$\phi_m(x) = -\phi_m(-x) \quad \text{for odd } m$$

(275)

Consequently, an output that is the mirror image of the input about $x = 0$ occurs at a distance $z = 3L_c$. For example, a 2×2 MMI couples light to the cross-port for $z = 3L_c$, and transmits to the bar-port at $z = 6L_c$. A simple device model can be made by assuming a large index contrast and slab modes with even and odd symmetry given by

$$\phi_m(x) \approx \begin{cases} \sin\left(\dfrac{\pi x(m+1)}{W}\right) & \text{for } m \text{ odd} \\[3mm] \cos\left(\dfrac{\pi x(m+1)}{W}\right) & \text{for } m \text{ even} \end{cases}$$

(276)

Multiple images of the input are formed at intermediate lengths. At a distance of $3L_c/N$, there are N self-images with the qth image located at x_q with phase ϕ_q given as follows [36]:

$$x_q = (2q - N)\frac{W_e}{N}$$

(277)

$$\phi_q = q(N - q)\frac{\pi}{N}$$

(278)

for $0 \leq q \leq N - 1$. The total field at $z = 3L_c/N$ is given by

$$\Phi\left(x, \frac{3L_c}{N}\right) = \frac{e^{-j\xi}}{\sqrt{N}}\sum_{q=0}^{N-1} \Phi(x - x_q)e^{j\phi_q}$$

(279)

where

$$\xi = \beta_0 \frac{3L_c}{N} + \pi\left(\frac{1}{N} + \frac{N-1}{4}\right).$$

A beam propagation simulation of a 1×2 MMI which functions as a 3dB splitter is shown in Figure 2-26b for $L = 3L_c/2$. The lengths for obtaining a splitting ratio of 3, 4 and 5 are also clearly visible. When an MMI is used as the splitter or combiner in an interferometer, the phase relationships between the output ports are needed. The phase relationships for the output images in terms of the input waveguide are given as follows [36]:

$$\phi_{ij} = -\xi + \frac{\pi}{4N}(j + i - 1)(2N - j - i + 1) \qquad \text{for odd } i + j$$

$$\phi_{ij} = -\xi + \frac{\pi}{4N}(2N - j + i)(j - i) + \pi \qquad \text{for even } i + j$$

(280)

where the inputs and outputs are numbered i and j, respectively, as shown in Figure 2-26a.

In general, larger refractive index differences are desirable for MMIs since more modes propagate for a given width. In constrast, higher index differences require smaller gaps for directional couplers, making them more difficult to fabricate. The MMI device length is proportional to the square of the width according to Eq. (272). The width is critical for achieving low excess loss. Wider input and output waveguides in conjunction with short device lengths help relax the fabrication tolerances on the index difference and width [35]. Special symmetries of the input and output waveguides can be exploited to restrict the excitation to a limited set of modes. Shorter devices can then be realized with restricted interference compared to the general mode excitation just described. Extremely short couplers 20–30 μm in length have been demonstrated in silica and 50–70 μm coupler lengths in InP, with excess losses less than 1 dB and uniformity between the outputs of 0.15 dB [37]. MMIs are advantageous for realizing splitters in high contrast waveguides, with short device lengths, and low excess loss compared to directional couplers. Polarization dependence is minimal in practical devices, a few tenths of a dB. The bandwidth is inversely proportional to the number of input and output ports. Bandwidths > 100 nm have been achieved for 2 × 2 3 db couplers assuming a 1 dB excess loss criterion [38]. Arbitrary splitting ratios have been reported using restricted interference [39] and butterfly configurations [40]. Applications for MMIs in switches [41] and wavelength routers [42] are discussed in Chapters 4 and 6.

2.5 MATERIAL PROPERTIES AND FABRICATION PROCESSES

Successful optical filter design and realization depend fundamentally on the properties of the materials and capabilities of the fabrication processes used. The basic issues include loss, temperature and polarization dependence, control over refractive indices and layer thicknesses, and methods for varying the refractive index or loss for tuning purposes. Temperature dependence is a practical concern because filters must operate over a broad temperature range, for example 0 to 80°C, while maintaining tight control over their center wavelength tolerances. For system applications where the filter operates on light which has propagated long distances through optical fibers, low polarization dependence is required since the input state of polarization is unpredictable. Although we shall focus on planar waveguides, these properties are important for other implementations, including optical fiber devices and thin film dielectric stacks. An overview of several materials used for planar waveguides is given, followed by a sketch of the fabrication processes for silica planar waveguides and their relevance to filter design.

2.5.1 Materials

Many materials are available for fabricating planar optical waveguides. These include glasses, semiconductors, and polymers. In addition to the properties listed above, the range of core-to-cladding refractive index difference, the wafer size, and methods to locally vary the refractive index in order to shift the optical phase are important for filter design. The material properties for silica, lithium niobate, InGaAsP, silicon and polymers are summarized in the following paragraphs.

The lowest planar waveguide losses have been demonstrated in silica waveguides. Straight waveguide propagation losses of 3 dB/m for $\Delta = 0.75\%$ [43] and 7 dB/m for $\Delta = 2\%$ [6] have been reported. Silicon substrates are in the range of 4–6 inches in diameter. The core is doped to raise its refractive index above the cladding. Common dopants are germanium, phosphorous, and titanium. Germanium is popular because of its large index and its UV photosensitivity. For $\Delta = 1\%$, approximately 10 mole-% of Ge is required. The maximum Ge concentration which can be deposited is around 25% to 30% before cracking occurs from tensile stresses in the doped core material [44]. A permanent index change is induced by irradiation of Ge-doped waveguides at 245 nm [45,46] and Ge- or P-doped waveguides at 193 nm [47]. The materials' photosensitivity can be enhanced by an order of magnitude by loading the samples with hydrogen or deuterium to a concentration of a few percent [48]. Refractive index changes between -3×10^{-4} [49] and $+2 \times 10^{-2}$ [48] have been reported in Ge-doped silica. Ultraviolet exposure of photosensitive glasses allows low-loss, low-polarization dependent Bragg grating filters to be fabricated in optical fibers as discussed in Chapter 5. In addition, long period grating filters, discussed in Chapter 4, are fabricated by UV exposure. Such a large photosensitivity also allows waveguides and couplers to be directly written [48,50], avoiding the lithography and etching steps discussed in Section 2.5.2. The temperature dependence of the refractive index of silica is $dn/dT = 1 \times 10^{-5}/°C$ [51]. The impact of this dn/dT is that filters realized in silica exhibit a temperature dependence of -1.5 GHz ($+0.0125$ nm)/°C in the 1550 nm range. The advantage of this temperature dependence is that phase shifters can be realized by thermo-optic tuning as discussed in section 2.5.2. The disadvantage is that temperature control has to be employed to keep the filter center wavelengths constant over a large ambient temperature range. Because of the latter issue, methods to reduce the temperature dependence have been explored by combining materials with different dn/dT, such as glass cores and lower cladding with a polymer upper cladding [52–54]. Silica waveguides grown on Si substrates have a birefringence of $B = 4 \times 10^{-4}$ [51]. For $\lambda = 1550$ nm, this implies a separation in peak wavelengths between the TE and TM polarization of 50 GHz (0.4 nm). The birefringence arises from compressive strain between the SiO_2 and Si due to the larger thermal expansion coefficient of Si. Equation (281) relates the stress-induced birefringence to the mechanical properties of the substrate and waveguide [55]:

$$B = KE(\alpha_{\text{sub}} - \alpha_{\text{cl}}) \, \Delta T \qquad (281)$$

where the photoelastic constant and Young's modulus for silica are $K = 3.5 \times 10^{-5}$ mm^2/kg and $E = 0.78 \times 10^4$ kg/mm^2 [51], respectively. The thermal coefficients of expansion for the substrate and cladding are roughly $\alpha_{Si} = 2.5 \times 10^{-6}/°C$ and $\alpha_{SiO2} \cong 0.5–1.0 \times 10^{-6}/°C$ [51]. The temperature difference ΔT is between the consolidation temperature and room temperature. Silica substrates offer a dramatic improvement by reducing the thermal expansion mismatch; a birefringence of $B = -9 \times 10^{-6}$ has been achieved [55]. Other approaches for reducing the polarization dependence include stress-relieving grooves [51], the use of a $\lambda/2$ waveplate to exchange the TE and TM polarizations in the middle of the device [56], and stress-inducing films [57,58] to compensate the birefringence.

Lithium niobate (LiNbO$_3$) waveguides are used for high speed modulators and switches. They are fabricated by several processes including ion exchange and thermal indiffusion. Core-to-cladding index differences are in the range of $\Delta n_e = 0.12$ ($\Delta n_o = -0.04$) for proton exchange and $\Delta n_e \cong \Delta n_o = 0.008$ for thermal indiffusion [59]. Since LiNbO$_3$ is a uniaxial crystal, it has a large birefringence. The indices for the extraordinary and ordinary rays are $n_e = 2.2$ and $n_o = 2.286$, and the temperature dependence of the refractive index is $dn_e/dT = +5.3 \times 10^{-5}/°C$ and $dn_o/dT = +0.56 \times 10^{-5} °C$ [60]. Losses of 0.1 dB/cm are obtained for $\Delta n = 0.04$, and wafer sizes are 3–4 inches [61]. LiNbO$_3$ has a large linear electro-optic coefficient of $r_{33} = 30.9 \times 10^{-10}$ cm/V [59]. The resulting index change from an applied field, known as the Pockels effect, is given by

$$\Delta n = \frac{r_{33}n^3}{2}\Gamma E$$

where E is the applied electric field and Γ is an overlap factor between the applied field and the mode. Modulators with bandwidths of 75 GHz [62] have been demonstrated. Another application of LiNbO$_3$ is for acousto-optic filters, which are discussed in Chapter 4. In this case, a change in the refractive index is caused by the stresses induced by an acoustic wave.

Since InGaAsP is used for lasers and detectors, it is attractive for fabricating filters so that all of these functions can be integrated on a single chip. Wafers are typically 2–3 inches in diameter. The refractive index of InP is 3.17 at $\lambda = 1.5$ μm. It can be increased to >3.5 using a quaternary alloy composition while keeping the lattice matched to InP; however, the increased index is accompanied by a decrease in the bandgap from 1.35 eV ($\lambda_g \sim 920$ nm) to less than 0.8 eV ($\lambda_g \sim 1550$ nm) [63] where the energy gap is related to the wavelength by

$$E = \frac{1.24}{\lambda_g}\left(\frac{ev}{\mu m}\right).$$

To avoid band-to-band absorption losses, the operating wavelength must be longer than the gap wavelength λ_g. Buried straight waveguides have losses less than 0.2 dB/cm, and a birefringence of 7×10^{-3} for rib-loaded slabs at 1550 nm [64]. Polarization independent waveguide designs have also been demonstrated [65]. A typical center wavelength shift with temperature is $d\lambda/dT \sim 0.11$ nm/°C [64], implying a $dn/dT \sim 2 \times 10^{-4}$. The linear electro-optic coefficient for InP is $r_{41} =$

-1.4×10^{-10} cm/V. Other mechanisms, which are polarization independent, are also available for varying the refractive index. By the free-carrier plasma effect, an increase in the free-carrier concentration increases the absorption loss and decreases the refractive index, as required by the Kramers-Kronig relations. Carriers may either be injected or depleted; however, carrier depletion is preferred since it is faster and requires less drive power [66]. At 1.52 μm, experimental data on n-doped InGaAsP yielded an index change of $\Delta n = 3.63 \times 10^{-21} N_d$ where N_d is the doping level [67]. By the Franz-Keldysh effect in bulk materials and the Stark effect in quantum wells, the absorption coefficient changes as a function of the reverse bias resulting in an index change. Data on InGaAsP with $\lambda_g = 1.2$ μm provides the following refractive index changes: $\Delta n = 32 \times 10^{-21} N_d$ at $\lambda = 1.30$ μm and $\Delta n = 15 \times 10^{-21} N_d$ at $\lambda = 1.55$ μm per cm^3 [67]. This latter band-filling effect dominates the free-carrier effect and is wavelength dependent. In practice, the overall index change is the sum of all the effects, including a small contribution from the Kerr effect, which is proportional to the square of the electric field. Phase modulators with an efficiency of 11°/V-mm and a loss of 1.6 dB/cm have been reported [67].

Waveguides have also been realized in silicon. For rib waveguides, a loss of 0.2 dB/cm at $\lambda = 1.3$ μm [68] and a birefringence of 9×10^{-5} have been achieved [69]. The index of silicon is 3.5 for $\lambda = 1.5$ μm. The temperature dependence of the refractive index is $1.86 \times 10^{-4}/°C$ [70], and its thermal expansion coefficient is $2.6 \times 10^{-6}/°C$. Its higher dn/dT offers an order of magnitude larger phase shift for the same amount of temperature change compared to silica. Current injection/depletion can be used to change the index on the order of $\Delta n = 10^{-3}$. A phase modulator with a figure of merit of 215°/V-mm has been demonstrated using the free-carrier plasma effect with a drive power of 43 mW, switching times shorter than 3.5 ns, and an amplitude modulation of 1.3 dB for a 1mm device [71].

Polymers also offer a unique set of features from a fabrication and functionality viewpoint. Fabrication involves low temperature processes such as spin coating polymers on substrates and etching or photo-bleaching techniques to form the waveguide structures. Waveguide applications require polymers with low loss and high thermal and environmental stability. Losses for singlemode waveguides at $\lambda = 1550$ nm have been reported for several different compositions, including 0.43 dB/cm for a deuterated polysiloxane [72], 0.25 dB/cm for a fluorinated polymer [73], and 0.06 dB/cm for a multifunctional acrylate monomer/oligomer combinations with various additives [74]. A range of refractive indices from 1.3 to 1.6 can be achieved by varying the polymer composition. The refractive index has a negative temperature dependence, typically in the range of -1×10^{-4} to $-4 \times 10^{-4}/°C$. Thermo-optic switches requiring only 20 mW of power for switching have been realized [75]. Nonlinear optical polymers are made by doping with a chromophore and poling at a temperature above the glass transition temperature to induce an electro-optic effect. Modulators with bandwidths over 100 GHz [76] have been demonstrated with polymers having linear electro-optic coefficients of $7–14 \times 10^{-10}$ cm/V. Very low polarization dependence can be achieved, as demonstrated by a 1×32 router with a polarization dependent wavelength shift of ≤ 0.01 nm [77] or $B \leq 9 \times 10^{-6}$.

2.5.2   Fabrication

For filter design, low loss and well-controlled coupling ratios are mandatory. Although larger index contrasts allow smaller devices to be made, they require tighter index and thickness control and are more demanding on the lithography resolution and etching reproducibility. Silica-based waveguides offer low loss and many years of experience at several laboratories [51, 78, 79]. The general steps in fabricating silica-based channel waveguides are described. Since the fabrication process determines the ultimate tolerances with which filters can be achieved, a basic understanding of the steps is helpful to the filter designer.

A mask is made with the desired waveguide pattern. A thin layer of chrome on a glass substrate is coated with resist and exposed using an electron-beam exposure system. Then, the resist is developed and the chrome is etched. The mask step size is on the order of 0.1 μm.

The fabrication steps are outlined in Figure 2-27. The substrate may be silicon or silica. The wafer thickness depends on the diameter and whether deposition is done on only one side or both, to keep the wafer from bowing. The temperature of subse-

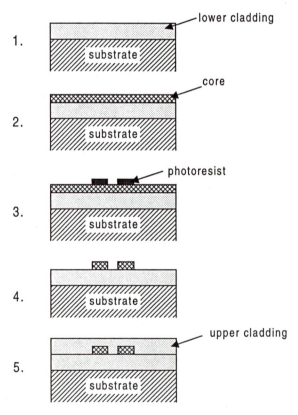

FIGURE 2-27   General channel waveguide fabrication processing steps showing two closely spaced waveguides.

quent processes can also be a factor in the tendency of the wafer to bow or not. Wafer thicknesses of 500 μm to 1 mm are typical for 4–6 inch diameters.

The first step is to deposit the lower cladding. Various deposition processes are used, including chemical vapor deposition (CVD), flame hydrolysis deposition (FHD), and electron beam (e-beam) deposition. Deposition is done on both sides of the wafer for CVD and only one side for FHD and e-beam. For a silicon substrate, the lower cladding is typically 10 to 20 μm thick. For FHD, silica soot is deposited to form the lower cladding. A doped layer of silica is then deposited to form the core. In the case of Ge-doped core, the flow of $SiCl_4$-$GeCl_4$ gas is controlled to achieve the desired elevation of refractive index over the cladding material. The soot is then consolidated at an elevated temperature of 1200 to 1300°C yielding a lower cladding thickness ~ 20μm [80]. CVD is a lower temperature process, 400° to 800°C for deposition and 900–1100°C for annealing, but has a slower deposition rate than FHD, 1 μm per hour compared to 1 μm per minute [81]. For any deposition process, the performance metrics include the minimum loss, the achievable uniformity of the layer thickness and index across the wafer, and the repeatability from wafer-to-wafer. The number of wafers that can be deposited at a time and the deposition rate are important parameters for production.

The next step is to define the waveguide pattern on the core. A photoresist is spun on top of the core layer and baked. The chrome mask previously discussed is used to expose the photoresist, typically by contact or projection photolithography. The resist is developed, which dissolves the resist where the core is to be removed (for positive resists). The resolution of the photolithography setup determines the minimum feature sizes, for example, minimum waveguide widths and gap sizes for directional couplers of 2–3 μm can be achieved with proximity printers.

The next step is to etch the core layer, leaving the desired waveguide pattern. Etching may be isotropic (typical of wet etching an amorphous material) or anisotropic (such as obtained by reactive ion etching (RIE)). The core height for doped-silica waveguides is typically several microns. Anisotropic etching is required to produce straight waveguide sidewalls. Smooth, vertical walls are needed for low loss. The feature size after etching may be smaller than the features defined in the photoresist, which is referred to as undercut. Knowledge of the expected undercut is critical to waveguide and filter design. For example, any variation in the gap between the waveguides of a directional coupler will change the coupling ratio from its nominal value. Any etching roughness or nonuniformity across the wafer contributes to variations in the waveguide core size that results in variations in the effective index.

After etching and removal of any remaining resist, the upper cladding is deposited. The required thickness depends on the core index. Typical thicknesses are 10 to 40 μm. The upper cladding must completely fill the areas where the core was removed. A particularly sensitive region is the gap in directional couplers. Incomplete filling can lead to loss and offset the coupling from the target value. Depending on the temperature of the upper cladding deposition process, the height and width of the waveguides, and the softening temperature of the core, the waveguides can warp. If the walls warp slightly towards each other, the coupling increases in a directional coupler.

For a successful filter demonstration, variations introduced by each process step must be considered. Post-fabrication tuning mechanisms are desirable to compensate for fabrication induced variations or for tuning a filter response. For silica waveguides, the thermo-optic effect is used to vary the local refractive index of a waveguide. A resistive material is deposited on the upper cladding over a section of the waveguide to be heated. Such heaters require two additional masks and associated processing steps, one for the resistive heater material and a second for low-resistance electrical contacts. Chromium and gold are typical materials used for the heaters and contacts. For $\lambda = 1550$ nm, a 5 mm long heater can change the optical path length by $\lambda/2$ (or the phase by π) with a temperature change of $<15°C$ and an applied power of 0.5 W. Switching times are typically 1–2 ms. An order of magnitude reduction in switching power can be obtained by etching underneath the lower cladding to create an insulative air gap around three sides of the waveguide being heated [82]. The reduction in switching power is achieved at the expense of the switching time, which increases by an order of magnitude.

REFERENCES

1. J. D. Jackson, *Classical Electrodynamics,* 2nd ed., New York: Wiley, 1975, p. 285.

2. I. Malitson, "Interspecimen comparison of the refractive index of fused silica," *J. Opt. Soc. Am.,* vol. 55, pp. 1205–1209, 1965.

3. C. Poole and J. Nagel, "Optical fiber components and devices," in Optical Fiber Telecommunications IIIA, I. Kaminow and T. Koch, Eds. San Diego: Academic Press, 1997, pp. 114–161.

4. R. Olshansky, "Propagation in glass optical waveguides," *Rev. Mod. Phys.,* vol. 51, no. 2, p. P341, 1979.

5. E. C. M. Pennings, J. Van Schoonhoven, J. W. M. Van Uffelen, and M. K. Smit, "Reduced Bending and scattering Losses in New Optical 'Double-Ridge' Waveguide," *Electron. Lett.,* vol. 25, no. 11, pp. 746–748, 1989.

6. S. Suzuki, M. Yanagisawa, Y. Hibino, and K. Oda, "High-Density Integrated Plannar Lightwave Circuits Using SiO_2-GeO2 Waveguides with a High Refractive Index Difference," *J. Lightw. Technol.,* vol. 12, no. 5, pp. 790–796, 1994.

7. M. Heiblum and J. H. Harris, "Analysis of curved optical waveguides by conformal transformation," *J. Quantum Elect.,* vol. 11, p. 186, 1975.

8. D. Marcuse, "Bending Losses of the Asymmetric Slab Waveguide," *Bell System Technical. J.,* vol. 50, no. 8, pp. 2551–1563, 1971.

9. W. A. Gambling, H. Matsumura, and C. M. Ragdale, "Field Deformation in Curved Single-mode Fiber," *Electron. Lett.,* vol. 14, no. 5, pp. 130–132, 1978.

10. E. A. J. Marcatili, "Bends in Optical Dielectric Guides," *Bell System Techn. J.,* vol. 48, no. 7, pp. 2103–2132, 1969.

11. Y. P. Li and C. H. Henry, "Silica-based optical integrated circuits," *IEE Proc.-Optoelectron.,* vol. 143, no. 5, 1996, pp. 263—280.

12. L. B. Jeunhomme, *Single-mode Fiber Optics: Principles and Applications,* New York: Marcel Dekker, 1983, p. 18.

13. F. Ladouceur and J. D. Love, *Silica-based Buried Channel Waveguides and Devices,* New York: Chapman and Hall, 1996.

14. Lord Rayleigh, "The problem of the whispering gallery," *Phil. Magn. S.,* vol. 20, no. 20, pp. 1001–1004, 1910.

15. M. K. Smit, E. C. M. Pennings and H. Blok, "A Normalized Approach to the Design of Low-Loss Optical Waveguide Bends," *J. Lightwave Technol.,* vol. 11, no. 11, pp. 1737–1742, 1993.

16. C. Madsen and J. Zhao, "Increasing The Free Spectral Range Of Silica Wave-Guide Rings For Filter Applications," *Optics Lett.,* vol. 23, no. 3, pp. 186–188, 1998.

17. E. A. J. Marcatili, "Dielectric Rectangular Waveguide and Directional Coupler for Integrated Optics," *Bell System Technical J.,* vol. 48, no. 7, pp. 2071–2102, 1969.

18. G. B. Hockcr and W. K. Burns, "Mode Dispersion in Diffused Channel Waveguides by the Effective Index Method," *Appl. Opt.,* vol. 16, pp. 113–118, 1977.

19. D. Marcuse, *Theory of Dielectric Optical Waveguides,* 2nd ed., San Diego: Academic Press, 1991, p. 49.

20. C. Pollock, *Fundamentals of Optoelectronics,* R. D. Irwin, 1995, p. 225.

21. N. Takato, K. Jinguji, M. Yasu, and H. Toba, and M. Kawachi, "Silica-Based Single-Mode Waveguides on Silicon and their Application to Guided-Wave Optical Interferometers," *J. Lightw. Technol.,* vol. 6, no. 6, pp. 1003–1010, 1988.

22. A. Kumar, A. K. Haul, and A. K. Ghatak, "Prediction of coupling length in a rectangular-core directional coupler: An accurate analysis," *Opt. Lett.,* vol. 10, pp. 86, 1985.

23. K. Jinguji, N. Takato, and M. Kawachi, "Analysis of rectangular-core directional couplers by the point matching method," Proc. Tech. Group Meeting, IECE of Japan, 1986, p. 59.

24. C. Dragone, "Efficient N×N star coupler based on Fourier optics," *Electron. Lett.,* vol. 24, no. 15, pp. 942–944, 1988.

25. C. Dragone, C. Henry, I. Kaminow, and R. Kistler, "Efficient multichannel integrated optics star coupler on silicon," *IEEE Photon. Technol. Lett.,* vol. 1, no. 8, pp. 241–243, 1989.

26. K. Okamoto, H. Okazaki, Y. Ohmori, and K. Kato, "Fabrication of large scale integrated-optic N×N star couplers," *IEEE Photon. Technol. Lett.,* vol. 4, no. 9, pp. 1032–1035, 1992.

27. K. Okamoto, H. Takahashi, M. Yasu, and Y. Hibino, "Fabrication of wavelength-insensitive 8×8 star coupler," *IEEE Photon. Technol. Lett.,* vol. 4, no. 1, pp. 61–63, 1992.

28. L. Soldano, F. Veerman, M. Smit, B. Verbeek, and E. Pennings, "Multimode interference couplers," Proc. Integrated Photonics Research Topical Meeting, Monterey, CA, April, 1991, Poster TuD1.

29. E. Pennings, R. Deri, A. Scherer, R. Bhat, T. Hayes, N. Andreadakis, M. Smit, and R. Hawkins, "Ultra-compact, low-loss directional coupler structures on InP for monolithic integration," Proc. Integrated Photonics Research Topical Meeting, Monterey, CA, April, 1991, Post-deadline PD2.

30. H. Talbot, "Facts relating to optical science," *Philos. Mag.,* vol. 9, no. 56, pp. 401–407, 1836.

31. M. Mansuripur, "The Talbot Effect," *Optics & Photonics News,* vol. 43, pp. 42–47, 1997.

32. O. Bryngdahl, "Image formation using self-imaging techniques," *J. Opt. Soc. Am.,* vol. 63, no. 4, pp. 416–419, 1973.

33. R. Ulrich, "Image formation by phase coincidences in optical waveguides," *Optics Communications,* vol. 13, no. 3, pp. 259–264, 1975.

34. T. Niemeier and R. Ulrich, "Quadrature outputs from fiber interferometer with 4×4 coupler," *Optics Lett.,* vol. 11, pp. 677–679, 1986.

35. L. Soldano and E. Pennings, "Optical Multi-Mode Interference Devices Based on Self-Imaging: Principles and Applications," *J. Lightw. Technol.,* vol. 13, no. 4, pp. 615–627, 1995.

36. M. Bachmann, P. Besse, and H. Melchior, "General self-imaging properties in N(N multimode interference couplers including phase relations," *Appl. Optics,* vol. 33, no. 18, pp. 3905–3911, 1994.

37. L. Soldano, M. Bouda, M. Smit, and B. Verbeek, "New small-size single-mode optical power splitter based on multi-mode interference," *Proc. European Conf. Opt. Commun. (ECOC).* Berlin, Germany, 1992, pp. 465–468, We. B10. 5.

38. P. Besse, M. Bachman, H. Helchior, L. Soldano, and M. Smit, "Optical bandwidth and fabrication tolerances of multimode interference couplers," *J. Lightw. Technol.,* pp. 1004–1009, 1994.

39. P. Besse, E. Gini, M. Bachmann, and H. Melchior, "New 1×2 multi-mode interference couplers with free selection of power splitting ratios," *Proc. European Conf. Opt. Commun. (ECOC).* Firenze, Italy, 1994, pp. We. C. 2. 4.

40. P. Besse, E. Gini, M. Bachmann, and H. Melchior, "New 2×2 and 1×3 multi-mode interference couplers with free selection of power splitting ratios," vol. 14, no. 10, *J. Lightw. Technol.,* pp. 2286–2292, 1996.

41. P. Besse, M. Bachmann, and H. Melchior, "Phase relations in multi-mode interference couplers and their application to generalized integrated Mach–Zehnder optical switches," 6th European Conf. on Integrated Optics, 1993.

42. L. Lierstuen and A. Sudbo, "8-channel wavelength division multiplexer based on multimode interference couplers," *IEEE Photon. Technol. Lett.,* vol. 7, no. 9, pp. 1034–1036, 1995.

43. Y. Hibino, H. Okazaki, Y. Hida, and Y. Ohmori, "Propagation Loss Characteristics of Long Silica-Based Optical Waveguides on 5 inch Si Wafers," *Electron. Lett.,* vol. 29, no. 21, pp. 1847–1848, 1993.

44. S. Suzuki, K. Shuto, H. Takahashi, and Y. Hibino, "Large-Scale and High Density Planar Lightwave Circuits with High-D GeO2-Doped Silica Waveguides," *Electron. Lett.,* vol. 28, no. 20, pp. 1863–1864, 1992.

45. G. Meltz and W. Morey, "Bragg grating formation and germanosilicate fiber photosensitivity," SPIE Vol. 1516 International Workshop on Photoinduced Self-Organization Effects in Optical Fiber, 185–199 (1991).

46. K. Hill, B. Malo, F. Bilodeau, D. Johnson, and J. Albert, "Bragg gratings fabricated in monomode photosensitive optical fiber by UV exposure through a phase mask," *Appl. Phys. Lett.,* vol. 62, no. 10, pp. 1035–1037, 1993.

47. B. Malo, J. Albert, F. Bilodeau, T. Kitagawa, D. Johnson, and K. Hill, "Photosensitivity in Phosphorous-Doped Silica Glass and Optical Waveguides," *Appl. Phys. Lett.,* vol. 65, no. 4, pp. 394–396, 1994.

48. V. Mizrahi, P. Lemaire, T. Erdogan, W. Reed, D. Digiovanni, and R. Atkins, "Ultraviolet Laser Fabrication of Ultrastrong Optical Fiber Gratings and of Germania-doped Channel Waveguides," *Appl. Phys. Lett.,* vol. 63, no. 13, pp. 1727–1729, 1993.

49. L. Dong, W. Liu, and Reekie L., "Negative Index Gratings Formed by a 193-nm Excimer Laser," *Optics Lett.,* vol. 21, no. 24, pp. 2032–2034, 1996.

50. G. Maxwell and B. Ainslie, "Demonstration of a directly written directional coupler using UV-induced photosensitivity in a planar silica waveguide," *Electron. Lett.,* vol. 31, no. 2, pp. 95–96, 1995.

51. M. Kawachi, "Silica Waveguides on Silicon and Their Application to Integrated-Optic Components," *Optical and Quantum Electron.,* vol. 22, pp. 391–416, 1990.

52. D. Bosc, B. Loisel, M. Moisan, and N. Devoldere, "Temperature And Polarization-insensitive Bragg Gratings Realized On Silica Waveguide On Silicon," *Electron. Lett.,* vol. 33, no. 2, pp. 134, 1997.

53. Y. Inoue, A. Kaneko, F. Hanawa, H. Takahashi, K. Hattori, and S. Sumida, "Athermal Silica-based Arrayed-wave-guide Grating Multiplexer," *Electron. Lett.,* vol. 33, no. 23, pp. 1945–1947, 1997.

54. Y. Kokubun, S. Yoneda, and S. Matsuura, "Temperature-insensitive optical filter at 1. 55 μm wavelength using a silica-based athermal waveguide," Integrated Photonics Research. Victoria, British Columbia, Canada, March 30-April 1, 1998, pp. 93–95.

55. S. Suzuki, Y. Inoue, and Y. Ohmori, "Polarization-insensitive arrayed-waveguide grating multiplexer with SiO_2-on-SiO_2 structure," *Electronic Lett.,* vol. 30, no. 8, pp. 642–643, 1994.

56. Y. Inoue, Y. Ohmori, M. Kawachi, S. Ando, T. Sawada, and H. Takahashi, "Polarization Mode Converter with Polyimide Half Waveplate in Silica-based Planar Lightwave Circuits," *IEEE Photonics Technol. Lett.,* vol. 6, no. 5, pp. 626–628, 1994.

57. M. Okuno, A. Sugita, K. Jinguji, and M. Kawachi, "Birefringence Control of Silica Waveguides on Si and its Application to a Polarization-Beam Splitters/Switch," *J. Lightw. Technol.,* vol. 12, no. 4, pp. 625–633, 1994.

58. H. Yaffe, C. Henry, R. Kazarinov, and M. Milbrodt, "Polarization Independent Silica-on-Silicon Mach–Zehnder Interferometers," *J. Lightw. Technol.,* vol. 12, no. 1, pp. 64–67, 1994.

59. R. Alferness, "Titanium-Diffused Lithium Niobate Waveguide Devices," in Guided-Wave Optoelectronics, T. Tamir, Ed. New York: Springer-Verlag, 1988, pp. 145–210.

60. H. Nishihara, M. Haruna, and T. Suhara, Optical Integrated Circuits. New York: McGraw-Hill, 1989.

61. Private communication with Ed Murphy of Lucent Technologies.

62. F. Leonberger and R. Ade, "Development and applications of commercial $LiNbO_3$ guided-wave devices," in Guided-Wave Optoelectron., T. Tamir, G. Griffel, and H. Bertoni, Eds. New York: Plenum Press, 1995, pp. 5–7.

63. F. Leonberger and J. Donnelly, "Semiconductor Integrated Optic Devices," in Guided-Wave Optoelectron., T. Tamir, Ed. New York: Springer-Verlag, 1988, pp. 317–396.

64. Private communication with Chris Doerr of Lucent Technologies.

65. B. Verbeek, A. Staring, E. Jansen, R. Van Roijen, J. Binsma, T. Van Dongen, M. Amersfoort, C. Van Dam, and M. Smit, "Large bandwidth polarisation independent and compact 8 channel PHASAR demultiplexer/filter," Optical Fiber Conference. San Jose, CA, February 20–25, 1994, pp. PD13–1.

66. A. Alping, X. Wu, T. Hausken, and L. Coldren, "Highly Efficient Waveguide Phase Modulator for Integrated Optoelectronics," *Appl. Phys. Lett.,* vol. 48, pp. 1243–1245, 1986.

67. J. Vinchant, J. Cavailles, M. Erman, P. Jarry, and M. Renaud, "InP/GaInAsP guided-wave phase modulators based on carrier-induced effects: theory and experiment," *J. Lightw. Technol.,* vol. 10, no. 1, pp. 63–69, 1992.

68. P. Trinh, S. Yegnanarayanan, and B. Jalali, "5x9 Integrated Optical Star Coupler in Silicon-on-Insulator Technology," *IEEE Photonics Technol. Lett.,* vol. 8, no. 6, pp. 794–796, 1996.

69. P. D. Trinh, S. Yegnanarayanan, F. Coppinger, and B. Jalali, "Silicon-on-insulator (SOI) Phased-array Wavelength Multi/demultiplexer With Extremely Low-polarization Sensitivity," *IEEE Photonics Technol. Lett.,* vol. 9, no. 7, pp. 940, 1997.

70. G. Cocorullo and I. Rendina, "Thermo-optical modulation at 1. 5 um in silicon etalon," *Electron. Lett.,* vol. 28, no. 1, pp. 83–85, 1992.

71. A. Cutolo, M. Iodice, P. Spirito, and L. Zeni, "Silicon Electrooptic Modulator Based On A 3-terminal Device Integrated In A Low-loss Single-mode SOI Wave-guide," *J. Lightwave Technol.,* vol. 15, no. 3, pp. 505, 1997.

72. M. Usui, M. Hikita, T. Watanabe, M. Amano, S. Sugawara, S. Hayashida, and S. Imamura, "Low-loss passive polymer optical waveguides with high environmental stability," *J. Lightw. Technol.,* vol. 14, no. 10, pp. 2338–2343, 1996.

73. G. Fischbeck, R. Moosburger, C. Kostrzewa, and A. Achen, "Singlemode Optical Waveguides Using A High-temperature Stable Polymer With Low Losses In The 1. 55 μm Range," *Electron. Lett.,* vol. 33, no. 6, pp. 518–519, 1997.

74. L. Eldada, L. Shacklette, R. Norwood, and J. Yardley, "Single-Mode Optical Interconnects in Ultra-Low-Loss Environmentally-Stable Polymers," in OSA Technical Digest Series Vol. 14. Organic Thin Films for Photonics Applications. Long Beach, CA: OSA, October 15–17, 1997, pp. 5–7.

75. N. Keil, H. Yao, and C. Zawadzki, "Polymer waveguide optical switch with <–40 dB polarisation independent crosstalk," *Electron. Lett.,* vol. 32, no. 7, pp. 655–657, 1996.

76. D. Chen, H. Fetterman, A. Chen, and W. Steier, "Demonstration of 110 GHz Electro-Optic Polymer Modulators," *Appl. Phys. Lett.,* vol. 70, no. 25, pp. 3335–3337, 1997.

77. S. Toyoda, A. Kaneko, N. Ooba, M. Hikita, H. Yamada, T. Kurihara, K. Okamoto, and S. Imamura, "Polarization-independent low crosstalk polymeric AWG-based tunable filter," European Conference on Optical Communication. Madrid, Spain, September 20–24, 1998, pp. postdeadline 103–104.

78. C. H. Henry, G. E. Blonder, and R. F. Kazarinov, "Glass Waveguides on Silicon for Hybrid Optical Packaging," *J. Lightw. Technol.,* vol. 7, pp. 1530–1539, 1989.

79. S. Vallette, S. Renard, H. Denis, J. P. Jadot, A. Fournier, P. Philippe, P. Gidon, A. M. Grouillet, and E. Desgranges, "Si-based Integrated Optics Technologies," Solid State Technol., pp. 69–75, 1989.

80. B. P. Pal, "Guided Wave Optics in Silicon: Physics, Technology and Status," Progress in Optics, vol. 32, pp. 3–59, 1993.

81. Y. Li and C. Henry, "Silicon optical bench waveguide technology," in Optical Fiber Telecommunications IIIB, I. Kaminow and T. Koch, Eds. San Diego: Academic Press, 1997, pp. 319–376.

82. M. Okuno, N. Takato, T. Kitoh, and A. Sugita, "Silica-based thermo-optic switches," NTT Review, vol. 7, no. 5, pp. 57–63, 1995.

PROBLEMS

1. Derive the expression for the phase change on reflection for parallel-polarized plane waves incident from medium n_1 to n_2. Show that the reflected wave leads the incident wave in phase.

2. Show that the characteristic equation for TE modes in a planar step index waveguide can be expressed as

$$\frac{1}{2}V\sqrt{1-b} = m\pi + \tan^{-1}\left(\sqrt{\frac{a+b}{1-b}}\right) + \tan^{-1}\left(\sqrt{\frac{b}{1-b}}\right) \quad m = 0, 1, 2, 3 \ldots$$

and the characteristic equation for TM modes can be written as

$$\frac{1}{2}V\sqrt{1-b} = m\pi + \tan^{-1}\left(\frac{1}{c-a(1-c)}\sqrt{\frac{a+b}{1-b}}\right) + \tan^{-1}\left(\frac{1}{c}\sqrt{\frac{b}{1-b}}\right)$$

$$m = 0, 1, 2, 3 \ldots$$

where $c = n_2^2/n_1^2$ and $V = k_0\dfrac{h}{2}\sqrt{n_1^2 - n_2^2}$.

3. For modes near cutoff, the wave can penetrate quite a distance into the cladding layer where the effective wave velocity is higher than the wave velocity for the lowest order mode. Use this argument to derive an intermodal dispersion expression. Discuss your result and compare with Eq. (187).

4. Using Eqs. (212)–(216) to derive the field expression in the four corners shown in Figure 2-18 and derive the characteristic Eqs. (222)–(229) by using proper boundary conditions.

5. Consider a symmetrical square waveguide of $h = w = 5.94$ μm. The core index of $n_{co} = 1.5$ is surrounded by a dielectric material of $n_s = 1.498$. Use the wave equation analytical method, effective index method and perturbation correction method to determine the lowest propagation mode in the waveguide. Discuss your results.

6. Derive Eq. (254).

CHAPTER THREE

Digital Filter Concepts for Optical Filters

In this chapter, we present digital filter concepts along with comparisons between digital and optical filters. The advantage of relating digital and optical filters is that numerous algorithms developed for digital filters can be used to design optical filters. Linear system theory is the foundation for analyzing signals in the time and frequency domains, and is reviewed first for both continuous and discrete signals. Then, we introduce the following three major filter classes: moving average (MA), autoregressive (AR) and autoregressive moving average (ARMA) filters. The filter magnitude, group delay and dispersion are given in terms of the Z-transform description. Single-stage optical filters are introduced next along with their Z-transform descriptions. Then, we turn out attention to multi-stage filters, which are required to closely approximate a desired magnitude or phase response. Examples of digital filter design techniques are given for each filter class, and an overview of multi-stage digital filter architectures is presented. The optical counterparts to these architectures are discussed in Chapters 4–6. For an in depth treatment of digital signal processing and digital filters, we refer the reader to texts such as [1–3,27–29].

3.1 LINEAR TIME-INVARIANT SYSTEMS

Filters, whether digital or optical, operate on an input signal to produce an output signal. The input and output signals may be functions of time or space and may be defined continuously in the independent variable or only at discrete points. The Fourier transform relates the time (or space) domain signal to its equivalent spectral domain counterpart which is a function of temporal (or spatial) frequency. Only signals with one independent variable will be considered, although the concepts are extendible to multiple variables. The filter is viewed as a system operating on a signal.

A linear time-invariant system is one which behaves linearly with respect to the input and whose parameters are stationary with time. The operation of a linear time-invariant system on an input signal is more simply described in the frequency domain than in the time domain. Linear system theory is presented for continuous signals first, and then discrete signals. Both are important for optical filters since the input optical signals are continuous in time, but optical filters are conveniently modeled as discrete time filters.

3.1.1   Continuous Signals

A continuous signal is defined by a function $x(t)$ over some continuous variable t which represents time or space. When a system operates on the signal, the output is given by $y(t) = S[x(t)]$ where $S[\ ]$ indicates the system operating on a signal enclosed in the brackets. If there are two inputs, $x_1(t)$ and $x_2(t)$, then the output for any system is given as $y(t) = S[x_1(t) + x_2(t)]$. For a linear system, denoted by $L[\ ]$, the output is equal to the superposition of the outputs for each individual input such that $y(t) = L[x_1(t)] + L[x_2(t)]$. If the input is multiplied by a constant factor, "a," the output of a linear system varies proportionally, $y(t) = L[ax(t)] = aL[x(t)]$. The condition that must be satisfied for a system to be linear is summarized as follows:

$$y(t) = L[a_1x_1(t) + a_2x_2(t)] = a_1L[x_1(t)] + a_2L[x_2(t)] \tag{1}$$

where a_1 and a_2 are constants. There are two flavors of linear optical systems. If the system is linear in the field amplitude, then the system is operating coherently. An incoherent system, on the other hand, is linear in intensity. The distinction arises from the coherence of the input signal relative to the system delay. The coherence time of the input signal, τ_c, is inversely proportional to its bandwidth, $\tau_c \approx 1/\Delta\nu$. Coherent optical systems are the main focus of this book; however, a brief description of incoherent optical processing is given in Section 3.3.4.

An important input function for describing linear systems is a very short pulse, called the impulse or delta function and denoted by $\delta(t)$. The impulse function cannot be defined in the simple manner used for most functions. Properties of the short pulse include infinite height at $t = 0$, zero width, and unit area as described in Eqs. (2)–(3).

$$\delta(t) = \begin{cases} \infty \text{ for } t = 0 \\ 0 \text{ for } t \neq 0 \end{cases} \tag{2}$$

$$\int_{-\infty}^{\infty} \delta(t)dt = 1 \tag{3}$$

The delta function is defined as the limit of a pulse as the width is decreased, its height increased and its area held constant. The definition is not restricted to a particular pulse shape. A convenient pulse shape is given by the rectangular function which is defined as follows:

$$\frac{1}{T}\Pi\!\left(\frac{t}{T}\right) = \begin{cases} \dfrac{1}{T} \text{ for } |t| \le \dfrac{T}{2} \\[2ex] 0 \text{ for } |t| > \dfrac{T}{2} \end{cases} \tag{4}$$

The delta function behavior in an integral is then defined as follows:

$$\lim_{T \to 0} \int_{-\infty}^{\infty} \frac{1}{T}\Pi\!\left(\frac{t}{T}\right)dt \tag{5}$$

The following sifting property of the delta function is important to linear system theory and is defined in the limit as $T \to 0$ of the integral of a pulse times the function $x(t)$:

$$\lim_{T \to 0} \frac{1}{T} \int_{-\infty}^{\infty} \Pi\!\left(\frac{t - t_0}{T}\right)x(t)dt = \int_{-\infty}^{\infty} \delta(t - t_0)x(t)dt = x(t_0) \tag{6}$$

The sifting property allows any input to be represented in the form of Eq. (6). The response of a linear system to the impulse function is given a special name, the impulse response. It is denoted by $h(t)$ and defined as follows:

$$h(t) \equiv L[\delta(t)] \tag{7}$$

A system is time invariant if

$$y(t - t_0) = L[x(t - t_0)] \tag{8}$$

so that the output is shifted in time but otherwise identical if the same input is applied at $t = 0$ or $t = t_0$. The output of a linear time invariant system is given by

$$y(t) = L\!\left[\int_{-\infty}^{\infty} \delta(t - t_0)x(t_0)dt_0\right] = \int_{-\infty}^{\infty} L[\delta(t - t_0)]x(t_0)dt_0 \tag{9}$$

The last equality in Eq. (9) can be written in terms of the impulse response as follows:

$$y(t) = \int_{-\infty}^{\infty} h(t - t_0)x(t_0)dt_0 = \int_{-\infty}^{\infty} h(t_0)x(t - t_0)dt_0 \tag{10}$$

The convolution operation defined by Eq. (10) is represented by $y(t) = x(t)*h(t)$. In the time domain, the output of a linear system is the convolution of the input with the system impulse response.

A signal can also be described in the frequency domain by taking the Fourier transform of the time domain function. The Fourier transform and its inverse are defined as follows [4]:

$$H(f) = \int_{-\infty}^{\infty} h(t)e^{-j2\pi ft}dt \tag{11}$$

$$h(t) = \int_{-\infty}^{\infty} H(f)e^{j2\pi ft}df \tag{12}$$

where t denotes a time variable and f denotes temporal frequencies. An alternate expression is given in terms of the angular frequency, $\Omega = 2\pi f$.

$$h(t) = \frac{1}{2\pi}\int_{-\infty}^{\infty} H(\Omega)e^{j\Omega t}d\Omega \tag{13}$$

The same transform pair relates the spatial variable x to spatial frequencies denoted by s. The Fourier transform exists if $h(t)$ is absolutely integrable,

$$\int_{-\infty}^{\infty} |h(t)|dt < \infty \tag{14}$$

and if $h(t)$ has only a finite number of discontinuities, maxima, and minima. Many important functions do not satisfy these criteria, such as $\cos(\Omega_0 t)$ and $\sin(\Omega_0 t)$. In these cases, a transform pair "in the limit" is found by using the delta function in the frequency response [4].

The Fourier transform of the impulse response is called the system frequency response and is denoted by $H(f)$. Let the system input be a complex sinusoidal function, $x(t) = \exp(j2\pi f_0 t)$, so the output is given by

$$y(t) = \int_{-\infty}^{\infty} h(t_0)e^{j2\pi f_0(t-t_0)}dt_0 \tag{15}$$

The response of the system is proportional to the transfer function at f_0.

$$y(t) = H(f_0)e^{j2\pi f_0 t} \tag{16}$$

Eq. (16) illustrates that sinusoidal inputs with a frequency f_0 are eigenfunctions of linear time-invariant systems with eigenvalues equal to $H(f_0)$. The transfer function is complex, $H(f) = |H(f)|e^{j\Phi(f)}$, so a linear system has both a magnitude response $|H(f)|$ and a phase response $\Phi(f)$ associated with it.

Several properties help to simplify calculating the Fourier transforms of functions. Three important properties are derived below. The first property relates the Fourier transform of a time shifted function to that of a non-shifted function. The Fourier transform of a shifted function is given as follows:

$$\int_{-\infty}^{\infty} h(t-t_0)e^{-j2\pi ft}dt = e^{-j2\pi ft_0}\int_{-\infty}^{\infty} h(x)e^{-j2\pi fx}dx = e^{-j2\pi ft_0}H(f) \tag{17}$$

where the substitution $x = t - t_0$ was made for the second equality. The magnitude of the Fourier transform remains the same; only a linear phase term proportional to the delay is added. The second property shows how scaling a function in the time domain transforms to the frequency domain. The scaling factor is denoted by "a", and the Fourier transform is given by:

$$\int_{-\infty}^{\infty} h(at)e^{-j2\pi ft}dt = \int_{-\infty}^{\infty} h(x)e^{-j2\pi fx/a}\frac{dx}{|a|} = \frac{1}{|a|}H\left(\frac{f}{a}\right) \tag{18}$$

where the substitution $x = at$ was made in the second equality. A real positive scaling factor larger than one compresses the time domain response, but expands the frequency domain response. The third property relates convolution in the time domain to multiplication in the frequency domain. The Fourier transform of Eq. (10) is written as follows:

$$Y(f) = \int_{-\infty}^{\infty}\left[\int_{-\infty}^{\infty} x(t_0)h(t - t_0)dt_0\right]e^{-j2\pi ft}dt \tag{19}$$

By changing the order of integration, we obtain

$$Y(f) = \int_{-\infty}^{\infty} x(t_0)\left[\int_{-\infty}^{\infty} h(t - t_0)e^{-j2\pi ft}dt\right]dt_0 = \int_{-\infty}^{\infty} x(t_0)e^{-j2\pi ft_0}H(f)dt_0 \tag{20}$$

The bracketed quantity in the middle term is the Fourier transform of the shifted impulse response, which yields $e^{-j2\pi ft_0}H(f)$. The expression on the right is the Fourier transform of $x(t)$; therefore, the convolution of $x(t)$ and $h(t)$ results in multiplication in the frequency domain as follows:

$$Y(f) = X(f)H(f) \tag{21}$$

TABLE 3-1   Fourier Transform Properties for Continous Signals

Property	Time	Fourier Transform
Linearity	$ax(t) + by(t)$	$aX(f) + bY(f)$
Time Shifting	$x(t - t_0)$	$e^{-j2\pi ft_0}X(f)$
Scaling	$x(at)$	$X(f/a)/|a|$
Time Reversal	$x(-t)$	$X(-f)$
Conjugation	$x^*(-t)$	$X^*(-f)$
Convolution	$\int_{-\infty}^{\infty} x(t_0)h(t - t_0)dt_0 = x(t)*h(t)$	$X(f)H(f)$
Autocorrelation & Power Spectral Density	$\int_{-\infty}^{\infty} x^*(t_0)x(t + t_0)dt_0 = x^*(t)*x(-t)$	$|X(f)|^2$
Integration	$\int_{-\infty}^{\infty} x(t)dt$	$X(0)$
Modulation	$e^{+j2\pi f_0 t}x(t)$	$X(f - f_0)$
Energy Conservation	$\int_{-\infty}^{\infty} |x(t)|^2 dt$	$\int_{-\infty}^{\infty} |X(f)|^2 df$

Other properties, such as time reversal and complex conjugation, are summarized in Table 3-1.

Fourier transforms are used in diffraction theory to relate the near field and far field as indicated in Figure 3-1a. Let a one dimensional field at $z = 0$ be expressed as $f(x)$. The distribution in the far field, $z = L$, is then given as follows [5]:

$$\frac{n\Phi}{\lambda L} F(-s) = \frac{n\Phi}{\lambda L} \int_{-\infty}^{\infty} f(x) e^{+j2\pi s x} dx \qquad (22)$$

where $F(s)$ is the Fourier transform of $f(x)$, $s = n \sin\theta/\lambda$, n is the refractive index in the medium, θ is the far field angle, and λ is the wavelength in vacuum. The phase factor is $\Phi = je^{-jk_0 nL} e^{-jk_0 nx'^2/2L}$. The far field is written in terms of the transverse coordinate x' using $\sin\theta \approx x'/L$, so that $s \approx x'n/\lambda L$. The distance required for the Fourier transform relationship to apply, known as the Fraunhofer approximation,

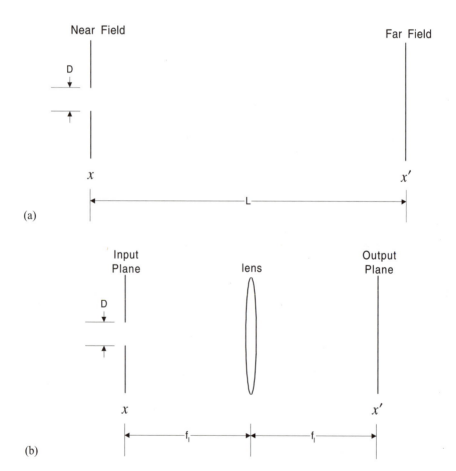

FIGURE 3-1   Fourier transform relationship (a) of the near and far field in the Fraunhofer approximation and (b) of the input and output focal planes in the Fresnel approximation.

must satisfy $L \gg nD^2/4\lambda$, where D is the aperture size. A lens positioned one focal distance away from the object creates a Fourier transform of the object at its focal plane as shown in Figure 3-1b. The focal distance is denoted by f_l. Equation (22) with f_l replacing L and $\Phi = je^{-j2k_0 n f_l}$ relates the object and image planes as long as the Fresnel approximation is valid, which is expressed by $\theta_{max}^4 nD/4\lambda \ll 1$.

To develop a better understanding, the Fourier transform is determined for three useful functions. First, the Fourier transform of the delta function is given by

$$\int_{-\infty}^{\infty} \delta(t) e^{-j2\pi f t} dt = 1 \tag{23}$$

It is a constant as a result of the sifting property. An impulse function contains all frequencies with equal magnitude and is a mathematical construct rather than a physical entity. The second example is a uniformly illuminated one-dimensional aperture represented by the rectangular function. Its Fourier transform is given by

$$\int_{-\infty}^{\infty} \prod(t) e^{-j2\pi f t} dt = \int_{-0.5}^{0.5} e^{-j2\pi f t} dt = \frac{\sin(\pi f)}{\pi f} \equiv \text{sinc}(f) \tag{24}$$

The resulting function is the sinc function. The third function is the Gaussian. It transforms into another Gaussian function as follows [4]:

$$\int_{-\infty}^{\infty} e^{-\pi t^2} e^{-j2\pi f t} dt = e^{-\pi f^2} \tag{25}$$

As discussed in Chapter 2, the Gaussian function is a useful approximation to the fundamental waveguide mode. Since the integral of a function is given by its Fourier transform evaluated at $f = 0$, the normalization constant for a Gaussian function can easily be determined using the scaling properties and Eq. (25). Transform pairs for various continuous functions are summarized in Table 3-2.

TABLE 3-2 Continuous Function Transform Pairs

Time Domain Function	Fourier Transform
$\delta(t)$	1
$\prod(t)$	$\text{sinc}(f)$
$\text{sinc}(t)$	$\prod(f)$
$u(t)$	$\dfrac{1}{2}\left[\delta(f) - \dfrac{j}{\pi f}\right]$
$\cos(2\pi f_0 t)$	$\dfrac{1}{2}[\delta(f-f_0) + \delta(f+f_0)]$
$\sin(2\pi f_0 t)$	$\dfrac{1}{2j}[\delta(f-f_0) - \delta(f+f_0)]$
$\exp(-\pi t^2)$	$\exp(-\pi f^2)$

The symmetry of a given function provides information about the symmetry of its Fourier transform. A function can be decomposed into even and odd parts, $h(t) = [h_e(t) + h_o(t)]$, which are defined as follows:

$$h_e(t) = \frac{1}{2}[h(t) + h(-t)] \quad h_o(t) = \frac{1}{2}[h(t) - h(-t)] \tag{26}$$

The Fourier transform kernel is described in terms of the sine and cosine functions using Euler's identity, $e^{j2\pi ft} = \cos(2\pi ft) + j\sin(2\pi ft)$, so that

$$H(f) = \int_{-\infty}^{\infty} \{h_e(t) + h_o(t)\}\{\cos(2\pi ft) - j\sin(2\pi ft)\} dt \tag{27}$$

Only the even functions in time produce a nonzero integral. The results are shown below for the surviving terms.

$$H_e(f) + jH_o(f) = \int_{-\infty}^{\infty} h_e(t)\cos(2\pi ft)dt - j\int_{-\infty}^{\infty} h_o(t)\sin(2\pi ft)dt \tag{28}$$

An even function transforms into an even function in f, while an odd function transforms into an odd function. For real valued impulse responses, $h_e(t)$ transforms into a real even function $H_e(f)$ and $h_o(t)$ transforms into an imaginary odd function $jH_o(f)$.

3.1.2 Discrete Signals

A similar set of properties applies for discrete signals. A discrete signal may be obtained by sampling a continuous time signal $x(t)$ at $t = nT$ where the sampling interval is T and n is the sample number. For a digital filter, T is the unit delay associated with the discrete impulse response. The impulse response of an optical filter, where each stage has a delay that is an integer multiple of the unit delay, is described by a discrete sequence. The Fourier transform of a sequence has a sum instead of an integral as follows:

$$X(f) = \sum_{n=-\infty}^{\infty} x(nT)e^{-j2\pi fnT} \tag{29}$$

where f denotes the absolute (unscaled) frequency. A normalized frequency is defined as $\nu \equiv fT = f/FSR$, where the free spectral range (FSR) is the period of the unscaled frequency response. The normalized angular frequency is given by $\omega = 2\pi\nu$, while the unscaled angular frequency is denoted by $\Omega = 2\pi f$. A discrete signal is often represented by $x(n)$, leaving T implied. The discrete-time Fourier transform (DTFT) is defined as

$$X(\nu) = \sum_{n=-\infty}^{\infty} x(n)e^{-j2\pi\nu n} \tag{30}$$

The infinite sum exists for functions that are absolutely summable, i.e.

$$\sum_{n=-\infty}^{\infty} |x(n)| < \infty \tag{31}$$

A system whose impulse response satisfies this criterion is stable since it can easily be shown that it has a bounded output for a bounded input [2]. The inverse DTFT is defined as follows:

$$x(n) = \int_{-1/2}^{1/2} X(\nu)e^{j2\pi\nu n} d\nu \tag{32}$$

Note that $X(\nu)$ is periodic with a period of 1, so

$$X(\nu) = X(\nu + m) \tag{33}$$

where m is an integer. A discrete function in one domain transforms to a periodic function in the other domain. Optical filters with discrete time delays have a periodic frequency response.

   If the Fourier transform of $x(t)$ is denoted by $X(f)$ and the Fourier transform of $x(nT)$ is denoted by $X_d(f)$, then $X_d(f)$ contains copies of $X(f)$ spaced by $\Delta f = 1/T$. If $X(f)$ is bandlimited such that $X(f) = 0$ for $|f| \geq 1/2T$, then there is no overlap of frequency components in $X_d(f)$ from adjacent copies. A signal that is bandlimited to $1/2T_2$ is sampled and the resulting $X_d(f)$ spectra are shown in Figure 3-2 for sampling intervals $T_3 < T_2 < T_1$. Aliasing occurs between frequencies from adjacent orders for T_3. Sampling at the Nyquist rate, $1/T = 2f_{max}$, or higher avoids aliasing.

   The delta function from the continuous-time case is replaced by the Kronecker delta function which is defined as follows:

$$\delta(n) = \begin{cases} 1, \, n = 0 \\ 0, \, n \neq 0 \end{cases} \tag{34}$$

Any discrete signal can be represented by a sum of constants times the delta function.

$$x(n) = \sum_{k=-\infty}^{\infty} x(k)\delta(n-k) \tag{35}$$

The response of a discrete time system to the Kronecker delta function input defines its impulse response. The response of a linear, time-invariant system to a shifted input is given as follows:

$$h(n-k) = L[\delta(n-k)] \tag{36}$$

Similar to the continuous case, the system output is the convolution of the input signal and the system impulse response.

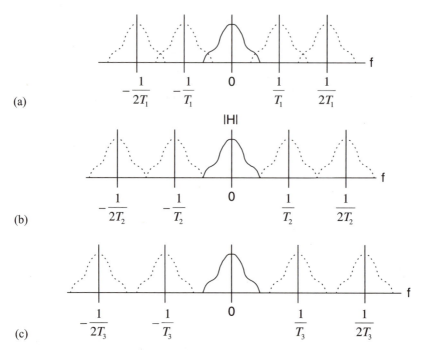

FIGURE 3-2 Frequency domain aliasing of bandlimited discrete time signals versus the sampling interval where $T_1 > T_2 > T_3$.

$$y(n) = L\left[\sum_{k=-\infty}^{\infty} x(k)\delta(n-k)\right] = \sum_{k=-\infty}^{\infty} h(n-k)x(k) \qquad (37)$$

The DTFT of this convolution operation simplifies to multiplication as follows:

$$Y(v) = \sum_{n=-\infty}^{\infty} \left[\sum_{k=-\infty}^{\infty} h(k)x(n-k)\right] e^{-j2\pi vn} \qquad (38)$$

$$= \sum_{k=-\infty}^{\infty} h(k)\left[\sum_{n=-\infty}^{\infty} x(n-k)e^{-j2\pi vn}\right]$$

$$= X(v)\sum_{k=-\infty}^{\infty} h(k)e^{-j2\pi vk} = X(v)H(v)$$

The DTFT properties for discrete signals are summarized in Table 3-3, and several discrete functions and their transform pairs are shown in Table 3-4.

Next, a discrete time signal is considered that is finite in extent such that $x(n) \neq 0$ for $n = 0, \ldots, N-1$ and $x(n) = 0$ otherwise. By shifting and repeating the nonzero

TABLE 3-3 DTFT and Z-Transform Properties for Discrete Signals

Property	Time	DTFT	Z Transform
Linearity	$ax(n) + by(n)$	$aX(\omega) + bY(\omega)$	$aX(z) + bY(z)$
Time Shifting	$x(n - k)$	$e^{-j\omega k}X(\omega)$	$z^{-k}X(z)$
Time Reversal	$x(-n)$	$X(-\omega)$	$X(z^{-1})$
Conjugation	$x^*(n)$	$X^*(-\omega)$	$X^*(z^*)$
Convolution	$x(n)*h(n)$	$X(\omega)H(\omega)$	$X(z)H(z)$
Autocorrelation & Power Spectral Density	$\displaystyle\sum_{n=-\infty}^{\infty} x^*(k)x(k+n)$ $= x^*(n)*x(-n)$	$\lvert X(\omega)\rvert^2$	$\lvert X(z)\rvert^2$

sequence, $x(n)$ can be made periodic in time such that $x(n) = x(n + N)$. Periodicity in time implies a discrete frequency domain representation. As previously seen, the sampling interval in one domain is inversely proportional to the periodicity in the opposite domain; therefore, the frequency domain sampling interval is given by one over the time domain periodicity, or $1/N$. The discrete Fourier transform and its inverse are given as follows:

$$X(k) = \sum_{n=0}^{N-1} x(n)e^{-j2\pi\frac{kn}{N}} \tag{39}$$

$$x(n) = \frac{1}{N}\sum_{k=0}^{N-1} X(k)e^{+j2\pi\frac{kn}{N}} \tag{40}$$

Both sequences have the same length and are periodic with period N. A property,

TABLE 3-4 Discrete Time Function Transform Pairs

Time Domain Function	Discrete-Time Fourier Transform
$\delta(n)$	1
$\Pi\left(\dfrac{n}{N}\right) = \begin{cases} 1 \text{ for } \lvert n\rvert \le N \\ 0 \text{ for } \lvert n\rvert > N \end{cases}$	$\dfrac{\sin\left(\omega\left(\dfrac{2N+1}{2}\right)\right)}{\sin\left(\dfrac{\omega}{2}\right)}$
$u(n)$	$U(\omega) = \pi\delta(\omega) + \dfrac{1}{2} - j\dfrac{1}{2}\cot\left(\dfrac{\omega}{2}\right)$
$a^n u(n)$	$\dfrac{1}{1 - ae^{-j\omega}}$

which can also be derived for the continuous time case, relates the energy in each domain. The energy is given by the square of the time domain sequence.

$$E = \sum_{n=0}^{N-1} |x(n)|^2 \tag{41}$$

The energy can be expressed in the frequency domain by substituting Eq. (40) into Eq. (41) as follows:

$$E = \sum_{n=0}^{N-1} \left[\frac{1}{N^2} \left\{ \sum_{k=0}^{N-1} X_k e^{+j2\pi\frac{kn}{N}} \right\} \left\{ \sum_{m=0}^{N-1} X_m^* e^{-j2\pi\frac{mn}{N}} \right\} \right] \tag{42}$$

where $X_k \equiv X(k)$. The bracketed expression can be written in matrix form where the sum of the matrix elements yields E.

$$E = \sum_{\text{matrix elements}} \left\{ \sum_{n=0}^{N-1} \begin{bmatrix} |X_0|^2 & X_0^* X_1 W & \cdots & X_0^* X_{N-1} W^{(N-1)} \\ X_0 X_1^* W^{-1} & |X_1|^2 & & \\ \vdots & & \ddots & \\ X_0 X_{N-1}^* W^{-(N-1)} & & & |X_{N-1}|^2 \end{bmatrix} \right\} \tag{43}$$

where $W \equiv \exp(+j2\pi/N)$. Two types of terms result, cross terms for $k \neq m$ and diagonal terms for $k = m$. When the sum over n is performed on each matrix element, the cross terms have the form

$$X_k X_m^* \sum_{n=0}^{N-1} e^{+j2\pi\frac{(k-m)n}{N}}$$

for $k \neq m$. The summation has the following solution:

$$\sum_{n=0}^{N-1} e^{+j2\pi\frac{(k-m)n}{N}} = \begin{cases} N \text{ for } k = m \\ 0 \text{ for } k \neq m \end{cases} \tag{44}$$

All the cross-terms are zero. Only the diagonal terms are left, which contribute $N|X_k|^2$ for each k. The result is that the coefficients in the time and frequency domains are related by

$$\sum_{n=0}^{N-1} |x(n)|^2 = \frac{1}{N} \sum_{k=0}^{N-1} |X(k)|^2 \tag{45}$$

Equation (45) is known as the Weiner–Klintchine theorem (or Parseval's identity). Both domains have the same number of coefficients, and the total energy is conserved. When the fields in an input and output array of waveguides are related by the discrete or continuous Fourier transform, power is conserved.

3.2   DIGITAL FILTERS

3.2.1   The *Z*-transform

The *Z*-transform is an analytic extension of the DTFT for discrete signals, similar to the relationship between the Laplace transform and the Fourier transform for continuous signals. The Laplace transform is used to describe analog filters (see the Appendix), while *Z*-transforms are used to describe digital filters. The *Z*-transform is defined for a discrete signal by substituting z for $e^{j\omega}$ in Eq. (30) as follows:

$$H(z) = \sum_{n=-\infty}^{\infty} h(n)z^{-n} \qquad (46)$$

where $h(n)$ is the impulse response of a filter or the values of a discrete signal, and z is a complex number that may have any magnitude. For the power series to be meaningful, a region of convergence must be specified, for example $r_{min} \leq |z| \leq r_{max}$ where r_{min} and r_{max} are radii. Of particular interest is $|z| = 1$, called the unit circle, because the filter's frequency response is found by evaluating $H(z)$ along $z = e^{j\omega}$. The inverse *Z*-transform is found by applying the Cauchy integral theorem to Eq. (46) to obtain:

$$h(n) = \frac{1}{2\pi j} \oint H(z)z^{n-1}dz \qquad (47)$$

The convolution resulting from filtering in the time domain

$$y(n) = x(n)*h(n) = \sum_{m=-\infty}^{\infty} x(m)h(n-m) \qquad (48)$$

reduces to multiplication in the *Z*-domain.

$$Y(z) = H(z)X(z) \qquad (49)$$

Equation (49) shows that a filter's transfer function, $H(z)$, can be obtained by dividing the output by the input in the *Z*-domain.

$$H(z) = \frac{Y(z)}{X(z)} \qquad (50)$$

A fundamental property of the *Z*-transform relates $h(n-1)$ to $H(z)$ as shown in Eq. (51).

$$\sum_{n=-\infty}^{\infty} h(n-1)z^{-n} = z^{-1}\sum_{n=-\infty}^{\infty} h(n)z^{-n} = z^{-1}H(z) \qquad (51)$$

The impulse response is assumed to be causal (discussed in Section 3.2.3) so that $h(n) = 0$ for $n < 0$. A delay of one results in multiplication by z^{-1} in the Z-domain, and a delay of N units results in multiplication by z^{N}. Table 3-3 summarizes other useful properties of the Z-transform.

The Z-transforms are determined for three functions to gain familiarity with this new domain. First, let the filter impulse response be the delta function shifted by N samples, $h(n) = \delta(n - N)$. It has the following Z-transform:

$$\sum_{n=-\infty}^{\infty} \delta(n - N)z^{-n} = z^{-N} \tag{52}$$

For this system, the output is defined by $y(n) = x(n - N)$. The Z-transform for a delay line is given by Eq. (52).

For the second example, let the output of a filter be the sum of the last $N + 1$ inputs: $y(n) = x(n) + x(n - 1) + \cdots + x(n - N)$. Such a filter contains N delays that are feedforward paths. The impulse response is $h(n) = \delta(n) + \delta(n - 1) + \cdots + \delta(n - N)$. The transfer function is $H(z) = 1 + z^{-1} + \cdots + z^{-N} = [1 - z^{-(N+1)}]/[1 - z^{-1}]$. There are N roots of $H(z)$ which all occur on the unit circle. The frequency response is given by

$$H(\omega) = e^{-j\omega\frac{N}{2}} \frac{\sin\left(\omega\frac{N+1}{2}\right)}{\sin(\omega/2)}$$

for $|\omega| \leq \pi$. The transmission is maximum at $\omega = 0$ and zero at N points given by $\omega_m = 2\pi m/(N + 1)$ for $m = 1, \ldots, N$. A diffraction grating, discussed in Chapter 4, has this frequency response. The passband width decreases as N increases, thereby providing a higher resolution filter response.

For the third example, let the output of a filter be given as follows: $y(n) = ay(n - 1) + x(n)$, where a is a real number satisfying $0 \leq |a| < 1$ and $n \geq 0$. This filter contains one delay, which is a feedback path. Its Z-transform is $Y(z) = az^{-1}Y(z) + X(z)$, which gives the transfer function $H(Z) = Y(z)/X(z) = 1/(1 - az^{-1})$. The transfer function is equivalent to the infinite sum

$$H(z) = \sum_{n=0}^{\infty} a^n z^{-n} = \frac{1}{1 - az^{-1}} \tag{53}$$

The region of convergence is $|z| > a$. A ring resonator and a Fabry–Perot cavity each have a response that can be expressed in the form of Eq. (53). The impulse response is defined in terms of the unit step function given by

$$u(n) = \begin{cases} 1 \text{ for } n \geq 0 \\ 0 \text{ for } n < 0 \end{cases} \tag{54}$$

The impulse response is $h(n) = a^n u(n)$ for $n \geq 0$, which has an infinite number of terms even though the filter has only one delay. The impulse response decays more slowly as $a \rightarrow 1$.

3.2.2 Poles and Zeros

As seen in the previous two examples, the filter input and output are related by weighted sums of inputs and previous outputs. This relationship is generalized for a discrete linear system with a discrete input signal as follows:

$$y(n) = b_0 x(n) + b_1 x(n-1) + \cdots + b_M x(n-M) - a_1 y(n-1) \tag{55}$$
$$- \cdots - a_N y(n-N)$$

The weights are given by the a and b coefficients. The Z-transform results in a transfer function that is a ratio of polynomials.

$$H(z) = \frac{\displaystyle\sum_{m=0}^{M} b_m z^{-m}}{1 + \displaystyle\sum_{n=1}^{N} a_n z^{-n}} = \frac{B(z)}{A(z)} \tag{56}$$

$A(z)$ and $B(z)$ are Mth and Nth-order polynomials, respectively. The expression for $H(z)$ can also be written in terms of the roots of the polynomials as follows [1]:

$$H(z) = \frac{\Gamma z^{N-M} \displaystyle\prod_{m=1}^{M} (z - z_m)}{\displaystyle\prod_{n=1}^{N} (z - p_n)} \tag{57}$$

The roots are complex numbers and are given different names depending on whether they are from the numerator or denominator polynomials. The zeros of the numerator, also called the zeros of $H(z)$, are represented by z_m. A zero that occurs on the unit circle, $|z_m| = 1$, results in zero transmission at that frequency. The roots of the denominator polynomial, designated by p_n, are called poles. The gain is set by Γ. For passive filters, the transfer function can never be greater than 1, so Γ has a maximum value determined by $\max\{|H(z)|_{z=e^{j\omega}}\} = 1$.

Digital filters are classified by the polynomials defined in Eq. (56). A Moving Average (MA) filter has only zeros and is also referred to as a Finite Impulse Response (FIR) filter. It consists only of feed-forward paths. A single stage MA digital filter is shown in Figure 3-3. An Autoregressive (AR) filter has only poles and contains one or more feedback paths as demonstrated by the single-stage digital filter shown in Figure 3-4. A pole produces an impulse response with an infinite number of terms in contrast to the finite number of terms for MA filters. Filters with both poles and zeros are referred to as Autoregressive Moving Average (ARMA) filters. An Infinite Impulse Response (IIR) filter must contain at least one pole. Since the

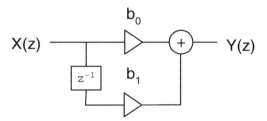

FIGURE 3-3   Single-stage MA digital filter.

IIR designation is ambiguous with respect to whether a filter has zeros or not, the MA/AR/ARMA terminology is used. The advantages, challenges, and tradeoffs between these different filter classes is explored for optical filters in Chapters 4–6.

3.2.3   Stability and Causality

There are two conditions that must be met for practical filters: stability and causality. A filter is stable if it has a bounded output response given a bounded input [2]. An absolutely summable impulse response satisfies this requirement.

$$|y(n)| \leq \sum_{k=-\infty}^{\infty} |h(k)||x(n-k)| \tag{58}$$

$$\leq M_x \sum_{k=-\infty}^{\infty} |h(k)|$$

where M_x represents the bounded input. For the Z-transform, the region of convergence must include the unit circle.

For a realizable filter, there can be no output before an input is applied. That is, if $x(n) = 0$ for $n < 0$, then $y(n) = 0$ for $n < 0$. This property is known as causality. In the convolution operation, causality changes the summation limits. The impact on $H(z)$ is that there are no positive powers of z in the sum; therefore, the region of convergence includes $|z| = \infty$. An important consequence of causality is that the real and

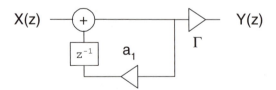

FIGURE 3-4   Single-stage AR digital filter.

imaginary parts of the frequency response, given by $H(\omega) = H_R(\omega) + jH_I(\omega)$, are related. To derive the relationship, we first define the impulse response in terms of even and odd functions. Note that $h(n) = 0$ for $n < 0$, so

$$h_e(n) = \frac{1}{2}[h(n) + h(-n)] = \frac{1}{2}h(n)$$ (59)

$$h_o(n) = \frac{1}{2}[h(n) - h(-n)] = \frac{1}{2}h(n)$$

for $n > 0$. The expression for $h(n)$ takes the following form:

$$h(n) = h_e(n) + h_o(n)$$ (60)

Note that at $n = 0$, the odd part gives no information since $h_0(0) = 0$. The even part suffices at $n = 0$ since $h_e(0) = h(0)$. For a causal filter, $h(n)$ is completely defined by either of the following expressions:

$$h(n) = 2h_e(n)u(n) - h_e(0)\delta(n)$$ (61)
$$= 2h_o(n)u(n) + h(0)\delta(n)$$

From the discrete case equivalent to Eq. (28), we know that the even part of a real impulse response transforms to the real part of the frequency response, $h_e(n) \leftrightarrow H_R(\omega)$, and the odd part to the imaginary portion, $h_o(n) \leftrightarrow jH_I(\omega)$. Using the transform of $u(n)$ from Table 3-4 results in the following expressions relating the real and imaginary parts of the frequency response.

$$H_R(\omega) = h(0) + \frac{1}{2\pi} \int_{-\pi}^{+\pi} H_I(\theta) \cot\left(\frac{\omega - \theta}{2}\right) d\theta$$ (62)

$$H_I(\omega) = -\frac{1}{2\pi} \int_{-\pi}^{+\pi} H_R(\theta) \cot\left(\frac{\omega - \theta}{2}\right) d\theta$$ (63)

The importance of this result is that $H(\omega)$ is completely defined by either $H_I(\omega)$ or $H_R(\omega)$. For causal filters, $H_I(\omega)$ and $H_R(\omega)$ form a Hilbert transform pair. An invaluable extension of this result, from an optical measurement perspective, is to relate the magnitude and phase in a similar fashion so that the phase response can be determined from the magnitude response. The natural logarithm of $H(\omega) = |H(\omega)|e^{j\Phi(\omega)}$ provides the following relationship between the magnitude and phase response [1,6]:

$$\tilde{H}(\omega) = \log_e|H(\omega)| + j\Phi(\omega) = \tilde{H}_R(\omega) + j\tilde{H}_I(\omega)$$ (64)

Equations (62)–(63) are the discrete time analog of the Kramers–Kronig relations which link the real and imaginary index of refraction, $n = n' - jn''$, given in Chapter 2. For $\tilde{H}_R(\omega)$ and $\tilde{H}_I(\omega)$ to be a Hilbert transform pair, the zeros of $H(z)$ must all have a magnitude less than one. Functions with this property are called minimum-phase functions and are discussed in Section 3.2.6. Hilbert transform relations for the magnitude and phase are given in Chapter 7.

Another important consequence of causality is that it limits our ability to realize ideal bandpass filters that have a rectangular frequency response. Their impulse response is the sinc function, which extends from $n = -\infty$ to $n = +\infty$, so it is not a causal filter. Truncation of the infinite impulse response leads to ringing around the transition frequencies, known as the Gibbs phenomenon. Besides infinitely sharp transitions between the passband and stopband, another property of ideal filters is that they have zero transmission over the stopband. Such a response cannot be realized with a causal filter. The Paley–Wiener theorem states that for a causal $h(n)$ with finite energy [2], the following relationship is true.

$$\int_{-\pi}^{\pi} \left| \ln |H(\omega)| \right| d\omega < \infty \tag{65}$$

Note that $|H(\omega)|$ can be zero at discrete frequencies but not over frequency bands because the above integral would no longer be bounded; therefore, ideal filters with $H(\omega) = 0$ over a continuous range of ω cannot be realized with a causal filter.

3.2.4 Magnitude Response

A filter's magnitude response is equal to the modulus of its transfer function, $|H(z)|$, evaluated at $z = e^{j\omega}$. Based on the pole/zero representation of $H(z)$, only the distance of each pole and zero from the unit circle, i.e. $|e^{j\omega} - z_m|$ or $|e^{j\omega} - p_n|$, affects the magnitude response. Consequently, a zero that is located at the mirror image position about the unit circle, i.e. $1/z_m^*$, cannot be differentiated from z_m based on the magnitude response. A convenient graphical method for estimating a filter's magnitude response is the pole-zero diagram as shown in Figure 3-5. The real and imaginary parts of each root are plotted. Zeros are designated by o's and poles by x's. One trip around the unit circle is equal to one FSR. Because there are two zero locations that yield the same magnitude response, a naming convention is used to distinguish them. Zeros with a magnitude >1 are called maximum-phase, and those with magnitudes <1 are called minimum-phase. A pole-zero diagram with a pair of zeros that are located reciprocally about the unit circle is shown in Figure 3-6.

For optical filters, the square magnitude response is most easily measured, since it requires only a square law detector and a means of sweeping the wavelength. The square magnitude response is given below in terms of the magnitude and phase of the transfer function's poles $p_i = r_{pi} \exp(j\phi_{pi})$ and zeros $z_i = r_{zi} \exp(j\phi_{zi})$ [7].

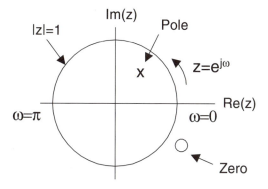

FIGURE 3-5   Pole-zero diagram showing the unit circle, one pole, and one zero.

$$|H(\omega)|^2 = \frac{|\Gamma|^2 \prod\limits_{i=1}^{M}\{1 - 2r_{zi}\cos(\omega - \phi_{zi}) + r_{zi}^2\}}{\prod\limits_{i=1}^{N}\{1 - 2r_{pi}\cos(\omega - \phi_{pi}) + r_{pi}^2\}}$$ (66)

Figure 3-7 compares the magnitude responses of single-stage AR filters having poles at $p_n = 0.8$, 0.9 and 0.95. Figure 3-8 shows the magnitude response for single-stage MA filters with zeros at $z_m = 0.8$, 0.9 and 0.95. The identical magnitude responses are obtained for the maximum-phase counterparts, i.e. $z_m = 1.25$, 1.11 and 1.05. The phase responses are different and will be discussed in section 3.2.5. The single-stage response for an AR filter is a narrow bandpass response. The single-stage response for an MA filter is sinusoidal, which acts as a notch filter by rejecting a narrow range of frequencies. The responses are plotted for one period as a function of the normalized frequency, ν. The optical frequency is obtained by setting $f = \nu/T$.

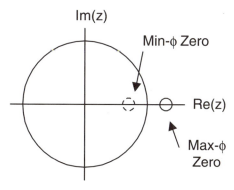

FIGURE 3-6   Pole-zero diagram showing a maximum and a minimum-phase zero.

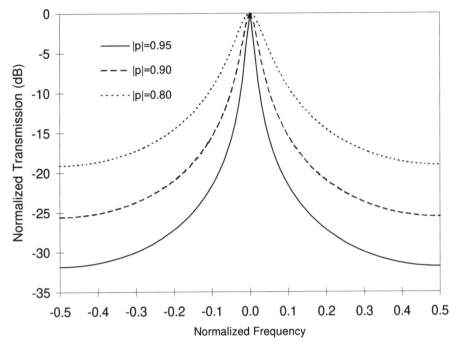

FIGURE 3-7 Magnitude response for a single-stage AR filter as a function of the root magnitude.

3.2.5 Group Delay and Dispersion

The filter's group delay is defined as the negative derivative of the phase of the transfer function with respect to the angular frequency as follows [1]:

$$\tau_n = -\frac{d}{d\omega} \tan^{-1} \left[\frac{\text{Im}\{H(z)\}}{\text{Re}\{H(z)\}} \right]_{z=e^{j\omega}} \tag{67}$$

where τ_n is normalized to the unit delay, T. The absolute group delay is given by $\tau_g = T\tau_n$. The delay is the slope of the phase at the frequency where it is being evaluated, equivalent to the definitions in electromagnetic theory. For example, a dispersionless delay line of unit length has a frequency response $H(\omega) = e^{-j\omega}$, a normalized group delay of 1, and an absolute group delay equal to T. If the phase is changing rapidly around ω, which can occur around a resonance, the derivative loses its meaning as a delay. In such cases, a negative delay may result from Eq. (67). This result should not be interpreted as a non-causal situation, but rather, that the delay for the signal after it has been filtered is no longer well defined. For example, the pulse may be so distorted that it is difficult to identify the center of the

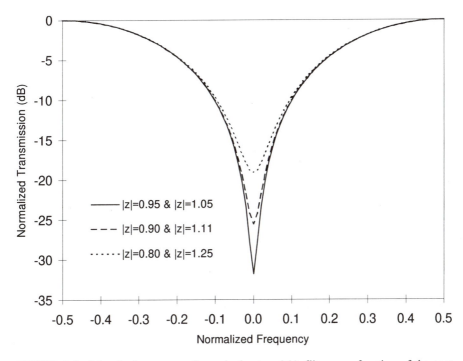

FIGURE 3-8   Magnitude response for a single-stage MA filter as a function of the root magnitude.

pulse after filtering, compared to before filtering. Negative group delay and definitions for signal velocity around a resonance are discussed in [8,9].

To obtain the group delay for a single zero, consider the transfer function $H_{1z}(z)$ $= 1 - re^{j\varphi}z^{-1}$ where r and φ are the magnitude and phase of the zero. By evaluating $H_{1z}(z)$ at $z = e^{j\omega}$ and defining the phase by $H_{1z}(\omega) = |H_{1z}(\omega)|e^{j\Phi_{1z}(\omega)}$, the phase is given explicitly in terms of r and ϕ as follows:

$$\Phi_{1z}(\omega) = \tan^{-1}\left[\frac{r\sin(\omega - \varphi)}{1 - r\cos(\omega - \varphi)}\right] \qquad (68)$$

Given that

$$\frac{d\tan^{-1}[g(x)]}{dx} = \frac{g'(x)}{1 + g^2(x)}$$

the group delay simplifies to

$$\tau_{1z}(r, \varphi) = \frac{r[r - \cos(\omega - \varphi)]}{1 - 2r\cos(\omega - \varphi) + r^2} \qquad (69)$$

Note that for a maximum-phase zero ($r > 1$) at $\omega = \phi$, $\tau_{1z} > 0$ and for a minimum-phase zero, $\tau_{1z} < 0$. For a zero that is a mirror image about the unit circle, i.e. $H_{1z}(z) = re^{-j\varphi} - z^{-1}$, the group delay is

$$\tau_{1z}\left(\frac{1}{r}, \varphi\right) = \frac{[1 - r\cos(\omega - \varphi)]}{1 - 2r\cos(\omega - \varphi) + r^2} \tag{70}$$

The group delay responses of two single-stage filters whose zeros are located at mirror image positions about the unit circle are related by

$$\tau_{1z}\left(\frac{1}{r}, \varphi\right) = 1 - \tau_{1z}(r, \varphi) \tag{71}$$

The group delay for the single zero case is easily extended to multiple zeros. Note that the phase of the overall transfer function is the sum of the phases for each root, i.e. $H(\omega) = |H_{1z}(\omega)| \cdots |H_{Mz}(\omega)|e^{j[\Phi_{1z}(\omega)+\cdots+\Phi_{Mz}(\omega)]}$. Since the group delay is proportional to the derivative of the phase, the group delays are additive.

The group delay for the single pole case can be found from the result for a single zero. Its transfer function is expressed as $H_{1p}(z) = \Gamma/(1 - re^{j\varphi}z^{-1})$ where Γ is a constant for normalizing the peak transmission. In the frequency domain, the magnitude and phase can be defined in terms of the previous magnitude and phase response as $H_{1p}(\omega) = \Gamma e^{-j\Phi_{1z}(\omega)}/|H_{1z}(\omega)|$. Therefore, $\Phi_{1p}(\omega) = -\Phi_{1z}(\omega)$ and $\tau_{1p}(\omega) = -\tau_{1z}(\omega)$. Since the pole magnitude r must always be less than one for stability, poles contribute positive group delay. The following equation expresses the group delay for N poles and M zeros [7].

$$\tau_n(\omega) = \sum_{i=1}^{N} \frac{r_{pi}\{\cos(\omega - \phi_{pi}) - r_{pi}\}}{\{1 - 2r_{pi}\cos(\omega - \phi_{pi}) + r_{pi}^2\}} + \sum_{i=1}^{M} \frac{r_{zi}\{r_{zi} - \cos(\omega - \phi_{zi})\}}{\{1 - 2r_{zi}\cos(\omega - \phi_{zi}) + r_{zi}^2\}} \tag{72}$$

The group delay for a zero at $\omega = \phi_{zi}$ is $-r/(1 - r)$. For a pole at $\omega = \phi_{pi}$, the group delay is $+r/(1 - r)$. The group delay response is shown in Figure 3-9 for poles with various magnitudes. As the magnitude of a pole or zero approaches the unit circle, the magnitude of the group delay increases, but the response becomes more localized in frequency as shown in the inset where the full width at half the maximum (FWHM) is equal to 0.016 times the FSR for a pole with a magnitude of 0.95. In Figure 3-10, the group delay for a minimum-phase and maximum-phase zero are shown. The sum of the group delays is a constant indicating that the filter has linear-phase. The constant group delay is equal to half the filter order. Ideal bandpass filters have linear-phase so that they do not distort signals in the passband. Linear-phase filters are discussed in more detail in Section 3.2.6.

The filter dispersion, in normalized units, is defined by $D_n \equiv d\tau_n/d\nu$. Figure 3-11 shows the normalized dispersion for a pole with a magnitude of 0.80 (or a maximum-phase zero with a magnitude of 1/0.80). To compare the filter response to the optical fiber transmission properties, we relate the normalized group delay and dispersion to the standard definitions for group delay and dispersion in terms of the

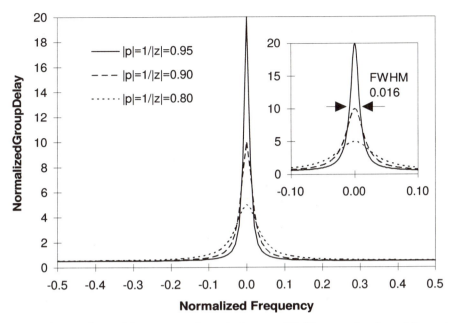

FIGURE 3-9   Group delay response for a single-stage AR filter as a function of the root magnitude.

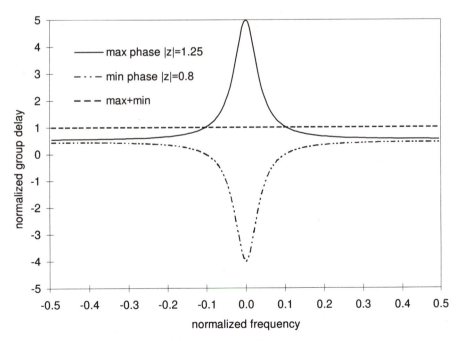

FIGURE 3-10   Constant group delay demonstrated by the sum of the group delay of a maximum- and minimum-phase zero with equal magnitudes.

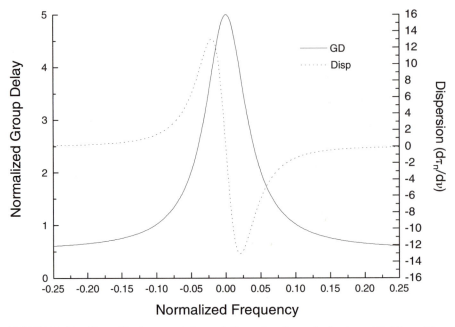

FIGURE 3-11. Normalized group delay and dispersion for a single-stage AR filter with a root magnitude equal to 0.80.

propagation constant. For a wave traveling in a bulk material, the frequency response for propagation a distance L is given by $e^{-\alpha(\Omega)L}e^{-j\beta(\Omega)L}$ where $\alpha(\Omega)$ is the absorption, the phase is $\beta(\Omega) = (\Omega/c)n(\Omega)$, and $n(\Omega)$ is the material's refractive index. For a guided wave, an effective index replaces $n(\Omega)$, and $\alpha(\Omega)$ includes scattering and other sources of loss, not just the material absorption. The propagation constant changes gradually with respect to frequency and is expanded in a Taylor series as follows:

$$\beta(\Omega) = \beta(\Omega_c + \Delta\Omega) = \beta(\Omega_c) + \beta'\,\Delta\Omega + \frac{1}{2!}\,\beta''\,\Delta\Omega^2 + \frac{1}{3!}\,\beta'''\,\Delta\Omega^3 + \cdots \quad (73)$$

where $\Delta\Omega = \Omega - \Omega_c$ and the primes indicate derivatives of β with respect to Ω evaluated at Ω_c. The first term is a constant which is neglected since the absolute phase is typically not important. The first derivative is defined as $\beta' = d\beta/d\Omega = \tau_g/L$. Note that $\tau_g = L(d\beta/d\Omega)$ for a waveguide and $\tau_g = -T(d\Phi/d\omega)$ for a filter. All higher order terms cause dispersion of a pulse. The second derivative of β is the quadratic dispersion term. For optical fibers, dispersion is typically defined as the derivative of the group delay with respect to wavelength and normalized with respect to length,

$$D = \frac{1}{L}\frac{d\tau_g}{d\lambda} = -\frac{1}{L}\frac{2\pi c}{\lambda^2}\beta''$$

so the units are ps/nm/km. The filter dispersion in absolute units is given by

$$D = -c\left(\frac{T}{\lambda}\right)^2 D_n \tag{74}$$

Large filter dispersions are achieved by increasing T; however, the FSR decreases accordingly. The third derivative of Eq. (73) gives the cubic dispersion, also referred to as dispersion slope and specified in units of ps/nm^2/km. For waveguides, $dD/d\lambda = 2\pi c[2\beta'' + (2\pi c/\lambda)\beta''']/\lambda^3$. The fourth derivative of β gives the quartic dispersion, and so forth for higher orders.

Standard singlemode fiber, i.e. non-dispersion shifted, has a typical dispersion of +17 ps/nm/km and a dispersion slope of +0.08 ps/nm^2/km for $\lambda = 1550$ nm. For $T = 100$ ps (FSR = 10 GHz), $D = -1250 D_n$ (ps/nm). For $D_n = -1$, D is roughly equal to 70 km of standard singlemode fiber. Filters with large dispersions can be designed by making the poles and zeros close to the unit circle. The filter phase can change rapidly with respect to frequency, particularly for filters with small FSRs, in contrast to the slow variation of the waveguide propagation constant. Any dispersion of the fiber's propagation constant can be approximated over a portion of a filter's FSR by manipulating the pole and zero locations. All-pass filters, discussed in Chapter 6, allow the phase response to be adjusted without affecting the filter's magnitude response.

3.2.6 Minimum-, Maximum-, and Linear-Phase Filters

An important class of filters are those having linear-phase. Their frequency response is described as follows: $H(\omega) = A(\omega)e^{-j(\omega\tau + \phi)}$ where $A(\omega)$ is real and ϕ is an arbitrary constant phase factor that does not change the linear-phase property. The group delay, τ, is a constant in this case, so the filter is dispersionless! No phase distortion is induced on the signal. From Eq. (71), it is clear that two zeros located at reciprocal points about the unit circle will cancel each other's frequency dependence and leave only a constant group delay. A pole and a maximum-phase zero similarly positioned about the unit circle will not exhibit a linear-phase response.

A unique property of MA filters is that they can be designed to have linear-phase. We will show that filters with linear-phase have an impulse response with Hermitian symmetry. To develop this property, the concept of a reverse polynomial is first introduced. The reverse polynomial appears in the Z-transform description of many optical filters. A single-stage MA filter with a zero at $z_1 = |z_1|\exp(j\phi_1)$ has a transfer function given by

$$H_1(z) = \Gamma(1 - z_1 z^{-1}) = h_0 + h_1 z^{-1} \tag{75}$$

A new transfer function is obtained when the zero is reflected about the unit circle, $z_1 \rightarrow 1/z_1^*$. It is written in a form that keeps the magnitude responses equal.

$$H_1^R(z) = \Gamma(-z_1^* + z^{-1}) = h_1^* + h_0^* z^{-1} \tag{76}$$

Both $H_1(z)$ and $H_1^R(z)$ have the following magnitude response.

$$|H_1(\omega)|^2 = |H_1^R(\omega)|^2 = 1 + |z_1|^2 + 2|z_1|\cos(\omega - \phi_1) \tag{77}$$

Furthermore, $H_1^R(z)$ can be obtained from $H_1(z)$ by the following linear operations: $H_1^R(z) = z^{-1}H_1^*(z^{*-1})$. The polynomial $H_1^R(z)$ is called the reverse polynomial of $H_1(z)$ because the coefficients are reversed in order. The polynomial $H_1(z)$ is the forward polynomial; however, to keep the notation simple, a superscript F will not be used.

Now consider a MA filter with a transfer function $H_N(z) = \Gamma\prod_{n=1}^{N}(1 - z_n z^{-1})$, which consists of a mixture of minimum and maximum-phase zeros. If Γ is complex, its argument adds a constant to the phase response but does not change the magnitude or group delay response of $H_N(z)$. Since only the relative phase is important for a filter response, we take Γ to be a real quantity for the following discussion. The reverse polynomial is obtained by reflecting each zero about the unit circle which yields

$$H_N^R(z) = \Gamma\prod_{n=1}^{N}(z_n^* - z^{-1}) \tag{78}$$

The reverse polynomial is defined in terms of $H_N(z)$ as follows:

$$H_N^*(z^{*-1}) = \Gamma\prod_{n=1}^{N}(1 - z_n^* z) \tag{79}$$

$$= \Gamma z^N\prod_{n=1}^{N}(-z_n^* + z^{-1}) = z^N H_N^R(z)$$

So, just as with the single stage case,

$$H_N^R(z) = z^{-N}H_N^*(z^{*-1}) \tag{80}$$

Using the properties in Table 3-3, these Z-domain operations are equivalent to time reversal, delaying N samples to make the new response causal, and complex conjugation of the impulse response. The coefficients of $H_N(z)$ define the impulse response $h(n)$.

$$H_N(z) = h_0 + h_1 z^{-1} + \cdots + h_N z^{-N} \tag{81}$$

The reverse polynomial is given by

$$H_N^R(z) = z^{-N}H_N^*(z^{*-1}) = h_N^* + h_{N-1}^* z^{-1} + \cdots + h_0^* z^{-N} \tag{82}$$

The frequency response of the reverse polynomial is the complex conjugate of the forward polynomial's frequency response times a delay, $H_N^R(\omega) = H_N^*(\omega)e^{-j\omega N}$. The product $H_N(z)H_N^R(z)$ evaluated on the unit circle is the square magnitude response times a delay, $|H_N(\omega)|^2 e^{-j\omega N}$. If the coefficients are real, then the impulse response of the reverse polynomial is $h^R(n) = h(N - n)$. The relationship is $h^R(n) = h*(N - n)$ for complex coefficients.

Reverse polynomials play an important role in optimal filter solutions in signal processing. A matched filter is the filter solution resulting from optimizing the signal-to-noise ratio (SNR) of a detected signal against a background of white noise [10], $y_{opt}(m) = [x(m) + n(m)]*h_{opt}(m)$ where the desired signal is $x(m)$ and the noise is $n(m)$. The matched filter's impulse response is equal to the complex conjugate of the time reversed, delayed signal.

$$h_{opt}(m) = x*(M - m) \qquad (83)$$

The reverse polynomial of the signal's Z-transform is the matched filter for the forward polynomial. Its frequency response is given by

$$H_{opt}(\omega) = X*(\omega)e^{-j\omega M} = X^R(\omega) \qquad (84)$$

Optical filters with one response equal to $H(z)$ and the other equal to $H^R(z)$ are common and will be discussed in Chapters 4–6.

Having defined the reverse polynomial for the general case, we now focus on relating the forward and reverse polynomials for a linear-phase filter. An Nth-order linear-phase filter contains some combination of zeros on the unit circle and pairs of zeros located at reciprocal points about the unit circle. For a general description, let there be K reciprocal pairs and P roots on the unit circle so that $N = 2K + P$. The transfer function is the product of these two terms.

$$H_N(z) = \Gamma \left\{ \prod_{k=1}^{K}(1 - z_k z^{-1})(-z_k^* + z^{-1}) \right\} \left\{ \prod_{p=1}^{P}(1 - e^{j\phi_p}z^{-1}) \right\} \qquad (85)$$

The reverse polynomial is then defined as follows:

$$H_N^R(z) = \Gamma \left\{ \prod_{k=1}^{K}(1 - z_k z^{-1})(-z_k^* + z^{-1}) \right\} \left\{ \prod_{p=1}^{P}(-e^{-j\phi_p} + z^{-1}) \right\} \qquad (86)$$

Now, we simplify the expression for the reverse polynomial in order to relate it to the forward polynomial. The terms in the first bracket are identical to those in the forward polynomial, so we only need to consider the terms in the second bracket. The pth root on the unit circle for the reverse polynomial is equal to $-e^{-j\phi_p}(1 - e^{j\phi_p}z^{-1})$ which is identical to the forward polynomial except for the phase factor. So, the coefficients of the reverse polynomial are identical to those in the forward polynomial multiplied by $(-1)^P e^{j\phi_P}$ where $\phi_P = \sum_{p=1}^{P}\phi_p$. Therefore, the reverse polynomial is given by

$$h^R(n) = (-1)^P e^{-j\phi_P}h(n) = h*(N - n) \qquad (87)$$

A linear-phase filter has the following symmetry for the impulse response.

$$h(n) = (-1)^P e^{+j\phi_P} h*(N-n) \qquad (88)$$

or $H_N^R(z) = (-1)^P e^{-j\phi_P} H_N(z)$. For real coefficients, $\phi_P = 0$ and the impulse response has even or odd symmetry, $h(n) = \pm h(N-n)$. A linear-phase filter with real coefficients and an odd number of roots at $z = 1$ has an odd impulse response.

An example of a linear-phase impulse response is given for an eighth-order filter, and a comparison is made to non-linear phase responses that have the same magnitude response. In general, for a filter with M zeros, there are 2^M different polynomials which have the same magnitude response, since each zero may be replaced with its reciprocal $1/z_m^*$ about the unit circle. If all of the zeros are inside the unit circle, a minimum-phase response results. Conversely, if all of the zeros are outside the unit circle, a maximum-phase response results. As an aside, all stable, causal AR filters are minimum-phase. The impulse responses for the linear-, minimum- and maximum-phase cases are given in Table 3-5, their roots are tabulated in Table 3-6, and the absolute value of their impulse responses are plotted in Figure 3-12. The naming convention arises because the minimum-phase roots give the least delay in the peak amplitude of the impulse response while the maximum-phase roots give the largest delay. Note that the linear-phase response is symmetric about the average delay. The group delay is shown for each implementation in Figure 3-13. The group delay changes sign depending on whether the zeros are minimum-phase or maximum-phase, but is constant for the linear-phase case.

As previously mentioned, the magnitude and phase response of a minimum-phase, causal filter form a Hilbert transform pair. Knowledge of one response completely specifies the other. Causality and the minimum-phase condition for continuous-time filters leads to the Kramers-Kronig equations, which relate the real and imaginary parts of the refractive index of a medium [11]. Resonant dielectric media are modeled as continuous AR filters [12]. Their magnitude response does not contain sharp nulls, indicative of zeros. For such a continuous time filter, the zeros of its Laplace transform determine whether it is minimum-phase or not. If the zeros

TABLE 3-5   Impulse Response for a Linear-, Minimum- and Maximum-Phase Eighth-Order MA Filter

n	$h_{\text{lin}}(n)$	$h_{\text{min}}(n)$	$h_{\text{max}}(n)$
0	−0.0041	0.0579	0.0003
1	0.0077	0.2037	−0.0021
2	0.0893	0.3226	−0.0058
3	0.2433	0.2763	0.0201
4	0.3276	0.1270	0.1270
5	0.2433	0.0201	0.2763
6	0.0893	−0.0058	0.3226
7	0.0077	−0.0021	0.2037
8	−0.0041	0.0003	0.0579

TABLE 3-6   Zeros of the Linear-, Maximum- and Minimum-Phase Eighth-Order MA Filter

n	$|z_{lin}|$	ϕ_{lin}	$|z_{min}|$	ϕ_{min}	$|z_{max}|$	ϕ_{max}
1	6.7328	0.0000	0.1485	0.0000	6.7328	0.0000
2	1.4506	2.4620	0.6894	2.4620	1.4506	2.4620
3	1.4506	−2.4620	0.6894	−2.4620	1.4506	−2.4620
4	1.0000	2.5597	1.0000	2.5597	1.0000	2.5597
5	1.0000	−2.5597	1.0000	−2.5597	1.0000	−2.5597
6	0.6894	2.4620	0.6894	2.4620	1.4506	2.4620
7	0.6894	−2.4620	0.6894	−2.4620	1.4506	2.4620
8	0.1485	0.0000	0.1485	0.0000	6.7328	0.0000

are in the left-half plane, then the filter is minimum-phase. The Kramers-Kronig relations, or equivalently, the Hilbert transform, tell us that a dispersive media is also lossy. For dielectric media, there is less dispersion farther away from an absorption peak. This behavior also characterizes discrete-time, minimum-phase systems. For minimum-phase systems, the phase response is proportional to the change in the magnitude response [13]. This behavior does not describe a mixed-phase system. In particular, linear-phase filters can have very sharp transitions in the magnitude response, although many stages are required, without introducing dispersion [14]. An-

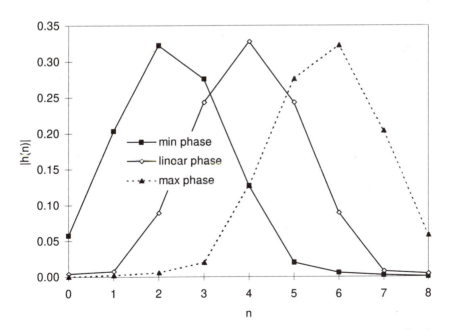

FIGURE 3-12   Impulse response for the minimum-, maximum-, and linear-phase filter implementations of an eighth-order MA filter.

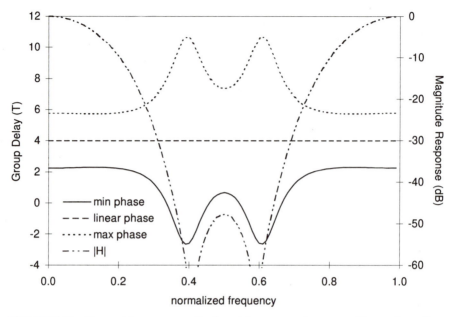

FIGURE 3-13   Group delay responses for the minimum-, maximum-, and linear-phase filter implementations and their common magnitude response.

other important mixed-phase system is an all-pass filter, where the amplitude response is constant and the phase response is non-linear.

The time dependence of a discrete filter with a continuous input is described next, since this is the case for periodic optical filters. The continuous impulse response for a discrete filter is given by

$$h(t) = \sum_{n=0}^{\infty} h(n)\delta(t - nT) \tag{89}$$

For a continuous input, the filter output is given by the convolution

$$y(t) = x(t)*h(t) \tag{90}$$

$$= \int_{-\infty}^{\infty} \sum_{n=0}^{\infty} x(t - t_0)h(n)\delta(t_0 - nT)dt_0$$

$$= \sum_{n=0}^{\infty} h(n)x(t - nT)$$

where the sifting property of the delta function was used between the second and third equalities. The optical frequency response is

$$Y(f) = X(f) \sum_{n=0}^{\infty} h(n)e^{-j2\pi fnT} = X(f)H(f) \tag{91}$$

and $H(f)$ is periodic with an FSR $= 1/T$. To prevent aliasing, the input pulse must have a bandwidth $\Delta f <$ FSR, which is equivalent to a pulse width $\Delta\tau > T$. Typically, $\Delta f \ll$ FSR and $\Delta\tau \gg T$, so the pulse spans many unit delays. The dispersion of an AR filter in the time domain is now investigated. A Gaussian input pulse propagating through a single-stage AR filter with various pole magnitudes is shown in Figure 3-14. The distortion produced by the non-linear phase response is evident in the extended tails of the pulse after passing through the filter. The dispersion increases as the pole magnitude increases.

There are two ways of compensating dispersion. First, a subsequent filter can be added which approximates the negative dispersion response of the dispersive element. All-pass filters are especially suited for this purpose. All-pass means that all frequencies are passed by the filter. The frequency response is given by

$$A(\omega) = e^{j\Phi(\omega)} \tag{92}$$

Note that the magnitude response is $|A(\omega)| = 1$. The Z-transform has the special form

$$A(z) = \frac{d_N^* + d_{N-1}^* z^{-1} + \cdots + z^{-N}}{1 + d_1 z^{-1} + \cdots + d_N z^{-N}} = \frac{\displaystyle\prod_{k=1}^{N}(-z_k^* + z^{-1})}{\displaystyle\prod_{k=1}^{N}(1 - z_k z^{-1})} = \frac{D_N^R(z)}{D_N(z)} \tag{93}$$

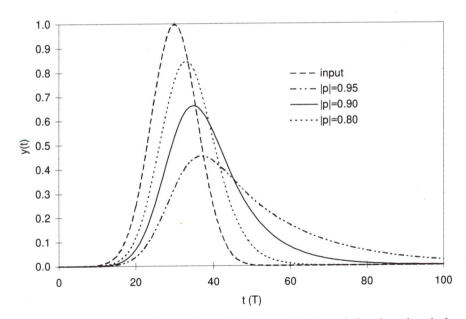

FIGURE 3-14 A Gaussian input pulse and the output after transmission through a single stage AR filter with different pole magnitudes.

The numerator polynomial is the reverse polynomial of the denominator. Optical allpass filters are discussed in Chapter 6.

A second approach is to compensate for nonlinear phase by time reversal of the output and transmission back through the filter. In this case, frequencies that see the least delay on the first pass are most delayed on the second pass. This time reversal, or frequency inversion, is equivalent to optical non-linear phase-conjugation [15]. This technique, called spectral inversion, has been explored for compensation of dispersion in optical systems [16–18].

3.3   SINGLE-STAGE OPTICAL FILTERS

The basic, single-stage MA and AR optical filters and their Z-transforms are developed in this section. The filter functions arise from the interference of two or more waves that are delayed relative to each other. The incoming signal is split into multiple paths by a division of the wavefront or amplitude. Diffraction gratings are an example of wavefront division, while directional couplers and partial reflectors are examples of amplitude division. After traveling along different paths, the fields are combined and interference occurs. For interference, the optical waves must have the same polarization, the same frequency and be temporally coherent over the longest delay length. When the signals are recombined, their relative phases determine whether they interfere constructively or destructively. The phase ϕ for each path is the product of the distance traveled L and the propagation constant $\beta = 2\pi n_e/\lambda$, which is expressed in terms of the refractive index n for a diffraction-based delay line or an effective index n_e for a waveguiding delay line. The key to analyzing optical filters using Z-transforms is that each delay be an integer multiple of a unit delay length L_U. The pase for each path is then expressed as a multiple of βL_U, so $\phi_p = p\beta L_U$ where p is an integer. The total transverse electric field is the sum over each optical path given by $E_{out} = E_0 e^{-j\phi_0} + E_1 e^{-j\phi_1} + \cdots + E_{N-1}e^{-j\phi_{N-1}}$ for N paths where the complex mode amplitude is denoted by E and the real-valued field is $e(t) = \Re e\{Ee^{j\Omega T}\}$. To obtain a Z transform of E_{out}, we first express the phase as a multiple of the unit delay T. Rewriting βL_U in terms of Ω yields $\beta L_U = \Omega[n_e(\Omega)L_U/c] = \Omega T(\Omega)$. The bracketed term defines a frequency-dependent unit delay. For a dispersionless delay line, T is a constant ad we obtain $E_{out}(z) = E_0 + E_1 z^{-1} + \cdots + E_{N-1}z^{-(N-1)}$ by substituting z^{-1} for $e^{-j\Omega T}$. The optical frequency response is periodic with a FSR of $1/T$. The normalized frequency $\nu = \omega/2\pi$ is related to the optical frequency by $\nu = (f - f_c)T = (\Omega - \Omega_c)T/2\pi$. The center frequency $f_c = c/\lambda_c$ is defined so that L_U is equal to an integer number of wavelengths, i.e. $m\lambda_c = n_e L_U$. At λ_c, the contributions from each path of length pL_U differ by 2π and thus add constructively in the absence of any other source of relative phase difference, such as the splitter and combiner.

Now, we examine the more realistic case of a delay line with dispersion. Let the period (or FSR) of the frequency response be denoted by $\Delta f = f_1 - f_2$. Then, the phases after propagation over a distance L_U differ by 2π so

$$\frac{\phi(f_1)}{2\pi} = f_1 \frac{n_e(f_1)L_U}{c} = m$$

and

$$\frac{\phi(f_2)}{2\pi} = f_2 \frac{n_e(f_2)L_U}{c} = (m-1)$$

where m is an integer. An expression for the FSR is obtained by redefining the frequencies $f_{1,2} = f_0 \pm \Delta f/2$ in terms of the center frequency f_0 and solving for Δf. The result is

$$FSR = \Delta f = \frac{c}{n_g L_U} = \frac{1}{T} \qquad (94)$$

where

$$n_g = n_e(f_0) + f_0 \frac{dn_e}{df}\bigg|_{f_0} = n_e(\lambda_0) - \lambda_0 \frac{dn_e}{d\lambda}\bigg|_{\lambda_0}$$

is the group index discussed in Chapter 2. Equation (94) defines a constant unit delay for delay lines with dispersion as long as the group index is approximately constant over the FSR. The FSR is expressed in terms of wavelength as $\Delta\lambda_{FSR} \cong \lambda^2/n_g L_U$. Assuming $\lambda = 1550$ nm and $n_g = 1.5$, FSRs from 100 nm (12.5 THz) to 0.1 nm (12.5 GHz) correspond to unit delay lengths of 16 μm to 16 mm. Note that $f_c T = mn_g/n_e$ so $f = (\nu + mn_g/n_e)$ FSR.

The formula for E_{out} is valid for coherent interference, i.e., when the longest delay time is much shorter than the coherence time of the source, denoted by τ_c. The coherence time is inversely proportional to the source spectral width, so the above condition is satisfied if $\tau_c \gg NT$ where N is the filter order. Given the desire to make compact circuits for higher levels of integration and the narrow source spectral widths required for transmitting high data rate signals over dispersive optical fibers, coherent interference of waves traveling along different paths is extremely important. A brief description of incoherent processing is given in Section 3.3.4, but the remainder of the book focuses exclusively on coherent optical signal processing. Propagation loss of a delay line is accounted for by multiplying z^{-1} by $\gamma = 10^{-\alpha L/20}$ where α is an average loss in dB per length and L is the delay path length.

The Z-transform relationships for basic optical circuit elements were first developed for fiber optic filters by Moslehi and co-workers [19,20]. The optical circuits are assumed to be linear and time-invariant. Other approaches besides Z-transforms may also be used to analyze optical circuits [21–23]. The basic optical filter circuits and their Z-transforms are described next using waveguide delays and directional couplers for splitting and combining the signals. A schematic diagram of a directional coupler is shown in Figure 3-15. The lines indicate waveguides of finite width and height. Two waveguides are brought close together so that their evanescent

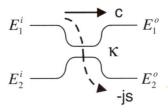

FIGURE 3-15 Transmission coefficients for the through- and cross-ports of an optical directional coupler.

fields overlap. A power coupling ratio, κ, is associated with each directional coupler. For an input on one port, the power coupled to the cross-port is κ times the input power. The length of the region where the waveguides are coupled determines the coupling ratio. For our present purposes, we need to represent the output fields in terms of the input fields. Using the results from Chapter 2, the relationships are expressed as a 2×2 transfer matrix, $\Phi_{cplr}(\kappa)$.

$$\begin{bmatrix} E_1^o \\ E_2^o \end{bmatrix} = \begin{bmatrix} c & -js \\ -js & c \end{bmatrix} \begin{bmatrix} E_1^i \\ E_2^i \end{bmatrix} = \Phi_{cplr}(\kappa) \begin{bmatrix} E_1^i \\ E_2^i \end{bmatrix} \qquad (95)$$

where κ is the power coupling ratio and E_n^i and E_n^o represent the coupler input and output field amplitudes for $n = 1,2$. The through and the cross-port transmission are designated by $c = \cos(\theta) = \sqrt{1 - \kappa}$ and $-js = -j\sin(\theta) = -j\sqrt{\kappa}$, respectively, where θ is equal to the coupling strength integrated over the coupling length as defined in Chapter 2. If the inputs and outputs are reversed, the new transfer matrix is the complex conjugate of Φ_{cplr}. The above transfer matrix assumes that the coupler has no excess loss so that the sum of the output powers equals the sum of the input powers. In addition, the coupling ratio is assumed to be wavelength independent, so the matrix elements are constants.

The basic filter structures require at least two paths for interference. The output is the sum of each optical path [24] as illustrated in Figure 3-16. The transfer function from any input port to any output port can be written by inspection using the transmission of each path segment. The directional coupler transmission is given by $-js$ for the cross port and c for the through-port. The transmission for each delay path is expressed in terms of the unit delay. A filter's transmission is then written by summing all paths between a particular input and output port. For a single stage device, shown in Figure 3-16b, the cross-port transmission has two terms. For a two stage device, shown in Figure 3-16c, there are four terms. The delays associated with common path lengths contribute to the overall filter delay, but are frequently neglected since they do not change the filter's amplitude or dispersion response.

Each stage is assumed to have a delay that is an integer multiple of the unit delay. The actual delay length may vary from the unit delay if there is a small change in the path length or effective group index from their nominal values. These variations are represented by

(a)

(b)

(c)

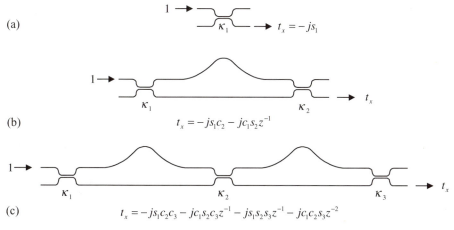

FIGURE 3-16   Principle of the sum of all optical paths.

$$n_g L = (L_U + \delta L)(\bar{n}_g + \delta n_g) \tag{96}$$
$$\approx \bar{n}_g L_U + L_U \delta n_g + \delta L \bar{n}_g = \bar{n}_g L_U + \delta(n_g L)$$

where $\bar{n}_g$ and L_U represent the nominal values and δL and δn_g are the deviations. In such a case, z^{-1} is replaced by $e^{-j\phi} z^{-1}$ where $\phi = (2\pi/\lambda)\delta(n_g L)$. The phase term may originate from fabrication variations or by design intent. The capability of adding phase terms by appropriately changing the length or group index of each delay allows complex filter coefficients to be realized. As an example, a phase change of π can be realized with $\delta n_g \approx 8 \times 10^{-4}$ for $L = 1$ mm and $\lambda = 1550$ nm.

3.3.1   A Single-Stage MA Filter

A fundamental building block in optical planar waveguide circuits is the Mach–Zehnder interferometer (MZI) shown in Figure 3-17. The MZI has two inputs and two outputs and consists of two couplers with power coupling ratios designated by κ_1 and κ_2. The waveguides connecting the couplers have lengths L_1 and L_2. To analyze its frequency response, the unit delay is set equal to the difference in path lengths, $L_U = \Delta L = L_1 - L_2$. The longer arm has a transfer function of z^{-1} relative to the shorter arm. A 2×2 delay matrix is written for both arms as follows:

$$\Phi_{delay} = \gamma e^{-j\beta L_2} \begin{bmatrix} z^{-1} & 0 \\ 0 & 1 \end{bmatrix} \tag{97}$$

where $-20 \log_{10}\gamma = \alpha L_2$ is the loss in dB for propagation through the common path length of both arms. Typically, $\Delta L \ll L_2$, so loss is neglected for the differential path

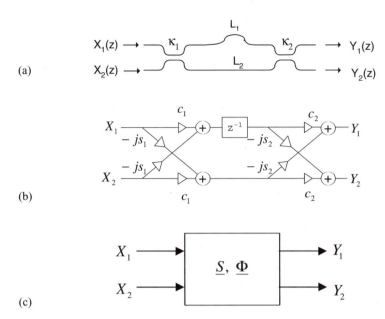

FIGURE 3-17 A Mach–Zehnder interferometer (a) waveguide layout, (b) Z-transform schematic, and (c) scattering ($\underline{S}$) and transfer ($\underline{\Phi}$) matrix representation.

length. In addition, the common path length, which contributes a linear phase (or constant delay) to the frequency response, is neglected below. The overall transfer matrix for the MZI is the product of matrices

$$\Phi_{MZI} = \Phi_{cplr}(\kappa_2)\Phi_{delay}\Phi_{cplr}(\kappa_1) \tag{98}$$

$$= \gamma \begin{bmatrix} c_2 & -js_2 \\ -js_2 & c_2 \end{bmatrix} \begin{bmatrix} z^{-1} & 0 \\ 0 & 1 \end{bmatrix} \begin{bmatrix} c_1 & -js_1 \\ -js_1 & c_1 \end{bmatrix}$$

$$= \gamma \begin{bmatrix} (-s_1s_2 + c_1c_2z^{-1}) & -j(s_1c_2 + c_1s_2z^{-1}) \\ -j(c_1s_2 + s_1c_2z^{-1}) & (c_1c_2 - s_1s_2z^{-1}) \end{bmatrix}$$

The order of the matrix multiplication is the reverse of the physical order. The MZI has one zero for each transfer function and is the simplest MA optical filter. Equation (99) defines four transfer functions for the MZI.

$$\begin{bmatrix} Y_1 \\ Y_2 \end{bmatrix} = \begin{bmatrix} H_{11}(z) & H_{12}(z) \\ H_{21}(z) & H_{22}(z) \end{bmatrix} \begin{bmatrix} X_1 \\ X_2 \end{bmatrix} = \begin{bmatrix} A(z) & B^R(z) \\ B(z) & A^R(z) \end{bmatrix} \begin{bmatrix} X_1 \\ X_2 \end{bmatrix} \tag{99}$$

Note that the coefficients of $H_{22}(z)$ are the reverse of those for $H_{11}(z)$. Likewise, $H_{12}(z)$ and $H_{21}(z)$ also form a forward and reverse polynomial pair. The two forward polynomials are labeled $A(z)$ and $B(z)$, and their relationship to the transfer functions is indicated in Eq. (99). The transfer functions in Eq. (98) are rewritten as follows assuming $\kappa = \kappa_1 = \kappa_2$:

$$\begin{bmatrix} Y_1 \\ Y_2 \end{bmatrix} = \gamma \begin{bmatrix} -s^2 + c^2 z^{-1} & -jcs(1 + z^{-1}) \\ -jcs(1 + z^{-1}) & c^2 - s^2 z^{-1} \end{bmatrix} \begin{bmatrix} X_1 \\ X_2 \end{bmatrix} \tag{100}$$

The coefficients for each transfer function are determined by the coupling ratios. When the couplers are identical, $H_{12}(z)$ and $H_{21}(z)$ always have a zero on the unit circle. For $H_{11}(z)$ and $H_{22}(z)$ to have a zero on the unit circle requires that $\kappa_1 = \kappa_2 = 0.5$; consequently, it is much easier to get a good rejection filter using the cross-port than for the through-port. For the special case $\kappa = 0.5$, the zero is on the unit circle for each transfer function, and the frequency responses are given by

$$|H_{11}(\Omega)|^2 = |H_{22}(\Omega)|^2 = \gamma^2 \sin^2\left(\frac{\Omega T}{2}\right) \tag{101}$$

$$|H_{12}(\Omega)|^2 = |H_{21}(\Omega)|^2 = \gamma^2 \cos^2\left(\frac{\Omega T}{2}\right)$$

For a lossless MZI, $\gamma = 1$ and the peak transmission for each port is 1. For other values of κ, the zeros are not on the unit circle for the through-ports and the ratio of maximum to minimum transmission over one FSR is reduced. In the above derivation, we have assumed that all the ports are matched so that there are no reflections to be considered.

For the MZI to be a general architecture, we must be able to place the zero anywhere on the Z-plane. We will consider $H_{11}(z)$ for the lossless case when the couplers are identical. The zero is given by $z_1 = c^2/s^2 = (1 - \kappa)/\kappa$. The coupling ratio has a value in the range $0 \le \kappa \le 1$, so $0 \le |z_1| \le \infty$. An arbitrary phase is obtained by adding a non-zero phase term as discussed above so that z^{-1} is replaced by $e^{-j\phi}z^{-1}$. The zero can then be located anywhere on the Z-plane.

3.3.2 A Single-Stage AR Filter

A ring resonator with two couplers, as shown in Figure 3-18a, is the simplest optical waveguide filter with a single-pole response. The unit delay is equal to $L_U = L_1 + L2 + L_{C1} + L_{C2}$ where L_{C1} and L_{C2} are the coupling region lengths for each coupler. Note that the feed-forward delay is the relative delay between two paths; whereas, the feedback delay is the total feedback path length. Using c and s as before, the sum of all optical paths is

$$Y_2(z) = -s_1 s_2 \sqrt{\gamma z^{-1}}\{1 + c_1 c_2 \gamma z^{-1} + \cdots\} X_1(z) \tag{102}$$

where $\gamma = 10^{-\alpha L/20}$. The common term, $-s_1 s_2 \sqrt{\gamma z^{-1}}$, is the transmission from the input to the output without the feedback path connected. Propagation once around the feedback path is given by $c_1 c_2 \gamma z^{-1}$. The infinite sum simplifies to the following expression for the ring's transfer function:

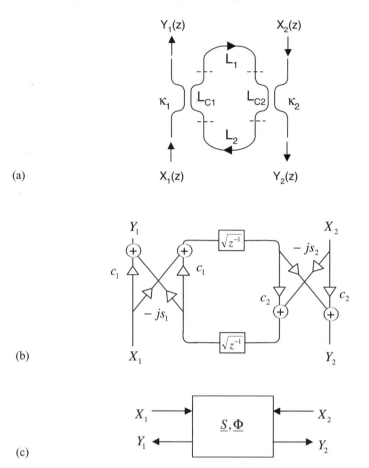

FIGURE 3-18 Ring resonator with two couplers (a) waveguide layout, (b) Z transform schematic, and (c) matrix representation.

$$H_{21}(z) = \frac{Y_2(z)}{X_1(z)} = \frac{-\sqrt{\kappa_1 \kappa_2 \gamma z^{-1}}}{1 - c_1 c_2 \gamma z^{-1}} = \frac{\sigma \sqrt{\gamma z^{-1}}}{A(z)} \qquad (103)$$

The location of the pole is determined by the coupling ratios and delay line attenuation. Since $c_1 c_2$ and γ are always less than unity for a passive waveguide, the resulting filter will always be stable. The maximum pole magnitude is given by $p_1 = \gamma$, so the loss limits the achievable pole magnitudes. A phase term can be added to move the pole location anywhere within the circle of radius γ in the Z-plane. If gain were introduced in the feedback path, it would be possible to create an unstable filter, i.e. a laser. The ring resonator provides a narrower passband with better stopband rejection as the pole moves closer to the unit circle, which is achieved by decreasing the coupling ratios and minimizing the ring loss. The FSR is inversely proportional to the ring radius assuming that $L_1 + L_2 = 2\pi R \gg L_{C1} + L_{C2}$. The minimum bend radius is a critical issue for waveguide AR ring filters, which are discussed in detail in Chapter 5. The square magnitude response is given by

$$|H_{21}(\omega)|^2 = \frac{\kappa^2 \gamma}{1 - 2c^2 \gamma \cos(\omega - \phi) + c^4 \gamma^2} \tag{104}$$

where the simplification $\kappa_1 = \kappa_2 = \kappa$ was made. The peak occurs at $\omega = \phi$. For a lossless filter, the peak transmission is 1 independent of κ.

Next, we investigate the other responses for the structure shown in Figure 3-18. The response for input X_1 to output Y_1 is obtained by adding the contributions from each path.

$$Y_1(z) = [c_1 - s_1^2 c_2 \gamma z^{-1} \{1 + c_1 c_2 \gamma z^{-1} + \cdots \}] X_1(z) \tag{105}$$

$$= \left[c_1 - \frac{s_1^2 c_2 \gamma z^{-1}}{1 - c_1 c_2 \gamma z^{-1}} \right] X_1(z)$$

$$= \left[\frac{c_1 - c_2 \gamma z^{-1}}{1 - c_1 c_2 \gamma z^{-1}} \right] X_1(z) = \frac{B(z)}{A(z)} X_1(z)$$

The transfer function $H_{11}(z) = Y_1(z)/X_1(z) = B(z)/A(z)$ has one pole and one zero, making it an ARMA response. The pole is identical to the pole for $H_{21}(z)$. Similarly, one can show that

$$H_{12}(z) = H_{21}(z) = \frac{\sigma\sqrt{\gamma}z^{-1}}{A(z)}$$

and

$$H_{22}(z) = \frac{c_2 - c_1 \gamma z^{-1}}{1 - c_1 c_2 \gamma z^{-1}} = \frac{-B^R(z)}{A(z)}$$

These results can be expressed in two different matrix forms. The first form, called the scattering matrix, relates the inputs to the outputs.

$$\begin{bmatrix} Y_1(z) \\ Y_2(z) \end{bmatrix} = S_{ring}(z) \begin{bmatrix} X_1(z) \\ X_2(z) \end{bmatrix} \text{ where } S_{ring}(z) = \begin{bmatrix} \dfrac{B(z)}{A(z)} & \dfrac{\sigma\sqrt{\gamma}z^{-1}}{A(z)} \\ \dfrac{\sigma\sqrt{\gamma}z^{-1}}{A(z)} & \dfrac{-B^R(z)}{A(z)} \end{bmatrix} \tag{106}$$

The second form relates the quantities in one plane to those in another as shown in Figure 3-18c. The resulting matrix is referred to as a transfer matrix and is given by

$$\begin{bmatrix} X_1(z) \\ Y_1(z) \end{bmatrix} = \Phi_{ring}(z) \begin{bmatrix} X_2(z) \\ Y_2(z) \end{bmatrix} \text{ where } \Phi_{ring}(z) = \frac{1}{\sigma\sqrt{\gamma}z^{-1}} \begin{bmatrix} A(z) & B^R(z) \\ B(z) & A^R(z) \end{bmatrix} \tag{107}$$

The scattering matrix form is convenient for expressing the implications of power conservation and reciprocity. Scattering matrices are used in many areas including microwave filters and quantum theory. The transfer matrix form, also called the chain matrix, is used extensively in Chapter 5 for analyzing multi-stage filters.

3.3.3   Power Conservation and Reciprocity

Now, we consider general properties of the transfer matrices for lossless filters by setting $\gamma = 1$. Let S denote the scattering matrix (for either a feed-forward or feedback optical circuit) that relates the inputs X to the outputs Y so that $Y = SX$. Using this definition, the transfer and scattering matrices are identical for feedforward structures such as the MZI. For an N×N (N input and output ports) lossless optical filter, the sum of the input powers must equal the sum of the output powers, $\sum_{n=1}^{N}|x_n|^2 = \sum_{n=1}^{N}|y_n|^2$. This power conservation condition is equivalent to requiring that the scattering matrix be unitary [25]. A matrix is unitary if $S^\dagger S = I$ where I is the identity matrix with ones on the diagonal and zeros for all the remaining terms and the Hermitian transpose, consisting of the transposition and conjugation operations, is indicated by the superscript †. The determinant of a unitary matrix has a magnitude of one, $|\det(S)| = 1$. Let the determinant be designated by $\det(S) = e^{j\theta_d}$. From the previous definitions of scattering matrices for optical filters, note that $\det(\Phi_{delay}) = e^{j\omega}$, $\det(\Phi_{cplr}) = 1$, and $|\det(S_{ring})| = 1$. Let the terms for a general 2×2 scattering matrix be denoted by

$$S = \begin{bmatrix} s_{11} & s_{12} \\ s_{21} & s_{22} \end{bmatrix}$$

Then, the unitary condition $S^{-1} = S^\dagger$ requires that

$$e^{-j\theta_d}\begin{bmatrix} s_{22} & -s_{12} \\ -s_{21} & s_{11} \end{bmatrix} = \begin{bmatrix} s_{11}^* & s_{21}^* \\ s_{12}^* & s_{22}^* \end{bmatrix} \tag{108}$$

The diagonal and off-diagonal terms must satisfy

$$s_{22}e^{-j\theta_d} = s_{11}^* \tag{109}$$

$$-s_{21}e^{-j\theta_d} = s_{12}^* \tag{110}$$

So, $|s_{11}| = |s_{22}|$ and $|s_{12}| = |s_{21}|$. For a unitary matrix, the sum of the square magnitudes of elements along any row is equal to one. In addition, the product of any column with the complex conjugate of any other column is zero. For a 2×2 matrix, the following relationships emerge:

$$|s_{11}|^2 + |s_{12}|^2 = 1 \tag{111}$$

$$|s_{21}|^2 + |s_{22}|^2 = 1$$

$$s_{11}s_{12}^* + s_{21}s_{22}^* = 0$$

From the first equation, note that $|s_{12}| = \sqrt{1 - |s_{11}|^2}$. By specifying the magnitude for one element, the magnitudes of all the remaining elements of a 2×2 unitary matrix are determined.

Next, we discuss the properties imposed by reciprocity on the scattering matrix. An N×N reciprocal filter has the property that if time is reversed, the output signal returns to the input from which it originated with its original input power level. In the frequency domain, time reversal is equivalent to phase conjugation. If the electric E and magnetic H fields are the solution of Maxwell's equations for a forward propagating wave, then E^* and H^* are solutions for the time reversed case as long as the dielectric constant satisfies $\varepsilon = \varepsilon^*$ (the material is lossless) [26]. To demonstrate time reversal using phase conjugation, consider a reciprocal device such as an MZI with arms having equal path lengths, folded about the middle with a mirror at the center as shown in Figure 3-19a. An input signal is introduced on the upper arm, and the coupler splits it evenly between the interferometer arms. The mirror reflects the light, and any phase shift introduced by the mirror is identical in both arms. The fields combine constructively in the lower arm and destructively in the upper arm, a result that does not appear to imply a reciprocal device. If a phase conjugate mirror is used instead, as indicated in Figure 3-19b, then the fields combine constructively in the input port as expected for a reciprocal device. The coupler's scattering matrix satisfies $S_{cplr}S^*_{cplr} = I$ which implies that $S_{cplr} = S^T_{cplr}$ using the unitary condition.

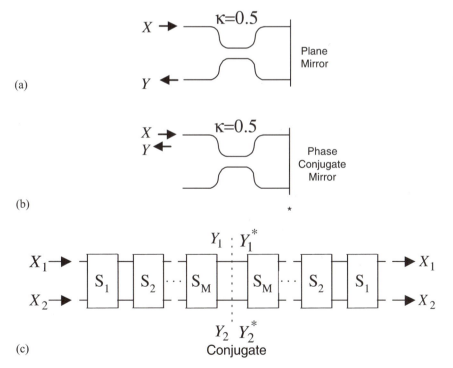

(a)

(b)

(c)

FIGURE 3-19 Demonstration of forward and backward propagation through a reciprocal coupler with the reflector being (a) a plane mirror and (b) a phase conjugate mirror. (c) Reciprocity illustrated for a cascade of 2×2 scattering matrices.

Now we generalize this example by considering a cascade of M unitary 2 × 2 elements $S_{tot} = S_M \cdots S_1$ such that $Y = S_{tot}X$. Forward propagation, phase conjugation, and reverse propagation through the cascade are illustrated in Figure 3-19c using an unfolded structure for clarity. The output is given by

$$S_1 \cdots S_M Y^* = S_M \cdots S_1 S_1^* \cdots S_M^* X^* \tag{112}$$

For a reciprocal device, the output is X^* if $S_m S_m^* = I$ for each matrix S_m where $m = 1, \ldots, M$. Since each matrix is unitary, $S_m = S_m^T$. Therefore, reciprocity requires that $s_{12} = s_{21}$ for each matrix S_m. This relationship also implies a condition on the relative phases of the scattering matrix elements. Let $s_{11} = se^{j\theta_{11}}$, $s_{22} = se^{j\theta_{22}}$, and $s_{12} = s_{21} = \sqrt{1-s^2}e^{j\theta_{12}}$. From Eq. (109), the phase of the determinant is given by $\theta_d = \theta_{11} + \theta_{22}$. From Eq. (110), the condition $\theta_{12} = \theta_d/2 + \pi(2m + 1)/2$ results where m is an integer. Thus, the off-diagonal terms differ in phase by an odd multiple of $\pi/2$ from the average phase of the diagonal terms. An example of nonreciprocal behavior is a lossy medium. The impact of loss on filter responses is investigated in Chapters 4–6.

For 2 × 2 filters which are lossless and reciprocal, the following conditions exist on their scattering matrix elements: $s_{12} = s_{21}$, $|s_{12}| = \sqrt{1 - |s_{11}|^2}$, $|s_{11}| = |s_{22}|$, and $\theta_{12} = \theta_d/2 + \pi(2m + 1)/2$. Therefore, the degrees of freedom for a 2 × 2 matrix are limited to one magnitude and two phases. For an $N \times N$ lossless, reciprocal device, there are $N(N + 1)/2$ degrees of freedom [25] for the scattering matrix coefficients. The degrees of freedom correspond to the number of coefficient magnitudes and phases that can be freely chosen. These restrictions are important for optical filter design since the filter coefficients depend on the matrix elements.

3.3.4   Incoherent Optical Signal Processing

For applications where the shortest relative delay is longer than the coherence length, incoherent signal processing occurs and the system responds linearly to the signal intensity instead of the fields. The filter operates on the modulated intensity, $I_{in}e^{j\Omega_m t}$ where Ω_m is the modulation angular frequency. Assuming that the highest practical modulation frequencies, limited by electrical modulators, are on the order of 10 GHz, it is interesting to note that incoherent processing operates on frequency ranges which are 4 orders of magnitude smaller than for coherent processing where the frequency of interest is on the order of 200 THz ($\lambda = 1500$ nm). For FSRs in the range of 10 MHz to 10 GHz, the unit delay length ranges from 21 m to 21 mm assuming $n_g = 1.45$. Practically, the minimum FSR is limited by the wafer size and loss for planar waveguides. For small FSRs, optical fiber circuits, which can easily contain long low-loss delay lengths, are more practical.

The incoherent transfer matrix for a directional coupler is

$$\begin{bmatrix} |E_1^o|^2 \\ |E_2^o|^2 \end{bmatrix} = \begin{bmatrix} 1 - \kappa & \kappa \\ \kappa & 1 - \kappa \end{bmatrix} \begin{bmatrix} |E_1^i|^2 \\ |E_2^i|^2 \end{bmatrix} \tag{113}$$

The incoherent transfer function for the single stage MZI shown in Figure 3-17 is

$$H_{11}(z) = (1 - \kappa_1)(1 - \kappa_2) + \kappa_1 \kappa_2 \gamma^2 z^{-1} \qquad (114)$$

where $z^{-1} = e^{-j\Omega_m T}$. Note that the resulting filter coefficients are real and have positive values. The zero is located at $z_1 = -\gamma^2 \kappa_1 \kappa_2 / [(1 - \kappa_1)(1 - \kappa_2)]$. Although the magnitude can take on any value by changing the coupling ratios, an arbitrary phase cannot be achieved as in the coherent processing case.

The incoherent transfer function for the two coupler ring shown in Figure 3-18 is given by

$$H_{11}(z) = \frac{\gamma \kappa_1 \kappa_2 \sqrt{z^{-1}}}{1 - (1 - \kappa_1)(1 - \kappa_2)\gamma^2 z^{-1}} \qquad (115)$$

The pole is located at $p_1 = \gamma^2 (1 - \kappa_1)(1 - \kappa_2)$. The pole magnitude is limited by γ^2 instead of γ as in the coherent case.

3.4 DIGITAL FILTER DESIGN

The field of digital filter design cannot be covered in a few pages; however, some examples are chosen to illustrate different design methods and to contrast the different filter classes (AR, MA and ARMA). Digital filter design starts with the desired response, which is typically the frequency response, but could be the impulse response. The output is the filter order and the coefficients that give a satisfactory approximation to the desired response. Approximating a desired response by a truncated Fourier series is discussed first. Design methods are then demonstrated for bandpass filters, which are important for multiplexing and demultiplexing channels in WDM systems. The design of filters approximating an arbitrary magnitude and/or phase response are considered next. Such filters find application as gain equalizers and dispersion compensators.

Once the filter coefficients for a multistage design are determined, a particular filter architecture must be chosen. Filter architectures impact the computation time and sensitivity to roundoff errors for digital filters. A review of several multi-stage digital filter architectures is provided for comparison to optical filter architectures in Chapters 4–6. Two additional steps are required for optical filter design. The coefficients must be translated into coupling ratios and path lengths, and the sensitivity to fabrication errors must be investigated to determine if the design is practical. These issues are investigated in Chapters 4–6 as well.

3.4.1 Approximating Functions

The goal of filter design is to approximate a desired function $H_d(\omega)$ with a realizable filter $H_N(\omega)$ having N stages. We have three choices for approximating functions that correspond to the filter classes: $B(z)$, $\Gamma/A(z)$, and $B(z)/A(z)$. The most efficient choice, i.e. requiring the fewest stages, depends on the desired response. The theory of approximating functions with a polynomial, i.e. $H_d(z) \approx B_N(z)$, is reviewed first.

Our goal is to minimize the error between the desired and approximate frequency response. The error for an Nth-order filter is defined as follows:

$$E_N(\omega) = |H_d(\omega) - H_N(\omega)| \tag{116}$$

where $H_N(\omega)$ is given by the truncated Fourier series

$$H_N(\omega) = \sum_{n=-N/2}^{N/2} h_d(n)e^{-j\omega n} \tag{117}$$

The error depends on the filter order. If $H_d(\omega)$ is continuous over the approximation interval, then the series converges uniformly [2] as follows:

$$\lim_{N \to \infty} E_N(\omega) = \lim_{N \to \infty} |H_d(\omega) - H_N(\omega)| = 0 \tag{118}$$

Another way to describe the condition for uniform convergence is that $h(n)$ must be absolutely summable,

$$\sum_{n=-\infty}^{\infty} |h_d(n)| < \infty \tag{119}$$

For functions which have discontinuities, a weaker convergence criterion is applicable if the series has finite energy.

$$\sum_{n=-\infty}^{\infty} |h_d(n)|^2 < \infty \tag{120}$$

One example is $H_d(\omega) = \prod(\omega/\omega_c)$ which has an impulse response equal to $h_d(n) = \sin(\omega_c n/2)/\pi n$. In this case, the approximate function converges in a mean-square sense defined as follows:

$$\lim_{N \to \infty} \int_{-\pi}^{\pi} |H_d(\omega) - H_N(\omega)|^2 \, d\omega = 0 \tag{121}$$

Mean-square convergence at discontinuities gives rise to the Gibbs phenomenon. An optimal mean-square design for MA filters is obtained by truncating the impulse response of the desired function, assuming that $h_d(n)$ is known exactly. This approach is the window method which is discussed along with the Gibbs phenomenon in Section 3.4.3. A major advantage for optimizing MA designs is that the frequency response depends linearly on the filter coefficients. For an ARMA filter, minimizing the mean square error results in a set of nonlinear equations. An efficient iterature algorithm for their solution is the Steiglitz-McBride method [27]. By minimizing a different error criterion, a set of linear equations for the coefficients of an AR filter are obtained as discussed in Section 3.4.5.

The mean-square error can be weighted as follows:

$$E_N = \int_{-\pi}^{\pi} w(\omega)|H_d(\omega) - H_N(\omega)|^2 \, d\omega \tag{122}$$

where $w(\omega)$ is a weighting function with positive or zero values. With a weighting function, the error can be reduced at frequencies which are most important such as the passband and stopband and relaxed in "don't care" regions such as the transition band. The error is typically determined for a sampled response as follows:

$$E_N = \sum_{k=0}^{N-1} w_k |H_d(\omega_k) - H_N(\omega_k)|^2 \tag{123}$$

where $w_k = w(\omega_k)$. An optimization algorithm is used to minimize E_N with respect to the filter coefficients.

In many instances, a more desirable criterion is to minimize the maximum deviation from the desired response.

$$E_N = \max_{\text{over } k}(w_k |H_d(\omega_k) - H_N(\omega_k)|) \tag{124}$$

This criterion describes the Chebyshev, or minimax, approximation problem. The optimal Chebyshev solution results in the maximum error being spread evenly over the interval, giving an equiripple design. Chebyshev's minimax theorem states that the optimal solution has an error function with $N + 2$ extrema of equal amplitude and alternating sign where N is the filter order, or the sum of the orders for the $A(z)$ and $B(z)$ polynomials for an ARMA filter. Routines can be found in standard filter design packages which find the minimax solution for MA filters. The Parks–McClellan algorithm combined with the Remez exchange algorithm is used to design linear-phase MA filters that have equiripple errors with respect to the desired frequency response. For AR and ARMA designs, numerical optimization packages are used to vary the filter coefficients of a starting solution and iterate until the best approximately equiripple solution is found. In addition, equiripple analog filter designs can be transformed to the Z-domain as discussed in Section 3.4.4.

3.4.2 Bandpass Filters

Digital filters are typically classified as lowpass, highpass, and bandpass designs. For optical filters, the frequency response of interest is over one FSR that is centered at an optical frequency much larger than the FSR. Bandpass optical filters can therefore be realized using lowpass digital filter designs.

The ideal bandpass response has a rectangular magnitude response with a passband width of ω_c and a linear-phase response. Three regions are specified, the passband, transition band, and stopband. Ideally, the passband has a magnitude of one, the stopband has a magnitude of zero, and the transition band occurs over the smallest possible frequency range. The linear-phase response is important only over the passband. Since the rectangular response cannot be realized with a causal filter, filter design involves approximating the ideal bandpass response. The particular error criterion over each band differentiates various optimum designs.

From the design perspective, the filter order can always be increased to reduce the error in the approximation. Increasing the filter order, however, is undesirable for optical filters since more stages are associated with added components and sensitivity to fabrication errors. In this respect, optical filters are more akin to analog electrical filters, whose parameters are limited by component tolerances, than to digital filters whose coefficients can be determined to double precision accuracy. Analog electrical filters typically have less than 10 stages; whereas, digital filters with hundreds of stages are easily implemented with double precision arithmetic.

The filter order is minimized if the desired filter response closely resembles the single-stage response (or sum of responses in dB) of one of our filter classes (MA, AR or ARMA). For example, if a sinusoidal response were desired, only one MA stage would suffice. The single-stage AR response more closely resembles a bandpass filter than the MA filter response; consequently, we anticipate that a bandpass response can be approximated with fewer AR stages than MA stages. This behavior has been quantified for bandpass filter designs with similar rolloff and stopband rejection [28]. The MA filters require a large number of stages compared to AR and ARMA filters; however, MA filters may be designed to have zero phase distortion, whereas AR and ARMA filters can only approximate a linear-phase response.

3.4.3   The Window Method for MA Bandpass Filters

The window method for MA filter design is straightforward. It requires that the desired response is fully specified in the frequency domain and that the impulse response can be calculated exactly using Eq. (32). The ideal bandpass response is given by

$$H_d(\nu) = \begin{cases} 1 \text{ for } |\nu| \le \tfrac{1}{2}\nu_c \\ 0 \text{ for } |\nu| > \tfrac{1}{2}\nu_c \end{cases} \tag{125}$$

where the cutoff frequency is denoted by ν_c. The impulse response is

$$h_d(n) = \frac{\sin(\pi\nu_c n)}{\pi n} \text{ for } -\infty < n < \infty \tag{126}$$

To form a causal filter, the impulse response is truncated and shifted to yield

$$h_N(n) = \frac{\sin[\pi\nu_c(n - N/2)]}{\pi(n - N/2)} \quad \text{for} \quad 0 \le n \le N \tag{127}$$

The truncation involves multiplying the impulse response by a rectangular function, or window, of width $N + 1$. This is equivalent to convolving the frequency response with the sinc function. The frequency response for $\nu_c = 0.1$ is shown in Figure 3-20

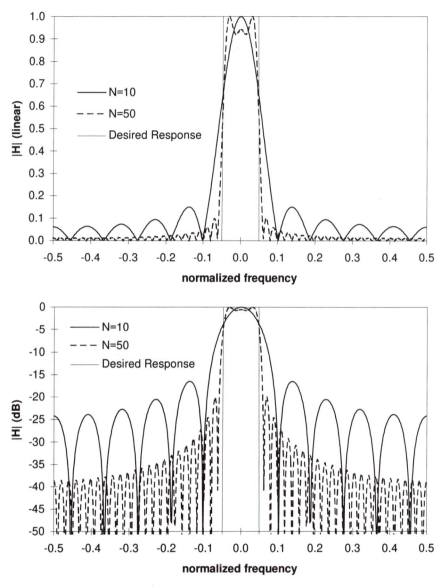

FIGURE 3-20 Uniform window design for a 10th and 50th-order MA bandpass filter.

for $N = 10$ and $N = 50$. The convolution produces ripples in the frequency response around the transition regions which do not decrease in amplitude when the filter order is increased. This behavior is known as Gibbs phenomenon. The impulse response is the Fourier series coefficients of the desired frequency response. The integral squared error is minimized when the desired response is approximated by the truncated Fourier series.

$$E = \frac{1}{2\pi} \int_{-\pi}^{\pi} |H_N(\omega) - H_d(\omega)|^2 d\omega \qquad (128)$$

The Gibbs phenomenon demonstrates that the maximum error can vary significantly, even when the integral squared error is minimized.

A simple way to reduce the ripples with the window method is to apply a tapered window function instead of a uniform one. For example, a Hanning window is defined as follows [29].

$$w(n) = \frac{1}{2}\left[1 - \cos\left(\frac{2\pi(n+1)}{N+1} \right) \right] \quad \text{for} \quad n = 0, \cdots, N-1 \qquad (129)$$

The new frequency response is shown in Figure 3-21 for the 10th- and 50th-order filters. The ripple amplitude is reduced; however, the transition region is more gradual. The stopband rejection has also been improved by tapering the window function. Figure 3-22 shows a plot of the zeros for the uniform and tapered windows. The uniform window has all of its zeros on the unit circle; whereas, the tapered window has two zeros which are not. The non-uniform window response is not optimal in the integral squared sense. Tapering functions to reduce unwanted sidelobes are widely used in optical filters also.

3.4.4 Classical Filter Designs for ARMA Bandpass Filters

Numerous methods for analog filter design are available in the signal processing literature. A technique to translate an analog design to the digital domain allows these

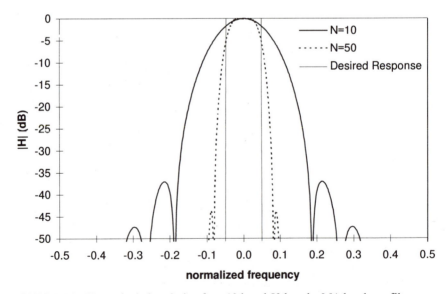

FIGURE 3-21   Tapered window design for a 10th and 50th-order MA bandpass filter.

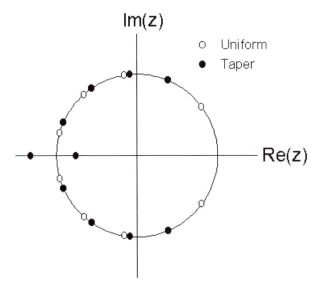

FIGURE 3-22 Zero plot comparing the uniform and tapered profiles for a 10th-order MA filter.

designs to be implemented. Analog filter designs are expressed as a ratio of polynomials, $B(s)/A(s)$, in the Laplace transform complex variable $s = \sigma + j\Omega$. The frequency response of an analog filter is found by setting $s = j\Omega$; whereas, the frequency response of a digital filter is determined by setting $z = e^{j\omega}$. The digital frequency response is periodic whereas the analog response is not. The bilinear transform, given by Eq. (130), provides a mapping between z and s so that designs in the Laplace-domain can be converted to the Z-domain.

$$s = \frac{2}{T}\frac{1 - z^{-1}}{1 + z^{-1}} \tag{130}$$

This relationship maps the left hand side of the s plane inside the unit circle and the right hand side outside the unit circle. The $j\Omega$ axis maps to the unit circle resulting in the following relationship between the digital angular frequency ω and analog angular frequency Ω:

$$\omega = 2 \tan^{-1}\frac{\Omega T}{2} \tag{131}$$

The following classical lowpass filter designs are described next in terms of their passband and stopband characteristics: 1) Butterworth, 2) Chebyshev Type I, 3) Chebyshev Type II, and 4) elliptic. As an aside, it is noted that the Bessel filter is an important analog filter since it is designed to have a linear-phase response over the passband; however, its phase properties are not retained through the bilinear transformation [2]. The complete design details for the four analog filters will not be given since the filter coefficients can be determined from commercially available filter design software packages. The translation between the Laplace transform description and the Z-transform description is outlined to highlight the origin of the poles and zeros for each case. A brief description of the Laplace transform is given in the appendix.

The Butterworth filter has a maximally flat passband response. The analog design employs a Taylor series approximation about $\Omega = 0$ and $\Omega = \infty$. Since the ideal passband response is a constant, the resulting approximate response is referred to as maximally flat. The square magnitude response is given by [3]

$$|H_N(s)|^2 = \frac{1}{1 + (s/j\Omega_0)^{2N}} = \frac{1}{A(s)A^*(-s)} \tag{132}$$

$$|H_N(\Omega)|^2 = \frac{1}{1 + (\Omega/\Omega_0)^{2N}} \tag{133}$$

It is an all-pole filter in the s-domain with a 3dB cutoff frequency of Ω_0. The magnitude response for the digital filter can be determined by substituting Eq. (131) into Eq. (133). The roots of the denominator polynomial of Eq. (132) are given by [2]

$$s_i = \Omega_0 e^{j\theta_i} \quad \text{where} \quad \theta_i = \frac{\pi}{2} + \frac{\pi(2i+1)}{2N} \quad \text{for} \quad i = 0, 1, \cdots, N-1 \tag{134}$$

The roots are evenly distributed over the left semicircle of radius Ω_0. Each root has a transfer function as follows:

$$H_i(s) = \frac{1}{A_i(s)} = \frac{1}{1 - (s/s_i)} \tag{135}$$

where $H(s) = H_0(s) \cdots H_{N-1}(s)$. The following Z-transform results after applying the bilinear transformation with $T = 2$:

$$H_i(z) = \frac{G_i(1 + z^{-1})}{1 - a_i z^{-1}} \tag{136}$$

where

$$G_i = \frac{\Omega_0 e^{j\theta_i}}{\Omega_0 e^{j\theta_i} - 1} \quad \text{and} \quad a_i = \frac{1 + \Omega_0 e^{j\theta_i}}{1 - \Omega_0 e^{j\theta_i}}$$

Note that the all-pole analog filter transforms to an ARMA digital filter with a zero at $z = -1$ for each pole. Thus, an Nth-order all-pole analog filter has N poles and N zeros in the Z-domain. The pole locations transform according to Eq. (130). The only design inputs for the Butterworth are the filter order and the cutoff frequency.

Chebyshev filters have a maximally flat response either in the stopband or passband, which distinguishes Type I from Type II, respectively. The opposite band has an equiripple response. The equiripple property allows steeper transitions to be realized for an equivalent filter order compared to the maximally flat approximation [3]. The Chebyshev Type I filter is an all-pole filter in the s domain with the following magnitude response:

$$|H_N(\Omega)|^2 = \frac{1}{1 + \varepsilon^2 T_N^2(\Omega/\Omega_c)} \qquad (137)$$

where $T_N(\Omega)$ is the Nth-order Chebyshev polynomial given by

$$T_N(x) = \begin{cases} \cos(N \cos^{-1} x) & \text{for } |x| \le 1 \\ \cosh(N \cosh^{-1} x) & \text{for } |x| > 1 \end{cases} \qquad (138)$$

The parameter ε determines the passband ripple.

The Chebyshev Type II magnitude response has both poles and zeros and is given by

$$|H_N(\Omega)|^2 = \frac{1}{1 + \varepsilon^2 [T_N^2(\Omega_s/\Omega_p)/T_N^2(\Omega_s/\Omega)]} \qquad (139)$$

where Ω_p and Ω_s demarcate the transition region on the passband and stopband sides, respectively. The stopband ripple is determined by ε. A zero in the analog domain transforms to the Z-domain as follows:

$$H_i(s) = 1 - s/s_i \qquad (140)$$

$$H_i(z) = \frac{-(1 - s_i) + (1 + s_i)z^{-1}}{s_i(1 + z^{-1})}$$

A Z-domain zero occurs at a position corresponding to the bilinear transformation of s_i. In addition, there is a pole at $z = -1$. The appearance of this unstable pole means that an all-zero analog filter transforms to an unstable digital filter. To avoid an unstable digital filter, the analog filter should have at least as many poles as zeros. Ripples in the stopband imply that the zeros of $H(z)$ are distributed about the unit circle instead of being located at $z = -1$ as with the Butterworth design.

The elliptic filter has an equiripple response in both the passband and stopband. For a given order, the elliptic filter has the smallest transition bandwidth. Its magnitude response is given by

$$|H_N(\Omega)|^2 = \frac{1}{1 + \varepsilon^2 U_N^2(\Omega/\Omega_p)} \qquad (141)$$

where $U_N(\Omega)$ is a Jacobian elliptic function of order N and ε is related to the passband ripple. For the elliptic filter, both the passband and stopband ripple are design inputs.

To compare these four filter types, a 10th-order filter of each type was designed with a cutoff of $\nu_c = 0.1$. The maximum stopband transmission was set to $R_s = 30$ dB and the maximum passband ripple to $R_p = 10 \log_{10}(1 - 10^{-R_s/10}) \approx 0.004$ dB. Since the outputs of a 2×2 optical filter are power complementary, this value for R_p provides a 30 dB stopband rejection for both ports. The magnitude responses are shown in Figure 3-23. An expanded view of the passband is given in Figure 3-24. The max-

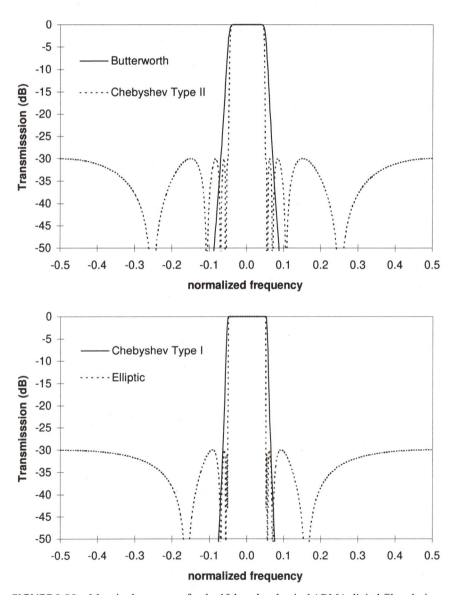

FIGURE 3-23   Magnitude response for the 10th-order classical ARMA digital filter designs.

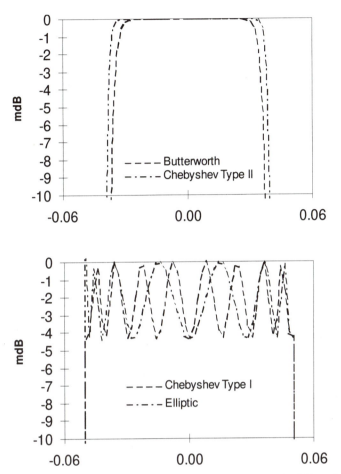

FIGURE 3-24 Passband magnitude response in mdB for the classical filter designs.

imally flat passband design of the Butterworth and Type II are evident along with the equiripple behavior of the Type I and Elliptic designs. Note that the equiripple designs have a slightly wider passband for the same input cutoff frequency. Pole-zero plots for each filter type are shown in Figure 3-25. All the filters are optimum with respect to either the maximally flat or equal ripple design criteria. The group delay responses are shown in Figure 3-26, and the dispersion in Figure 3-27. Both the Butterworth and Chebyshev Type I have no ripples in the stopband. By allowing ripples in the stopband, a sharper transition can be realized as evidenced by the Chebyshev Type II and elliptic filter examples. The sharper transitions of the latter filters lead to larger group delays and dispersion at the passband edges. Note that all of the zeros are on the unit circle. As a consequence, the numerator polynomials are linear-phase. Thus, the filter dispersion arises strictly from the contributions of the poles.

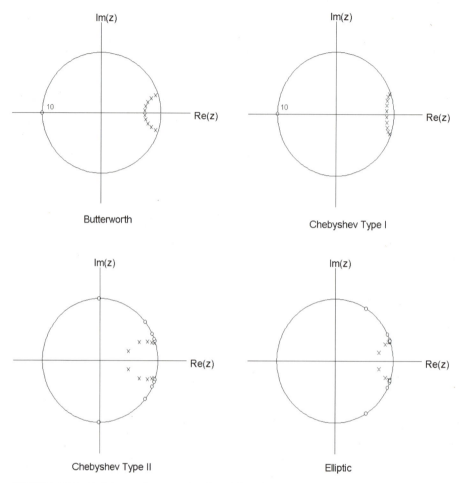

FIGURE 3-25 Pole/zero diagrams for the 10th-order Butterworth, Chebyshev Type I and Type II, and Elliptic filters.

3.4.5 The Least Squares Method for AR Filter Design

A method for designing AR filters, known as the least squares or autocorrelation method of linear prediction, is outlined next. The approximation is defined as follows:

$$H_d(z) \approx H_N(z) = \frac{1}{A_N(z)} \qquad (142)$$

The coefficients of $A_N(z)$ are determined by minimizing the mean square error between the desired function and $H_N(z)$. In the time domain, the relationship is [27]

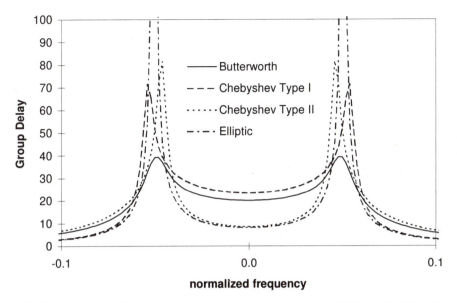

FIGURE 3-26 Normalized group delay response for the classical ARMA digital filter designs.

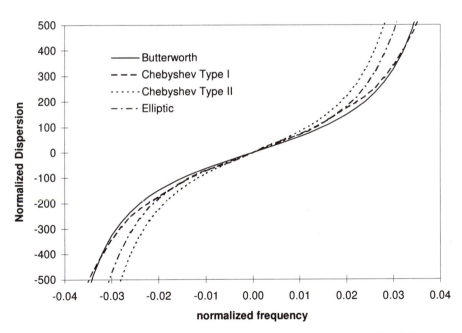

FIGURE 3-27 Passband normalized dispersion for the classical ARMA digital filter designs.

$$h_d(n)*a_N(n) = \delta(n) + e(n) \tag{143}$$

where $e(n)$ is the error sequence and the $a_N(n)$ are the coefficients of $A_N(z)$ which we are trying to find. The error term arises if $H_d(z)$ is not an all-pole function, or if it is all-pole but has an order larger than N. Assuming that $h_d(n)$ is defined for $0 \le n \le L$ where $L > N$, the convolution can be written in matrix form as [27]

$$
\begin{bmatrix}
h_{d0} & 0 & \cdots & 0 \\
h_{d1} & h_{d0} & \ddots & \vdots \\
\vdots & & \ddots & 0 \\
h_{dN} & & & h_{d0} \\
\vdots & \ddots & & \\
h_{dL} & & & \\
0 & h_{dL} & & \\
\vdots & \ddots & \ddots & \\
0 & \cdots & 0 & h_{dL}
\end{bmatrix}
\begin{bmatrix}
a_{0N} \\
a_{1N} \\
\vdots \\
a_{NN}
\end{bmatrix}
=
\begin{bmatrix}
1 \\
0 \\
\vdots \\
0
\end{bmatrix}
+
\begin{bmatrix}
e_0 \\
e_1 \\
\vdots \\
e_L
\end{bmatrix}
\tag{144}
$$

where $a_{nN} = a_N(n)$. A shorthand notation is given by $\underline{H_d}\underline{a} = \underline{\delta} + \underline{e}$, where the underscores indicate a matrix or vector. The mean square error is defined as

$$E_N = \underline{e}^\dagger \underline{e} = \sum_{n=0}^{L} |e(n)|^2 \tag{145}$$

Note that the error being minimized in the mean square sense is $e(n) = h_d(n)*a_N(n) - \delta(n)$, not the difference in the desired and approximating impulse or frequency responses. By substituting $\underline{e} = \underline{H_d}\underline{a_N} - \underline{\delta}$ into Eq. (145), the error is given in terms of the coefficients as follows:

$$E_N = \underline{a}_N^\dagger \underline{H}_d^\dagger \underline{H}_d \underline{a}_N - \underline{\delta}^\dagger \underline{H}_d \underline{a}_N - \underline{a}_N^\dagger \underline{H}_d^\dagger \underline{\delta} + \underline{\delta}^T \underline{\delta} \tag{146}$$

It is straightforward to show that $\underline{\delta}^\dagger \underline{\delta} = 1$, $\underline{\delta}^\dagger \underline{H}_d \underline{a}_N = h_{d0} a_{0N}$, and $\underline{a}_N^\dagger \underline{H}_d^\dagger \underline{\delta} = h_{d0}^* a_{0N}^*$. To further simplify E_N, let $\underline{R} = \underline{H}_d^\dagger \underline{H}_d$ where $R_m = \sum_{n=0}^{L-m} h_d^*(n+m)h_d^*(n)$ for $m = 0, \cdots,$ N. Note that $R_{-m} = R_m^*$, and that R_0 is real valued. As $L \to \infty$, R_m converges to the autocorrelation sequence of $h_d(n)$. For a finite L, R_m is an approximation to the autocorrelation sequence. With these simplifications, the error is given by $E_n = \underline{a}_N^\dagger \underline{R} \underline{a}_N - [h_{d0} a_{0N} + h_{d0}^* a_{0N}^*] + 1$. To minimize the error, the derivative with respect to each coefficient is found and set equal to zero, i.e. $\partial E_N / \partial a_{nN} = 0$ for $n = 0, \ldots, N$. Alternatively, $\partial E_N / \partial a_{nN}^* = 0$ may be used. In either case, the solution is referred to as the normal equations given by $\underline{R}\underline{a}_N = h_{d0}^* \underline{\delta}$, or in expanded form by

$$
\begin{bmatrix}
R_0 & R_1^* & \cdots & R_N^* \\
R_1 & R_0 & & \\
\vdots & & \ddots & \vdots \\
R_N & R_{N-1} & \cdots & R_0
\end{bmatrix}
\begin{bmatrix}
a_{0N} \\
a_{1N} \\
\vdots \\
a_{NN}
\end{bmatrix}
= h_{d0}^*
\begin{bmatrix}
1 \\
0 \\
\vdots \\
0
\end{bmatrix}
\tag{147}
$$

Solving for $\underline{a}_N$ gives the best Nth-order all-pole filter to approximate $H_d(\omega)$ in the least-squares sense. As an aside, multiplying $\underline{e}$ by $\underline{H}_d^\dagger$ and substituting into Eq. (147) yields

$$\underline{H}_d^\dagger \underline{e} = R\underline{a}_N - \underline{H}_d^\dagger \underline{\delta} = 0 \tag{148}$$

The minimum error is orthogonal to the subspace spanned by the columns of $\underline{H}_d^\dagger$, which is the well known orthogonality principle. The minimum square error is $E_N = 1 - h_{d0}a_{0N}$. As $N \to \infty$, the error goes to zero and $a_{0N} \to 1/h_{d0}$. In previous definitions of AR filters, we included a gain factor, Γ, in the numerator and assumed that $a_{0N} = 1$. A new set of coefficients is defined that follows this convention. Let $\hat{a}_{nN} = a_{nN}/a_{0N}$ and $\Gamma_N = 1/a_{0N}$ so that

$$H_N(z) = \frac{1}{a_{0N} + a_{1N}z^{-1} + \cdots + a_{NN}z^{-N}} = \frac{\Gamma_N}{1 + \hat{a}_{1N}z^{-1} + \cdots + \hat{a}_{NN}z^{-N}} \tag{149}$$

For a passive optical filter, the transmission satisfies $|H_d(\omega)| \leq 1$ which implies a maximum gain factor $\Gamma_N^{\max} = \min\{|1 + \hat{a}_{1N}e^{-j\omega} + \cdots + \hat{a}_{NN}e^{-j\omega N}|\}$. Two extensions of the above approach for the design of ARMA filters are Prony's and Shank's methods [2]. The input is either $h_d(n)$ or its autocorrelation, and the error is $e(n) = h_d(n)*a(n) - b(n)$.

We now focus on solving the normal equations for AR filter design, which are rewritten as

$$h_{d0}\Gamma_N\underline{\delta} = \underline{R}\hat{\underline{a}}_N \tag{150}$$

Since $h_d(n)$ can be multiplied by a constant phase term without changing the frequency response, h_{d0} is taken to be real without loss of generality. The terms along each diagonal of $\underline{R}$ are identical, so $\underline{R}$ is a Toeplitz matrix. The Levinson-Durbin algorithm exploits this symmetry to efficiently solve the normal equations. The importance of this algorithm to optical filters is twofold. First, an alternative filter description is given in terms of reflection coefficients, denoted by γ_n. As the name implies, the γ_n's describe the reflectance between layers of a multi-stack dielectric filter, which will be developed in Chapter 5. Second, the digital lattice filter architecture is intricately linked to this new description. The design and analysis of digital lattice filters establishes a framework for the study of optical lattice filters in Chapters 4–6.

The Levinson-Durbin algorithm solves the normal equations recursively by obtaining the solution for $n = 0$, then $n = 1$, and so forth until the Nth-order solution is found. The matrix equations for $n = 0$, $n = 1$, and $n = 2$ are subsets of Eq. (147) and are shown in expanded form below [30].

$$n = 0: [R_0][1] = h_{d0}[\Gamma_0] \tag{151}$$

$$n = 1: \qquad \begin{bmatrix} R_0 & R_1^* \\ R_1 & R_0 \end{bmatrix} \begin{bmatrix} 1 \\ \hat{a}_{11} \end{bmatrix} = h_{d0} \begin{bmatrix} \Gamma_1 \\ 0 \end{bmatrix} \qquad (152)$$

$$n = 2: \qquad \begin{bmatrix} R_0 & R_1^* & R_2^* \\ R_1 & R_0 & R_1^* \\ R_2 & R_1 & R_0 \end{bmatrix} \begin{bmatrix} 1 \\ \hat{a}_{12} \\ \hat{a}_{22} \end{bmatrix} = h_{d0} \begin{bmatrix} \Gamma_2 \\ 0 \\ 0 \end{bmatrix} \qquad (153)$$

For $n = 0$, the solution is $\underline{\hat{a}}_0 = \hat{a}_{00} = 1$ and $\Gamma_0 = R_0/h_{d0}$. The Nth-order equation is $\underline{R}\underline{\hat{a}}_n = h_{d0}\Gamma_n\underline{\delta}$ where the dimension of $\underline{R}$ and $\underline{\delta}$ vary with n. It is convenient to define a second matrix equation in terms of $\underline{R}$ and the reverse polynomials of $\underline{\hat{a}}_n$ and $h_{d0}\Gamma_n\underline{\delta}$. A property of $\underline{R}$ is that it is invariant with respect to interchanging the rows and columns, so $\underline{J}\underline{R}\underline{J}^{-1} = \underline{R}$ where the "reversing" matrix $\underline{J}$ is defined as follows [30]:

$$\underline{J} \equiv \begin{bmatrix} 0 & 0 & \cdots & 1 \\ \vdots & & \cdot^{\cdot^{\cdot}} & \vdots \\ 0 & 1 & 0 & 0 \\ 1 & 0 & \cdots & 0 \end{bmatrix} \qquad (154)$$

Using this property, it can be shown that $\underline{R}\underline{\hat{a}}_n^{*R} = h_{d0}\Gamma_n^*\underline{\delta}^R$, which is given in extended form by

$$\begin{bmatrix} R_0 & R_1^* & \cdots & R_n^* \\ R_1 & R_0 & & \\ \vdots & & \ddots & \vdots \\ R_n & R_{n-1} & \cdots & R_0 \end{bmatrix} \begin{bmatrix} \hat{a}_{nn}^* \\ \vdots \\ \hat{a}_{1n}^* \\ 1 \end{bmatrix} = h_{d0}\Gamma_n^* \begin{bmatrix} 0 \\ \vdots \\ 0 \\ 1 \end{bmatrix} \qquad (155)$$

The reverse polynomials are obtained by reversing the coefficient order and taking the complex conjugate. The solution $\underline{\hat{a}}_1$ is written in terms of $\underline{\hat{a}}_0$, $\underline{\hat{a}}_2$ in terms of $\underline{\hat{a}}_1$, and so forth until the solution $\underline{\hat{a}}_N$ is obtained. To express the next higher order polynomial in terms of the previous order polynomial, we assume that a constant γ_n exists such that the next higher order polynomial is a linear combination of the lower one and its reverse polynomial. The dimension of the lower order polynomial is extended by adding a zero as the last coefficient. For $n = 1$,

$$\underline{a}_1 = \begin{bmatrix} 1 \\ \hat{a}_{11} \end{bmatrix} = \begin{bmatrix} 1 \\ 0 \end{bmatrix} - \gamma_1 \begin{bmatrix} 1 \\ 0 \end{bmatrix} = \begin{bmatrix} \hat{a}_0 \\ 0 \end{bmatrix} - \gamma_1 \begin{bmatrix} 0 \\ \hat{\underline{a}}_0^{*R} \end{bmatrix} \qquad (156)$$

The general form is

$$\underline{\hat{a}}_n = \begin{bmatrix} \hat{\underline{a}}_{n-1} \\ 0 \end{bmatrix} - \gamma_1 \begin{bmatrix} 0 \\ \hat{\underline{a}}_{n-1}^{*R} \end{bmatrix} \qquad (157)$$

Equation (157) is a step-up recursion relation since it defines the next higher order polynomial in terms of the previous one. For convenience, one more term is defined, Δ_n, as follows:

$$\underline{R}\begin{bmatrix} \hat{\underline{a}}_n \\ 0 \end{bmatrix} = \begin{bmatrix} h_{d0}\Gamma_n \\ 0 \\ \vdots \\ \Delta_n^* \end{bmatrix} \text{ and } \underline{R}\begin{bmatrix} 0 \\ \hat{\underline{a}}_n^{*R} \end{bmatrix} = \begin{bmatrix} \Delta_n \\ 0 \\ \vdots \\ h_{d0}\Gamma_n^* \end{bmatrix} \quad (158)$$

The right-hand expression results from the property demonstrated in Eq. (155). So,

$$\Delta_n^* = R_{n+1} + \hat{a}_{1n}R_n + \cdots + \hat{a}_{nn}R_1 \quad (159)$$

for $1 \le n < N$. For the $n = 0$ case,

$$\begin{bmatrix} R_0 & R_1^* \\ R_1 & R_0 \end{bmatrix}\begin{bmatrix} 1 \\ 0 \end{bmatrix} = \begin{bmatrix} h_{d0}\Gamma_0 \\ \Delta_0^* \end{bmatrix} \quad (160)$$

So, $\Delta_0^* = R_1$. Now we are ready to solve for the nth solution. Substituting Eq. (157) into Eq. (150) yields

$$\underline{R}\hat{\underline{a}}_n = h_{d0}\Gamma_n\underline{\delta} = \underline{R}\begin{bmatrix} \hat{\underline{a}}_{n-1} \\ 0 \end{bmatrix} - \gamma_n\underline{R}\begin{bmatrix} 0 \\ \hat{\underline{a}}_{n-1}^{*R} \end{bmatrix} \quad (161)$$

Substituting in Eq. (158), we obtain

$$\begin{bmatrix} h_{d0}\Gamma_n \\ 0 \\ \vdots \\ 0 \end{bmatrix} = \begin{bmatrix} h_{d0}\Gamma_{n-1} \\ 0 \\ \vdots \\ \Delta_{n-1}^* \end{bmatrix} - \gamma_n\begin{bmatrix} \Delta_{n-1}^* \\ 0 \\ \vdots \\ h_{d0}\Gamma_{n-1}^* \end{bmatrix} \quad (162)$$

The bottom row gives $\Delta_{n-1}^* = \gamma_n h_{d0}\Gamma_{n-1}^*$, which is rearranged to solve for the reflection coefficient.

$$\gamma_n = \frac{1}{h_{d0}}\left(\frac{\Delta_{n-1}}{\Gamma_{n-1}}\right)^* \quad (163)$$

The top row of Eq. (162) combined with Eq. (163) gives an expression for the gain factor.

$$\Gamma_n = (1 - |\gamma_n|^2)\Gamma_{n-1} \quad (164)$$

Note that Γ_n is always less than Γ_{n-1} provided that the reflection coefficient $|\gamma_n| > 0$. The steps in the Levinson-Durbin algorithm are outlined below.

1. Input h_{d0} and R_m for $m = 0, \ldots, N$
2. Set $\hat{\underline{a}}_0 = \hat{a}_{00} = 1$, $\Gamma_0 = R_0/h_{d0}$, and $\Delta_0 = R_1^*$

3. For $n = 1$ to N
   a. calculate γ_n using Eq. (163)
   b. calculate Γ_n using Eq. (164) and $\hat{a}_n$ using Eq. (157)
   c. calculate Δ_n using Eq. (159) (skip for $n = N$)

The outputs are Γ_n, $\hat{a}_n$ and γ_n for $0 \leq n \leq N$. Note that $\hat{a}_{nn} = -\gamma_n$ according to Eq. (157). In summary, the coefficients of $A_N(z)$ and the reflection coefficients are equivalent ways of describing the same filter. The Levinson-Durbin algorithm takes the autocorrelation sequence as an input and finds both representations.

Since the square magnitude response of the desired function is typically specified, one procedure for obtaining R_m and h_{d0} is outlined below. The input h_{d0} is a scale factor that only affects Γ_n and not $\hat{a}_n$ or γ_n. So, we let $h_{d0} = 1$ for the algorithm. Then, Γ_N is set equal to the minimum value of $|A_N(\omega)|$ so that the filter's peak transmission is unity. The steps are

1. Sample input $|H_d(\omega_k)|^2$ for $k = 0, \ldots, 2K$ where $K > N$
2. Inverse DTFT$\{|H_d(\omega_k)|^2\} = R_k$, note $k = -K, \ldots, -N, \ldots, -1, 0, 1, \ldots, N, \ldots, K$
3. Apply the Levinson-Durbin algorithm to obtain $\hat{a}_N$ and γ_n
4. Calculate $\Gamma_N^{\max} = \min\{|1 + \hat{a}_{1N}e^{-j\omega} + \cdots + \hat{a}_{NN}e^{-j\omega N}|\}$

In step 2, the approximate autocorrelation sequence is noncausal, having both nonzero positive and negative delays. For input to the Levinson–Durbin algorithm, R_k is needed for $k = 0, \ldots, N$. Therefore, the number of samples in step 1 must be $2K + 1$ where $K > N$.

The reflection coefficient description is efficiently implemented in a digital lattice structure. The Z-transform for the step-up recursion of Eq. (157) is given as follows:

$$A_{n+1}(z) = A_n(z) - \gamma_{n+1}z^{-1}A_n^R(z) \qquad (165)$$

A similar equation for the reverse polynomial is

$$A_{n+1}^R(z) = -\gamma_{n+1}^* A_n(z) + z^{-1}A_n^R(z) \qquad (166)$$

A single stage digital filter which performs these operations is shown in Figure 3-28. Note the similarity in structure to the Mach–Zehnder diagram in Figure 3-17. An overview of digital multi-stage lattice architectures for MA and AR filters is given in Section 3.4.6.

3.4.6   Multi-Stage Filter Architectures

Multi-stage architectures are required for closely approximating a desired function. In this section, we start with a list of desired properties for optical multi-stage fil-

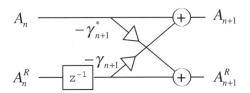

FIGURE 3-28   Single-stage digital lattice filter.

ters. Then, various digital architectures and their tradeoffs are discussed as a prelude to the optical architectures presented in Chapters 4–6. A list of desired characteristics for optical filter architectures is given below.

1. Capable of realizing general filter functions, which includes arbitrary zero and/or pole placement and being extendible to higher orders (multiple stages).
2. Minimal passband loss.
3. Many inputs and outputs.
4. Insensitive to fabrication variations.
5. Minimum number of stages.

These characteristics are applicable to MA, AR, and ARMA filters.

The first characteristic gives the most versatile filter since any function can be approximated over a finite frequency range to within a given error by increasing the number of stages. For bandpass filters, a general filter affords full design control over the passband width and flatness, stopband rejection and transition bandwidth. Gain equalization and dispersion compensating filters also require arbitrary placement of poles and zeros. If the poles and zeros are restricted to specific locations, then only a limited set of functions are available with which to approximate a desired response. We shall refer to this case as a restricted architecture, to contrast it with a general architecture. Important examples of restricted optical architectures are those employing vernier operation, discussed in Chapters 4–5.

The second desired characteristic is to minimize the passband loss. For digital filters, any gain is easily obtained by multiplying by a constant. If overflow occurs, the coefficients can be scaled. A passive optical filter, on the other hand, can achieve 100% transmission at most. It is preferable to avoid the need for optical amplifiers since they are typically expensive relative to a passive filter and add noise over a broad spectrum.

Many of the optical filter architectures are based on directional couplers and have two inputs and two outputs, forming a 2×2 device. Since the number of channels carried by WDM systems may be quite large and the filters must operate on a per channel basis at some point in the system, an important capability is to extend the 2×2 device architectures to $1 \times N$, $N \times 1$ and $N \times N$ operation.

Finally, a filter architecture that is most insensitive to fabrication variations is desirable. This is analogous to digital filters where the sensitivity to roundoff errors depends on the specific implementation. For optical filters, the uncertainty in component values is on the order of 10^{-2}. For digital filters, quantization errors of the filter coefficients are limited by the number of bits n in the representation of each number. For fixed point arithmetic with a resolution of only 10 bits, the quantization error is already an order of magnitude smaller than the uncertainty in optical filter component values. There are several reasons why the simplest optical filter is desirable. Filters with fewer stages are typically less sensitive to fabrication variations and require a smaller chip area for planar waveguide implementations. Another consideration is that the source coherence length must be longer than the longest relative delay, so a smaller number of stages relaxes the required source coherence length. Other considerations which differentiate optical filters from their digital counterparts include loss, polarization dependence, temperature dependence, and long term stability of the component (coefficient) values. Table 3-7 compares various properties of digital and optical filters. For digital filters, we shall only discuss general architectures.

There are many ways of implementing multi-stage digital filters given the $A(z)$ and $B(z)$ polynomials. If the filter coefficients were perfectly implemented by infinite precision arithmetic, the digital architectures would be equivalent. In practice, some architectures are more robust to finite precision implementations than others. To provide a background for understanding the sensitivity of various optical filters to fabrication variations, it is instructive to review the effects of quantization errors

TABLE 3-7   Comparison of Digital and Optical Multi-Stage Filters

Digital Filters	Optical Filters
• MA, AR & ARMA general architectures	• MA, AR & ARMA general architectures • Restricted architectures, e.g. Vernier operation for increasing FSR
• Any gain, scale the coefficients if overflow problems arise	• Power conservation determines maximum transmission • Low passband loss is critical • Loss changes the filter response
• Sensitivity to quantization errors	• Sensitivity to fabrication variations • Test methods for postfabrication optimization
• Minimize number of stages to reduce computation time • Use real coefficients to avoid complex number computations	• Minimize number of stages to reduce complexity, device size, and sensitivity to fabrication variations • Complex coefficients are easily realized
• Independent of environmental conditions	• Polarization dependence • Temperature dependence • Long-term stability

on digital filters. Another consideration for digital filter architectures is computation time and memory requirements. A concern for digital AR and ARMA filters is that quantization errors can lead to unstable filters if the pole magnitudes become equal to or larger than 1. For passive optical filters, the poles will always have a magnitude <1; however, this is a relevant concern for optical filters which employ gain in the feedback paths.

First, we consider the direct implementation of the $A(z)$ or $B(z)$ polynomial in a tapped delay line, or transversal, architecture as shown in Figures 3-29 and 3-30 for a MA and AR filter, respectively.

$$H_{MA}(z) = B(z) = \sum_{n=0}^{N} b_n z^{-n} \tag{167}$$

$$H_{AR}(z) = \frac{\Gamma}{A(z)} = \frac{\Gamma}{1 + \sum_{n=1}^{N} a_n z^{-n}}$$

The incoming signal is split into $N + 1$ paths where each path experiences a different delay and tap coefficient. A variation in one of the tap coefficients affects all of the pole or zero locations. The variation of the ith pole to changes in the a_n values is given by [2]

$$\Delta p_i = -\sum_{n=1}^{N} \frac{p_i^{N-n}}{\prod_{j=1, j\neq i}^{N} (p_i - p_j)} \Delta a_n \tag{168}$$

Poles that are clustered, such as for narrowband filters, are particularly sensitive to uncertainties in the a_n values since the denominator becomes large. By reducing the number of coefficients and maximizing the distance between the poles, the overall error can be reduced. A similar expression can be obtained for the zeros. The sensi-

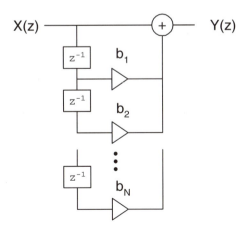

FIGURE 3-29 Transversal MA digital filter with $b_0 = 1$.

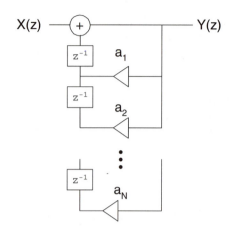

FIGURE 3-30   Transversal AR digital filter with $\Gamma = 1$.

tivity of the roots to coefficient accuracy increases with the order. For MA filters, the error in the requency response, $E(\omega) = H(\omega) - H_d(\omega)$, due to quantization of the filter coefficients has a standard derivation of $\sigma_E = 2^{-(b+1)}/\sqrt{M/3}$ where b is the number of bits and M is the filter order [2]. The overall filter response is typically broken into first- and second-order sections that are then cascaded. Second-order sections are preferred for digital filters when conjugate poles and zeros can be combined to avoid complex multiplications. For optical filters, no distinction is required between real and complex operations.

A straightforward multi-stage architecture is to cascade single stages as shown in Figure 3-31 for an MA and AR digital filter. The zeros and poles determine the coefficient for each stage.

$$H_{MA}(z) = \Gamma \prod_{m=1}^{M} (1 - z_m z^{-1}) \tag{169}$$

$$H_{AR}(z) = \frac{\Gamma}{\displaystyle\prod_{n=1}^{N} (1 - p_n z^{-1})}$$

For each stage, the incoming signal is split into two paths. One path has unity transmission and the other experiences a delay and multiplication by a coefficient. The advantage of the cascade architecture is that variations in one of the tap coefficients does not change the response of the other stages. The cascade structure is superior to the direct realization with respect to coefficient quantization effects. Cascade structures also display low stopband sensitivity. For optical filters, passband loss is a major issue for cascade architectures as explained in Chapters 4 and 5.

A third multi-stage architecture uses the equivalent expression for a polynomial in terms of a set of reflection coefficients as shown in Section 3.4.5. A multi-stage

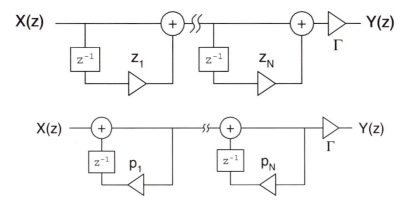

FIGURE 3-31   Cascade architecture MA (top) and AR (bottom) digital filters.

MA lattice filter is shown in Figure 3-32, and a multi-stage AR lattice filter is shown in Figure 3-33. The outputs of one stage are the inputs to the next stage, which is an important property for realizing low-loss optical filters. Digital lattice structures are capable of operating with significantly higher component variation than cascade structures, 4 to 5 bits are sufficient for coefficient quantization [31]. Focusing on the MA filter first, the following transfer matrix is defined relating the inputs of one stage to its outputs:

$$\begin{bmatrix} B_n(z) \\ B_n^R(z) \end{bmatrix} = \begin{bmatrix} 1 & -\gamma_n z^{-1} \\ -\gamma_n^* & z^{-1} \end{bmatrix} \begin{bmatrix} B_{n-1}(z) \\ B_{n-1}^R(z) \end{bmatrix} \tag{170}$$

The zeroth order polynomials are $B_0 = B_0^R = 1$.

To design a lattice filter having an Nth-order MA transfer function $H(z)$, one sets $B_N(z) = H(z)$ and $B_N^R(z) = z^{-n}H^*(z^{*-1})$. The last coefficient of $B_N(z)$ is equal to $-\gamma_N$. By taking the inverse of Equation (170), the following step-down relations are obtained:

$$\begin{bmatrix} B_{n-1}(z) \\ B_{n-1}^R(z) \end{bmatrix} = \frac{z}{1 - |\gamma_n^2|} \begin{bmatrix} z^{-1} & -\gamma_n^* \\ -\gamma_n z^{-1} & 1 \end{bmatrix} \begin{bmatrix} B_n(z) \\ B_n^R(z) \end{bmatrix} \tag{171}$$

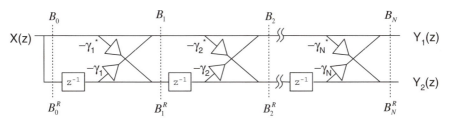

FIGURE 3-32   Multi-stage MA digital lattice filter with complex coefficients.

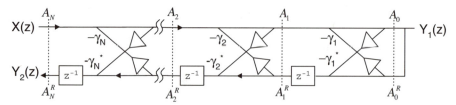

FIGURE 3-33   Multi-stage AR digital lattice filter.

for $n = N, \ldots, 1$. The last coefficient of B_{n-1} defines $-\gamma_{n-1}$ and so forth until all the reflection coefficients are found. This procedure is known as the Levinson algorithm.

The AR lattice filter is similar to the MA lattice, except with the inputs and outputs reversed. The transfer functions are: $H_1(z) = Y_1(z)/X(z) = 1/A_N(z)$, an AR filter, and $H_2(z) = Y_2(z)/X(z) = A_N^R(z)/A_N(z)$, an all-pass ARMA filter since the magnitude responses of the forward and reverse polynomials are the same. The step-up recursion relations are as follows:

$$\begin{bmatrix} A_n(z) \\ A_n^R(z) \end{bmatrix} = \begin{bmatrix} 1 & -\gamma_n \\ -\gamma_n^* z^{-1} & z^{-1} \end{bmatrix} \begin{bmatrix} A_{n-1}(z) \\ A_{n-1}^R(z) \end{bmatrix} \tag{172}$$

The magnitude of the reflection coefficients is related to the roots of the direct-form polynomial. If $|\gamma_n| < 1$ for $1 \leq n \leq N$, then the roots have a magnitude less than 1. Thus for the MA filter, this condition assures the roots are minimum-phase, and for the AR filter, it assures that the filter is stable. This test is known as the Schur-Cohn stability test. All-pole lattice filters are important in speech processing to model the vocal tract, in geology to model the layers of the earth and for optical filters as discussed in Chapter 5. The design of an AR filter given by $H(z) = 1/A_N(z)$ proceeds by

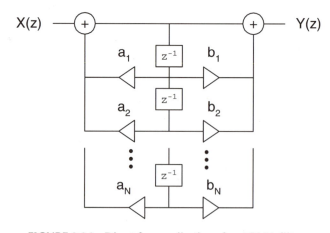

FIGURE 3-34   Direct form realization of an ARMA filter.

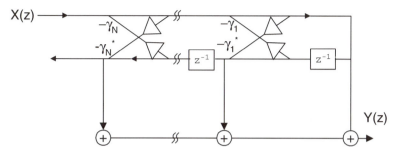

FIGURE 3-35   A lattice-ladder ARMA architecture.

inverting Eq. (172) to obtain step-down recursion relations for $A_n(z)$ and $A_n^R(z)$. For sharp bandpass responses, poles cluster near the unit circle. The step-down recursion includes the term $1/(1 - |\gamma|^2)$ which becomes very large as $\gamma \rightarrow 1$; consequently, poles near the unit circle combined with roundoff errors in the recursion can lead to unstable lattice designs.

The most straightforward implementation of ARMA filters is to cascade multistage AR and MA filters. For optical filters, architectures that combine the implementation of the poles and zeros are required to minimize the passband loss. A direct form digital implementation that uses the same delays for the poles and zeros is shown in Figure 3-34. Alternatively, zeros can be realized by any linear combination of multiple feedback paths. An architecture that uses a lattice filter to realize the poles and a linear combination of sampling from the lattice to realize the zeros is shown in Figure 3-35. Finally, an ARMA filter can be realized by decomposing the desired function into the sum or difference of two all-pass filters. The background for this approach and its optical implementation are discussed in Chapter 6.

APPENDIX

The Laplace transform for continuous signals is analogous to the Z-transform for discrete signals. The Laplace transform of a continuous function $h(t)$ is defined by [1–4]

$$H(s) = \int_{-\infty}^{\infty} h(t)\, e^{-st} dt \qquad (173)$$

where s is a complex number given by $s = \sigma + j\Omega$. We are interested in causal functions which are zero for $t < 0$; therefore, the lower integration limit is replaced by 0. The frequency response of $h(t)$ is obtained by evaluating $H(s)$ along the imaginary axis $s = j\Omega$. With this substitution, the Laplace transform reduces to the Fourier transform. The inverse Laplace transform is defined by

$$h(t) = \frac{1}{2\pi j} \int_{c-j\infty}^{c+j\infty} H(s)\, e^{st} ds \qquad (174)$$

where $c \geq 0$. The Laplace transforms for differentiation and integration of a function are given by

$$h'(t) \rightarrow sH(s) - h(0) \tag{175}$$

$$\int_0^t h(x)dx \rightarrow \frac{H(s)}{s} \tag{176}$$

For analog filter applications, the initial conditions that appear in the Laplace transform of the first and higher order derivatives of $h(t)$ are set equal to zero, i.e. $h(0) = h'(0) = \cdots = 0$. Thus, integrodifferential equations with constant coefficients can be solved algebraically using the Laplace transform. Consider a linear system whose output $y(t)$ is given in terms of the input $x(t)$ by

$$b_0 x(t) + b_1 x'(t) + \cdots + b_N x^{(N)}(t) = y(t) + a_1 y'(t) + \cdots + a_N y^{(N)}(t) \tag{177}$$

where the superscripts denote derivatives with respect to t. For electrical filters, the coefficients a_n and b_n depend on the circuit's lumped elements such as capacitance, inductance and resistance. The Laplace transform of the above equation is

$$[b_0 + b_1 s + \cdots + b_N s^N]X(s) = [1 + a_1 s + \cdots + a_N s^N]Y(s) \tag{178}$$

The filter's transfer function is then defined by

$$H(s) = \frac{Y(s)}{X(s)} = \frac{b_0 + b_1 s + \cdots + b_N s^N}{1 + a_1 s + \cdots + a_N s^N} = \frac{B(s)}{A(s)} \tag{179}$$

The roots of $B(s)$ are the zeros and the roots of $A(s)$ are the poles of the transfer function. The system is stable if the poles occur on the left-hand side of the s-plane, i.e. $\sigma < 0$.

REFERENCES

1. A. Oppenheim and R. Schafer, *Digital Signal Processing*, 2nd. Ed., Englewood, N.J.: Prentice-Hall, Inc., 1975.

2. J. Proakis and D. Manolakis, *Digital Signal Processing: Principles, Algorithms, and Applications*, 3rd. Ed., Upper Saddle River, NJ: Prentice-Hall, 1996.

3. S. Orfanidis, *Introduction to Signal Processing*, Englewood Cliffs, NJ: Prentice-Hall, 1996.

4. R. Bracewell, *The Fourier Transform and Its Applications*, New York: McGraw-Hill, 1986.

5. J. Goodman, *Introduction to Fourier Optics*, San Francisco: McGraw-Hill, 1968.

6. B. Gold, A. Oppenheim, and C. Rader, "Theory and implementation of the discrete Hilbert transform," in *Papers on Digital Signal Processing*, A. Oppenheim, Ed. Cambridge, MA: MIT Press, 1969, pp. 14–42.

7. A. Deczky, "Synthesis of Recursive Digital Filters Using the Minimum p-Error Criterion," *IEEE Trans. Audio Electroacoustics,* vol. 20, no. 4, pp. 257–263, 1972.

8. J. Stratton, *Electromagnetic Theory,* New York: McGraw-Hill, 1941.

9. L. Brillouin, *Wave Propagation and Group Velocity,* New York: Academic Press, 1960.

10. S. Haykin, *Communication Systems,* New York: Wiley, 1994.

11. P. Yeh, *Optical Waves in Layered Media,* New York: Wiley, 1988.

12. B. Saleh and M. Teich, *Fundamentals of Photonics,* New York: Wiley-Interscience, 1991, p. 266.

13. H. Bode, *Network Analysis and Feedback Amplifier Design,* New York: Van Nostrand, 1945.

14. G. Lenz, B. Eggleton, C. Giles, C. Henry, R. Slusher, and C. Madsen, "Dispersive properties of optical filters for WDM systems," *J. Quantum. Electron.,* vol. 34, pp. 1390–1402, 1998.

15. D. Pepper, "Applications of Optical Phase Conjugation," *Scientific American,* vol. 254, pp. 74–83, 1986.

16. A. Yariv, D. Fekete, and D. Pepper, "Compensation for channel dispersion by nonlinear optical phase conjugation," *Opt. Lett.,* vol. 4, pp. 52–54, 1979.

17. R. Jopson, A. Gnauck, and R. Derosier, "Compensation of fibre chromatic dispersion by spectral inversion," *Electron. Lett.,* vol. 29, pp. 576–578, 1993.

18. S. Watanabe, T. Naito, and T. Chikama, "Compensation of chromatic dispersion in a single-mode fiber by optical phase conjugation," *IEEE Photon. Technol. Lett.,* vol. 5, pp. 92–95, 1993.

19. B. Moslehi, J. Goodman, M. Tur, and H. Shaw, "Fiber-Optic Lattice Signal Processing," *Proc. IEEE,* vol. 72, no. 7, pp. 909–930, 1984.

20. K. Jackson, S. Newton, B. Moslehi, M. Tur, C. Cutler, J. Goodman, and H. Shaw, "Optical Fiber Delay-Line Signal Processing," *IEEE Trans. Microwave Theory Techniques,* vol. 33, no. 3, pp. 193–209, 1985.

21. J. Capmany and M. Muriel, "A New Transfer Matrix Formalism for the Analysis of Fiber Ring Resonators: Compound Coupled Structures for FDMA Demultiplexing," *J. Lightw. Technol.,* vol. 8, no. 12, pp. 1904–1919, 1990.

22. E. Heyde and R. Minasian, "A Solution to the Synthesis Problem of Recirculating Optical Delay Line Filters," *IEEE Photon. Technol. Lett.,* vol. 6, no. 7, pp. 833–835, 1994.

23. N. Ngo and L. Binh, "Novel Realisation of Monotonic Butterworth Type Lowpass, Highpass, and Bandpass Optical Filters Using Phase-Modulated Fiber-Optic Interferometers and Ring Resonators," *J. Lightw. Technol.,* vol. 12, no. 5, pp. 827–841, 1994.

24. Y. Li and C. Henry, "Silica-based Optical Integrated Circuits," *IEE Proc.-Optoelectron.,* vol. 143, no. 5, pp. 263–280, 1996.

25. R. Collin, *Foundations for Microwave Engineering,* Singapore: McGraw-Hill Book Co., 1966.

26. M. Mansuripur, "Reciprocity in classical linear optics," *Optics & Photon. News,* July, pp. 53–58, 1998.

27. L. Jackson, *Digital Filters and Signal Processing, Boston,* MA: Kluwer Academic, 1996.

28. L. Rabiner and B. Gold, *Theory and Application of Digital Signal Processing,* Englewood Cliffs, N.J.: Prentice-Hall, Inc., 1974, p. 284.

29. T. Parks and C. Burrus, *Digital Filter Design,* New York: Wiley, 1987, p. 224.

30. S. Orfanidis, *Optimum Signal Processing: An Introduction,* 2nd Ed., New York: Mc-Graw-Hill, 1988, pp. 195–246.

31. R. Crochiere, "Digital ladder filter structures and coefficient sensitivity," Res. Lab. *Electron. Mass. Inst. Technol.,* vol. 103, October 15, 1971.

PROBLEMS

1. Assuming a diffraction-based delay-line in fused silica, calculate the difference in FSRs for $\lambda = 1550$ nm and $L_U = 2$ mm if n is used instead of n_g. Use the material dispersion for silica given by the Sellmeir equation in Chapter 2. In addition, calculate the wavelength range for which the phase error caused by neglecting the wavelength dependence of the refractive index is $\delta\phi/\phi < 0.1\%$.

2. Design an AR filter to approximate a linear power response $|H(\nu)|^2 = \nu$ for $0.05 \leq \nu \leq 0.95$ using the AR least squares synthesis technique. Compare to the design of an MA filter with the same number of stages using the window technique.

3. Find the AR model for $h(n) = r^n u(n)$ using $L = 2$ and $L = \infty$.

4. Calculate the reflection coefficients for $\Gamma/A_2(z)$ where $\Gamma = 0.02419$ and $A_2(z) = (1 + 0.9e^{j(2\pi/50)}z^{-1})(1 + 0.9e^{-j(2\pi/50)}z^{-1})$ by taking the inverse discrete Fourier transform of $|A_2(\omega)|^2$ when sampled at $N = 64$ and $N = 128$ points as the autocorrelation input to the Levinson algorithm.

CHAPTER FOUR

Multi-Stage MA Architectures

This chapter focuses on multi-stage optical MA (all-zero) filters. Before plunging into multi-stage architectures, it is instructive to begin with an in depth look at a single-stage MZI. Several practical issues are addressed including the impact of loss, refractive index variations, coupling ratio variations and wavelength dependence on the filter response. The discussion of multi-stage architectures begins with two structures that closely resemble their digital filter counterparts, optical cascade and transversal architectures. Then, multi-port architectures, which are advantageous for multiplexing and demultiplexing numerous channels simultaneously, are introduced. Two examples are the diffraction grating and waveguide grating router (WGR). Their principles of operation are very similar and are discussed in Section 4.4. Both of these filters break the wavefront into discrete sections that are then delayed and recombined to obtain very narrowband filters. Finally, optical MA lattice filters are developed. These 2×2 filters (2 inputs and 2 outputs) have power complementary outputs and can approximate any desired function. Examples are given for dispersion compensation and gain equalization applications. As the delay is made smaller and the number of stages larger for a lattice filter, we obtain filters whose frequency response is efficiently described by coupled-mode theory. Two examples of forward-propagating coupled-mode devices are presented, long period gratings and acousto-optic filters. The coupled-mode equations are derived, and a matrix formalism is presented for describing devices with arbitrary mode coupling strength profiles.

4.1 SINGLE-STAGE MZI DESIGN

Since all of the filters in this chapter are based on the MZI, design details and practical issues are addressed on a single stage before proceeding to multi-stage architectures. A schematic is shown in Figure 4-1. As discussed in Chapter 3, there are four transfer functions defined. For a lossless filter, they are

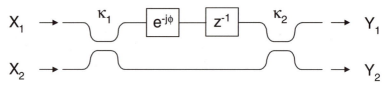

FIGURE 4-1   A single-stage MZI schematic.

$$\begin{bmatrix} Y_1(z) \\ Y_2(z) \end{bmatrix} = \begin{bmatrix} H_{11}(z) & H_{12}(z) \\ H_{21}(z) & H_{22}(z) \end{bmatrix} \begin{bmatrix} X_1(z) \\ X_2(z) \end{bmatrix} \tag{1}$$

where

$$\begin{bmatrix} H_{11}(z) & H_{12}(z) \\ H_{21}(z) & H_{22}(z) \end{bmatrix} = \begin{bmatrix} c_1 c_2 e^{-j\phi} z^{-1} - s_1 s_2 & -j(s_1 c_2 e^{-j\phi} z^{-1} + c_1 s_2) \\ -j(c_1 s_2 e^{-j\phi} z^{-1} + s_1 c_2) & -s_1 s_2 e^{-j\phi} z^{-1} + c_1 c_2 \end{bmatrix} \tag{2}$$

where $s_n = \sqrt{\kappa_n}$ and $c_n = \sqrt{1 - \kappa_n}$. The diagonal transfer functions, H_{11} and H_{22}, apply to the through-ports while the off-diagonal transfer functions, H_{12} and H_{21}, apply to the cross-ports. The zero locations in each case are given in Table 4-1. The relationship between the zeros for $H_{11}(z)$ and $H_{22}(z)$ is $z_{11} = 1/z_{22}^*$ and similarly for $H_{12}(z)$ and $H_{21}(z)$. Thus, $H_{11}(z)$ and $H_{22}(z)$ form a forward and reverse polynomial pair, as do $H_{12}(z)$ and $H_{21}(z)$. Conservation of power for a lossless filter requires that $|H_{11}(z)|^2 + |H_{21}(z)|^2 = |H_{12}(z)|^2 + |H_{22}(z)|^2 = 1$. A zero on the unit circle for either the through- or cross-port response implies 100% transmission for the other response. If the coupling ratios are identical ($\kappa_1 = \kappa_2$), the zero of the cross-port response occurs on the unit circle. For the through-port, the zero lies on the unit circle if $\kappa_1 = 1 - \kappa_2$. A special case arises for $\kappa_1 = \kappa_2 = 0.5$, where the zero for both the through-port and cross-port are on the unit circle, as shown in Figure 4-2a. Both output responses have transmission minimum that are zero in this case.

4.1.1   Loss and Fabrication Induced Variations

Loss is included in the transfer functions by replacing z^{-1} with γz^{-1} where the loss in dB for transmission along the differential path length ΔL is given by -20

TABLE 4-1   Zeros for a Single-Stage MZI

$H_{ij}(z)$	z_{ij}	$H_{ij}(z)$	z_{ij}
H_{11}	$e^{-j\phi} \sqrt{\dfrac{(1 - \kappa_1)(1 - \kappa_2)}{\kappa_1 \kappa_2}}$	H_{12}	$-e^{-j\phi} \sqrt{\dfrac{\kappa_1 (1 - \kappa_2)}{(1 - \kappa_1)\kappa_2}}$
H_{22}	$e^{-j\phi} \sqrt{\dfrac{\kappa_1 \kappa_2}{(1 - \kappa_1)(1 - \kappa_2)}}$	H_{21}	$-e^{-j\phi} \sqrt{\dfrac{(1 - \kappa_1)\kappa_2}{\kappa_1 (1 - \kappa_2)}}$

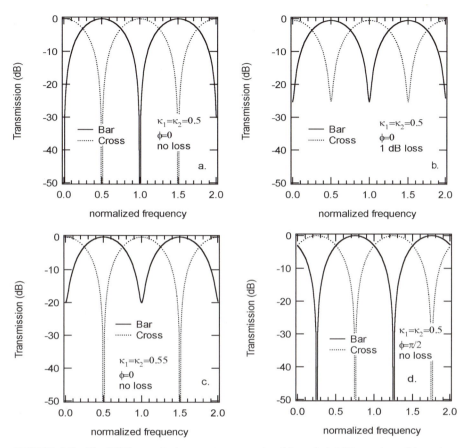

FIGURE 4-2 The MZI magnitude response assuming (a) perfect 3dB couplers, (b) a relative path loss of 1 dB, (c) coupler variation, and (d) a phase shift.

$\log_{10}(\gamma)$. Losses for silica waveguides are typically less than 0.1 dB/cm, so the differential path length must be many centimeters long to introduce significant loss. A family of MZIs have been demonstrated in doped-silica buried waveguides with FSRs from 8 pm to 250 nm [1]. The corresponding relative delay lengths range from 100 mm to 2.7 μm. For waveguides with $\Delta = 0.75\%$ and a minimum bend radius of 5 mm, the chip sizes range from 5 × 5 to 3 × 0.5 cm². The impact of a 1 dB loss (10 cm × 0.1 dB/cm) is shown in Figure 4-2b. The transmission nulls are shallower than the lossless case since the zeros are no longer on the unit circle for $\kappa_1 = \kappa_2 = 0.5$.

The fabricated coupling ratios will vary from their design value due to variations in the core-to-cladding refractive index, core height and width, and gap separation. A systematic variation, such as an increase in the index difference, will result in all coupling ratios being higher or lower than their design values. The response for identical couplers with coupling ratios equal to 0.55 is shown in Figure 4-2c. The null of the through-port is washed out; whereas, the cross-port null remains zero. Random variations lead to non-identical couplers. In such a case, the cross-port null would also be shallower.

The absolute wavelength of the passband must be accurately controlled. Constructive interference occurs for wavelengths that are integer multiples of the optical path length difference between the paths. For paths with a relative length difference of ΔL, the center wavelength of the passband satisfies $m\lambda_c = n_e\Delta L$ where m is the order. For a single stage MZI, variations in the optical path length difference from its nominal value translate the response in frequency causing the passband to be offset from its nominal center wavelength as shown in Figure 4-2d for $\phi = \pi/2$. Variations may arise from changes in the effective index, either systematic or random, or changes in the path length from the design intent. An example of the latter is quantization of the waveguide path during mask layout. Typically, this effect can be avoided by decreasing the mask grid size. Accumulated phase errors from random effective index variations scale with path length. Such variations are important for devices with small FSRs. A statistical treatment is given in Chapter 5. A practical solution is to compensate for the phase errors after fabrication, for example by using the thermo-optic effect and locally heating one arm relative to the other. An alternative is to permanently compensate for index variations by UV exposure or similar tuning mechanism.

4.1.2   A Tunable Coupler

A single-stage MZI with nominally identical arm lengths is called a symmetric MZI and can be used to make a coupler with any coupling ratio by changing the relative phase between the arms. A symmetric MZI may also be called a zeroth-order MZI since $\Delta L = 0$ and m is the order defined by $m = n_e\Delta L/\lambda$. The conditions for achieving a desired coupling ratio are explored by making both coupling ratios identical to simplify the problem, i.e. $\kappa_1 = \kappa_2 = \kappa$. This simplification is a realistic assumption from a fabrication viewpoint since the couplers are typically closely spaced. Assuming $X_1 = 1$ and $X_2 = 0$ in Eq. (1), the outputs are given by

$$Y_1 = (1 - \kappa)e^{-j\phi} - \kappa \tag{3}$$

$$Y_2 = -j\sqrt{(1 - \kappa)\kappa}(e^{-j\phi} + 1)$$

The square magnitude responses are

$$|Y_1|^2 = (1 - \kappa)^2 + \kappa^2 - 2\kappa(1 - \kappa)\cos(\phi) \tag{4}$$

$$|Y_2|^2 = 4\kappa(1 - \kappa)\cos^2\left(\frac{\phi}{2}\right)$$

For a lossless MZI, the effective coupling ratio of the MZI is given by

$$\kappa_{\text{MZI}} = \frac{|Y_2|^2}{|Y_1|^2 + |Y_2|^2} = |Y_2|^2 \tag{5}$$

First, the maximum value for $|Y_2|^2$ is one only if $\kappa = 0.5$. By changing the phase to $\phi = \pi$, $|Y_2|^2 = 0$ (and $|Y_1|^2 = 1$) independent of κ. Symmetric MZIs are used with phase shifters, such as heaters, to provide a tunable coupling ratio. The largest tuning range, 0–1, and the ability to use the device as a switch are achieved for $\kappa = 0.5$. Another important application of MZIs, discussed in Chapter 5, involves placing identical reflective filters in each arm so that the reflected signal exits in a separate port from the input signal. The best cross-talk performance is achieved for a power split of exactly 50%. A symmetric MZI can be used to achieve an effective coupling ratio of 50%. The range of κ's satisfying this condition are found by setting $|Y_2|^2 = 0.5$, which results in the following condition on κ:

$$\kappa(1 - \kappa) \geq \frac{1}{8} \tag{6}$$

If the coupling ratios are identical and $0.1464 \leq \kappa \leq 0.8536$, then a 50% split ($\kappa_{\text{MZI}} = 0.5$) can be achieved with the proper phase setting. If κ_1 and κ_2 are in this range but $\kappa_1 \neq \kappa_2$, a phase ϕ can still be found to achieve the 50% power split. A plot of ϕ versus κ_2 for several values of κ_1 is shown in Figure 4-3 for a 50% power splitter. The values for κ_1 and κ_2 can be reversed without changing the results. One difference between the symmetric MZI and a directional coupler is that the relative phase between the outputs is different and depends on κ_1 and κ_2 instead of being a constant equal to $\pi/2$. The relative phase, defined as $\phi_{\text{out}}^{\text{rel}} = \arg[Y_2/Y_1]$ for an input on X_1 (or X_2), is important if the symmetric MZI is used in an interferometer.

When employing a Z-transform description of a filter, we assume that the coupling ratios are constant over the wavelength range of interest. For small wavelength ranges, this approximation is good; however, over several hundred nanometers, there is a significant change in the filter response as demonstrated in Figure 4-4 for four periods of a MZI with a FSR = 125 nm. The nulls of the cross-port are preserved since $\kappa_1 = \kappa_2$; however, the nulls of the through-port are severely diminished for adjacent FSRs from the design center. When a filter response over a large wavelength range is to be designed, the wavelength dependence of the coupling ratios is important. In such a case, a symmetric MZI is advantageous over a directional coupler [2]. The wavelength dependence of a single directional coupler designed for a nominal 50% splitting ratio at 1550 nm was simulated using the following values: $\Delta = 0.64\%$, 5×5 μm^2 core, and a 9.5 μm waveguide center-to-center separation in the coupling region. The results are shown in Figure 4-5. Also plotted is the wavelength dependence of the cross-port response for a symmetric MZI. The coupling ratios and phase for the MZI were chosen to minimize the wavelength dependence over a 100 nm range. The resulting values were $\kappa_1 = 64.5\%$, $\kappa_2 = 19.0\%$ and $\phi = 1.206$ radians. The MZI response varies by only $\pm 3\%$ over a 200 nm range while the directional coupler varies by $\pm 17\%$.

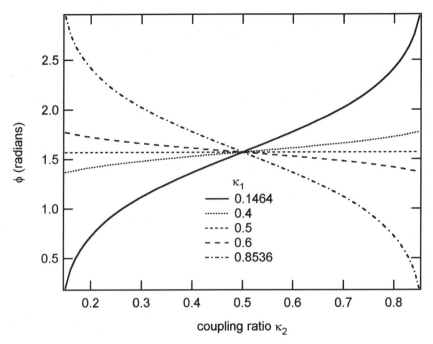

FIGURE 4-3 The required phase difference between the symmetric MZI arms for a 50% power split as a function of κ_1 and κ_2.

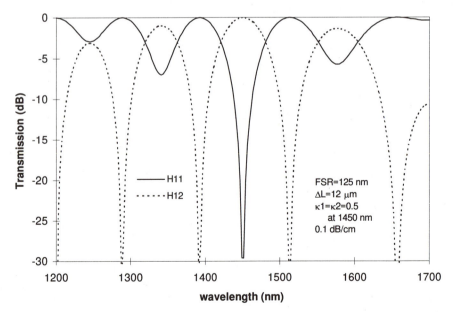

FIGURE 4-4 Four periods of an asymmetric MZI response showing the impact of the couplers' wavelength dependence.

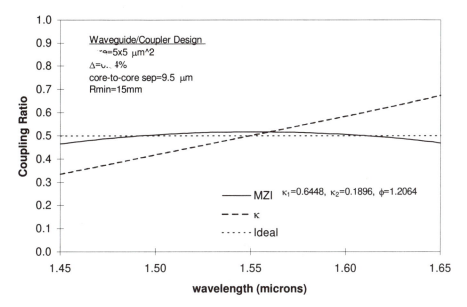

FIGURE 4-5   Wavelength dependence of a symmetric MZI and a directional coupler (κ) designed for a 50% power splitting ratio.

4.2   CASCADE FILTERS

A multi-stage MA filter can be implemented by cascading MZIs, as shown in Figure 4-6. Each stage implements a single zero; so, a total of 2N couplers are required for an Nth-order filter. Each stage is independent from the other stages since changes in the parameters of one stage do not change the zero locations of the other stages. Any combination of an input and output port can be chosen for each stage. Suppose that we want to realize a second-order filter with zeros at the complex conjugate positions $z_{1,2} = e^{\pm j\phi_d}$ using the X_1 input and Y_2 output for each MZI. The overall transfer function is

$$H(z) = \Gamma(1 - z_1 z^{-1})(1 - z_2 z^{-1}) = \frac{1}{4}[1 - 2\cos(\phi_d)z^{-1} + z^{-2}] = H_{21}^1(z)H_{21}^2(z) \quad (7)$$

where the cross-port transfer functions from Eq. (2) are used in the right-hand expression and the superscripts 1 and 2 denote the first and second MZI stages. Using the zero location given in Table 4-1 for H_{21}, one solution is $\kappa_{1a} = \kappa_{1b} = \kappa_{2a} = \kappa_{2b} = 0.5$ and $\phi_{1,2} = \pi \pm \phi_d$. In this example, the passband is centered at $\omega = \pi$. The peak transmission is unity for $\phi_d = 0$, but decreases as ϕ_d increases. The square magnitude responses of $H(\omega)$, $H_{21}^1(\omega)$, and $H_{21}^2(\omega)$ are shown in Figure 4-7 for $\phi_d = \pi/6$ radians. Even though the individual responses have a unity passband response, they are staggered in frequency giving rise to an overall loss of 0.6 dB at the center of the

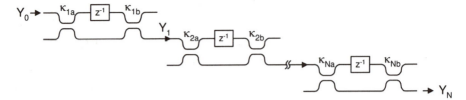

FIGURE 4-6   Cascaded MZI multi-stage filter.

passband. For a higher order filter, the passband loss is even larger. When the 10th-order bandpass filter designs from Chapter 3 are implemented using this architecture, the passband loss is 37 dB and 18 dB for the uniform and tapered windows, respectively. The tapered window response has a smaller passband loss because the zeros are located further from the passband. Because of its large passband loss, this architecture by itself is not practical for realizing higher order functions with arbitrary zero locations.

A specific cascade architecture with low passband loss consists of stages with delays that follow a geometric progression of L_U, $2L_U$, $4L_U$, etc. Figure 4-8 shows two applications that have been demonstrated [3,4], a demultiplexer and a channel

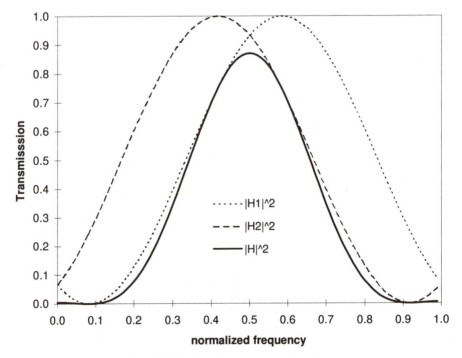

FIGURE 4-7   Two-stage cascade MA filter.

selector. The demultiplexer can also be used as a multiplexer since it is a reciprocal device. For a demultiplexer with M stages, there are 2^M outputs. The architecture is referred to as a $\log_2 N$ filter since the number of stages $M = \log_2 N$ where N is the number of channels. The frequency selector is similar to the de/multiplexer except that only one output port is used. The transfer function for any input and output port combination is the product of transfer functions for the individual stages. A three-stage, 1×8 demultiplexer is shown in Figure 4-9 along with the idealized spectra at each stage. The goal is to send each channel to a different output port. The transfer function for output port 1 is

$$H_1(z) = H_{21}^A(z)H_{21}^B(z)H_{21}^C(z) = (1 + z^{-4})(1 + z^{-2})(1 + z^{-1})/8 \qquad (8)$$

where the superscripts A, B, and C indicate the various stages. Constant multiplicative factors of $-j$ were neglected from the cross-port transfer functions in Eq. (8).

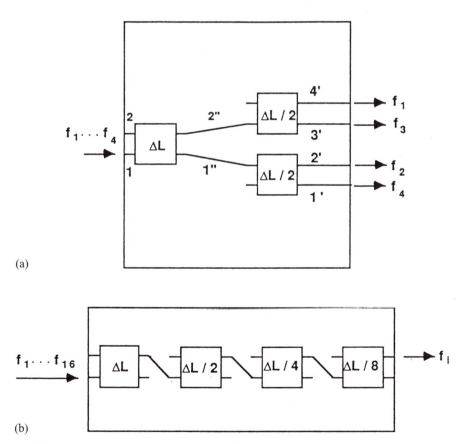

(a)

(b)

FIGURE 4-8   Schematic for (a) a 1×4 demultiplexer and (b) a 16 channel frequency selector [1]. Reprinted by permission. Copyright © 1990 IEEE.

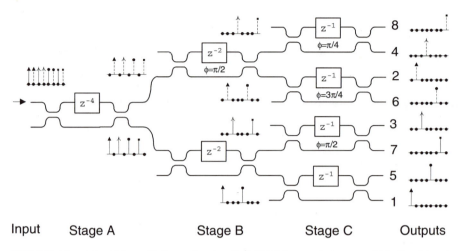

FIGURE 4-9   A 1 × 8 demultiplexer showing the MZI delays and phases and idealized input versus output spectra using the $\log_2 N$ architecture.

The resulting impulse response is $h(n) = 1/8$ for $0 \leq n \leq 7$. For 100% transmission of one wavelength into a given port, there must be zero transmission into the remaining ports by power conservation. A plot of the zeros for output port 1 is given in Figure 4-10a. To minimize passband loss, the coupling ratios must equal 0.5 so that the zeros are on the unit circle for both the cross-port and through-port transmission of each MZI. The magnitude response is shown in Figure 4-11 for each stage independently and for the overall response for output 1. The remaining outputs are found by shifting the zeros by $2\pi n/N$ where $N = 8$, which is realized by adding appropriate phase shifts to each MZI. The zero locations for output 2 are shown in Figure 4-10b, and the transfer function is $H_2(z) = (1 - z^{-4})(1 + jz^{-2})(1 - e^{-j3\pi/4}z^{-1})/8$. The phase shift for each stage is indicated in Figure 4-9. The phase oc-

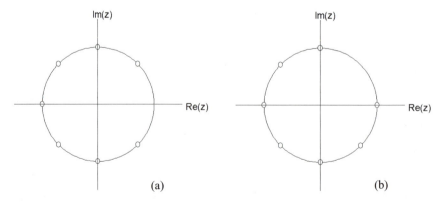

FIGURE 4-10   Zero diagram of a 1 × 8 demultiplexer for (a) output 1 and (b) output 2.

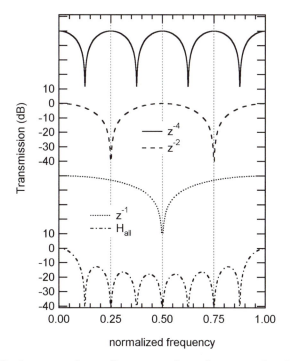

FIGURE 4-11   Single-stage and overall responses for a three-stage demultiplexer using MZIs with delays of $4L_u$, $2L_u$, and L_u. The canceling of extra maxima in the top response by zeros in the remaining stages is an example of Vernier operation.

curs on the delay side, so $z^{-1} \rightarrow e^{-j\phi}z^{-1}$. The transfer function for the nth output is generalized for a $1 \times N$ device as follows:

$$H_n(z) = \frac{1}{N}\sum_{k=0}^{N-1} e^{+j2\pi\frac{nk}{N}}z^{-k} \tag{9}$$

So, the frequency response is the discrete inverse Fourier transform of $h(k) = 1/N$ for $0 \le k \le N-1$, multiplied by a phase factor that determines the output port.

A 1×4 demultiplexer was demonstrated in P-doped SiO_2 waveguides using the $\log_2 N$ cascade architecture with a 2.6 dB passband loss and an average cross-talk loss of 16 dB [4]. The cross-talk loss was limited by fabrication variations in the coupling ratios and phases. Using thin film heaters, the phase variation can be compensated after fabrication. Later, a 1×16 demultiplexer with a 160 GHz FSR was demonstrated in Ge-doped SiO_2 with $\Delta = 1.5\%$ [5]. A minimum bend radius of 2 mm was used, and the overall chip size was 30×70 mm². The average passband loss for the packaged device was 6.1 dB and the average cross-talk loss was 24.3 dB.

A 16 channel frequency selector with the waveguide layout shown in Figure 4-12

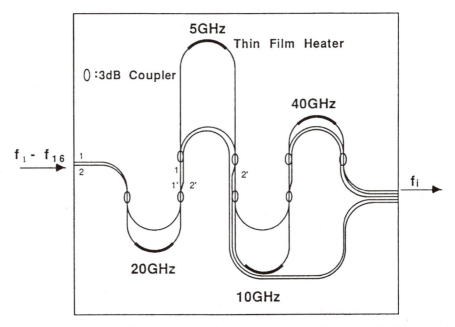

FIGURE 4-12   Waveguide layout for a 16 channel frequency selector with 5 GHz channel spacing [1]. Reprinted by permission. Copyright © 1990 IEEE.

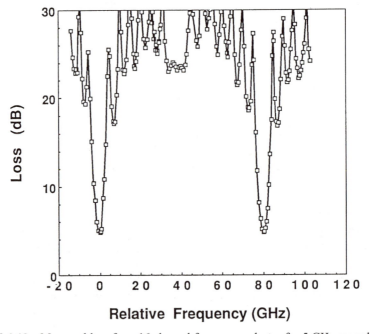

FIGURE 4-13   Measured loss for a 16 channel frequency selector for 5 GHz spaced channels [1]. Reprinted by permission. Copyright © 1990 IEEE.

was demonstrated [1]. On the output, one port was used for the selected frequency, and the remaining four ports were used to monitor the heater settings to optimize the output for the desired channel. This circuit was fabricated for a 5 GHz channel spacing on a 5.2×4 cm^2 chip with differential path lengths of 20, 10, 5 and 2.5 mm for the 5, 10, 20 and 40 GHz MZIs, respectively. The experimental results are shown in Figure 4-13. The passband loss is 5.2 dB, the adjacent channel cross-talk loss is >13 dB, and the nonadjacent channel cross-talk loss is >20 dB.

The three-stage architecture in Figure 4-9 yields a seventh-order filter. In general, the number of stages required to produce an Nth-order filter response can be significantly reduced if the desired polynomial contains a factor of the form $(1 + z^{-p} r_p e^{j\phi_p})$, where $1 < p \le N$. Then, the delay of one stage is set to p times the unit delay, and the total number of stages decreases accordingly. For such a factorization, the p roots are spread uniformly around a circle of radius $r^{1/p}$ with phases satisfying $\phi_m = \phi_p/p + 2\pi m/p$ for $m = 0, 1, \ldots, p - 1$.

4.3 TRANSVERSAL FILTERS

Transversal optical planar waveguide filters [6–8], shown schematically in Figure 4-14 with a unit delay indicated by τ, are analogous to the direct form architectures for digital filters. The magnitudes of the tap coefficients $[a_0, \ldots, a_{N-1}]$ are realized by varying the splitting ratios, for example, by using tunable couplers as discussed in Section 4.1.2. A phase shifter is included in each arm to set the tap coefficient phase. To derive the frequency response, let the combiner consist of couplers, Y-branches, or a star coupler such that each tap experiences the same loss through the

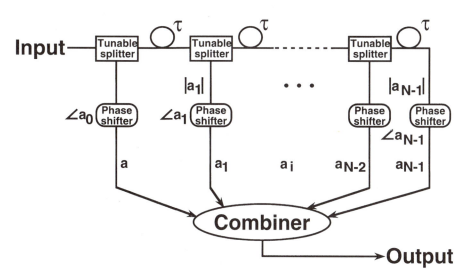

FIGURE 4-14 A transversal MA filter architecture [7]. Reprinted by permission. Copyright © 1991 IEEE.

combiner. By summing all the optical paths, the Z-transform for a lossless transversal filter is written as follows:

$$H_{N-1}(z) = \frac{1}{\sqrt{N}} [s_0 e^{-j\Phi_0} + c_0 s_1 e^{-j\Phi_1} + c_0 c_1 s_2 e^{-j\Phi_2} z^{-2} + \cdots$$

$$+ c_0 c_1 \cdots c_{N-2} s_{N-1} e^{-j\Phi_{N-1}} z^{-(N-1)}] \tag{10}$$

where each combiner has an amplitude transmission of $1/\sqrt{N}$. The phase for the nth path Φ_n includes the phase shifter contribution ϕ_n and any relative phase between paths introduced by the combiner. To design a filter with $H_d(z) = b_0 + b_1 z^{-1} + \cdots + b_{N-1} z^{-N-1}$, we must find N coupling ratios and phases. A small fraction of light is lost at the through-port of the last splitter, so $H_d(z)$ is normalized to obtain a peak transmission less than unity to account for this loss. The coupling ratios may be found iteratively beginning with the first coefficient, κ_0, as follows:

$$\kappa_0 = N|b_0|^2, \ \kappa_1 = \frac{|b_1|^2 N}{(1-\kappa_0)}, \ldots, \ \kappa_{N-1} = \frac{|b_{N-1}|^2 N}{\displaystyle\prod_{n=0}^{N-2}(1-\kappa_n)} \tag{11}$$

The phases in the tapped arms, ϕ_n, are set so that $\Phi_n = -\arg[b_n]$.

In practice, tunable couplers are used for the splitters so that the tap coefficients can be measured and set to their desired values after fabrication [7]. An optical pulse with a duration shorter than the unit delay is transmitted through the filter, and the received pulses as a function of delay give the tap coefficients. The measurement setup is discussed in Chapter 7. An 8-tap filter with a unit delay of 50 ps (20 GHz FSR) was demonstrated by Sasayama [7]. The measured and calculated spectral responses, as well as the corresponding tap coefficients, are shown in Figure 4-15. A passband loss of 9.4 dB and a maximum stopband-to-passband rejection of 10 dB were obtained. The close agreement between the calculated and measured results show that any arbitrary frequency response can be approximated.

The transversal filter has also been used to demonstrate a frequency selector that can add or drop one or more channels [8] in contrast to the previous selector and demultiplexer that could select only one channel for a given output port. The desired frequency response is either 1 or 0 for each channel. A discrete sequence of 1's and 0's with length N results, where N is equal to the number of taps. The filter coefficients are then determined from the inverse DTFT of this sequence, and the coupling ratios are calculated. A 16-tap filter with a 15 dB fiber-to-fiber insertion loss and a FSR = 20 GHz was demonstrated [8]. The frequency response for single channel selection, 3 channel and 7 channel selection along with the tap coefficients are shown in Figure 4-16. The minimum cross-talk loss is 13 dB for the single channel, 10 dB for the 3 channel selector, and 7 dB for the 7 channel selector. A small nonuniformity of 0.3 dB is evident between the passbands. The advantage of this architecture is the straightforward design algorithm for adding and dropping one or

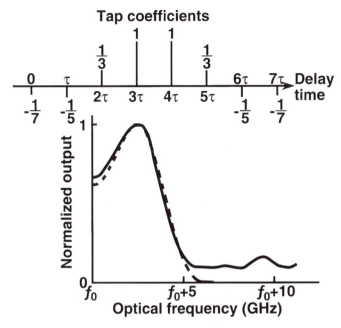

FIGURE 4-15 Coefficients for an eight-tap filter (top) and the measured (solid) and predicted (dashed) spectral response (bottom) [7]. Reprinted by permission. Copyright © 1991 IEEE.

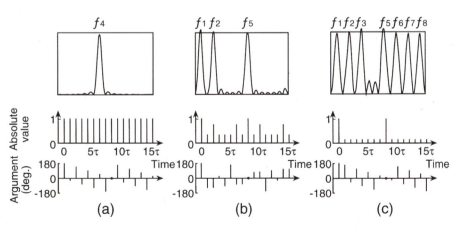

FIGURE 4-16 A 16-tap transversal filter as a multi-channel selector for (a) 1 channel, (b) 3 channels, and (c) 7 channels [8]. Reprinted by permission. Copyright © 1994 IEEE.

more channels. In practice, a larger cross-talk loss and a wider and flatter passband are desirable for dropping channels. These characteristics can be improved by increasing the number of taps.

4.4   MULTI-PORT FILTERS

For multiplexers and demultiplexers in WDM systems, a device with many input and output ports is highly desirable. In section 4.2, a $1 \times N$ architecture with $\log_2 N$ stages was discussed. Both the coupling ratios and relative phases had to be well controlled to achieve the desired performance. For small N, the overall circuit size and complexity are practical; however, it becomes too cumbersome for large N. In this section, we consider devices that use other approaches for splitting, delaying, and combining so that the number of ports can be increased significantly while minimizing the added complexity.

4.4.1   Diffraction Grating Filters

We begin the discussion of multi-port MA filters with the diffraction grating. The operating principles are very familiar, but are not typically discussed in digital filter terminology. There are two types of diffraction gratings, transmission and reflection. A transmission grating can be viewed as a multi-slit aperture, as shown in Figure 4-17a. Interference occurs between waves arising from each slit. Let the slits be separated by a distance d, the surrounding index be n_s, and the source be located at infinity so that the incoming beam is collimated. In the far field of the grating, we sum the diffracted fields from each slit. An interference pattern results because the rays from each slit experience different optical path lengths. The path length difference between an adjacent pair of slits, as drawn in Figure 4-17b, is $\Delta L \equiv CD - AB = d[\sin\theta_m - \sin\theta_i]$. The angles of incidence and transmission are denoted by θ_i and θ_m,

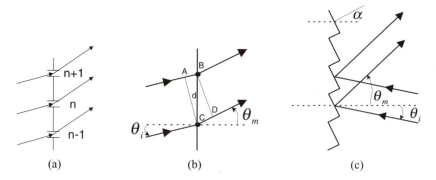

(a)   (b)   (c)

FIGURE 4-17   (a) A multi-slit transmission grating, (b) an enlarged view showing the path length differences between two adjacent slits, and (c) a reflection grating with a blaze angle of α.

respectively, measured counter-clockwise from the normal to the grating. At wavelengths where the difference in optical path length between each slit is a multiple of λ, constructive interference occurs. This condition is described by the grating equation

$$m\lambda = n_s d(\sin\theta_m - \sin\theta_i) = n_s \Delta L \tag{12}$$

where m is the diffraction order. The grating acts like a prism since different wavelengths have different output angles. Unlike a prism, however, there are several diffraction orders for a given wavelength. The transmission angle for each diffraction order is given by θ_m. By differentiating the grating equation, the angular dispersion is obtained.

$$D_\alpha \equiv \frac{d\theta_m}{d\lambda} = \frac{m}{n_s d \cos\theta_m} \tag{13}$$

The angular dispersion is proportional to m. For a positive m, the diffraction angle increases as the wavelength increases. The grating equation also applies to reflection gratings. A reflection grating with a blaze angle of α is shown in Figure 4-17c. It can be folded about the grating centerline and viewed as a transmission grating so that the angles are defined as before.

Now, we examine the far field in more detail, assuming that N identical slits are uniformly illuminated by a plane wave incident at an angle θ_i. The field just after the grating is

$$E_g(x) = -\sum_{n=0}^{N-1} u(x - nd)e^{-jk_0 n_s x \sin\theta_i} \tag{14}$$

where $u(x)$ is the transmission function for one slit or reflecting facet, and the amplitude of the incoming plane wave is assumed to be one. Let $U^{env}(s)$ denote the far field of a single slit where $s = n_s \sin\theta/\lambda$ as defined in Eq. (22) of Chapter 3 and θ is the far field angle. Then, the far field of $E_g(x)$ is given by

$$E_f(s) = U^{env}(s - s_i)\sum_{n=0}^{N-1} e^{+j2\pi(s-s_i)nd} = U^{env}(s - s_i)U^{array}(s - s_i) \tag{15}$$

where $s_i = n_s \sin\theta_i/\lambda$. There are two contributions to the far field, a slowly varying envelope $U^{env}(s - s_i)$ equal to the diffraction pattern of a single slit $U^{array}(s - s_i)$ and an array factor that depends on the input and output angles as well as the slit spacing and number of slits. To simplify the array factor, let $\theta_i = 0$. Then,

$$U^{array}(\theta) = \frac{1}{N}\sum_{n=0}^{N-1} e^{+jnk_0 n_s d\sin\theta} \approx \frac{1}{N}\sum_{n=0}^{N-1} e^{+j2\pi n \sin\theta/\Delta\theta_p} \tag{16}$$

The array factor is periodic with a period $\Delta\theta_p \approx \lambda/n_s d\cos\theta_m$, assuming a far field angle centered around θ_m. This angular periodicity results from sampling the incoming field with the slits. The array factor can also be written as a function of s with a period $\Delta s_p = 1/d$ or as a Z-transform,

$$U^{\text{array}}(z) = \frac{1}{N} \sum_{n=0}^{N-1} z^{-n} \tag{17}$$

where $z^{-1} = e^{+j2\pi sd}$. Since the aperture must have a finite width for a practical device, the grating has a finite impulse response that is $h(n) = 1/N$ for $n = 0, \ldots, N-1$. Instead of uniform illumination, a Gaussian beam may be incident on the grating. Then, the impulse response would have non-uniform coefficients. As long as the impulse response remains symmetric, the grating's frequency response will have linear phase and exhibit no temporal dispersion. Assuming uniform illumination, the spatial frequency response is

$$U^{\text{array}}(s) = \frac{e^{+j\pi(N-1)sd}}{N} \frac{\sin[\pi N sd]}{\sin[\pi sd]} \tag{18}$$

The peak value is unity and the half width, defined from the peak to the first zero, is $s_{1/2} = 1/dN = \Delta s_p/N$ (or $\Delta\theta_{1/2} = \Delta\theta_p/N$). Thus, a higher resolution (or narrower passband) is obtained by increasing the number of slits that are illuminated. The slit pitch can also be increased; however, it will decrease the angular periodicity, which may be undesirable. For an input at $\theta_i \neq 0$, Eq. (15) shows that the array and envelope factors are shifted by θ_i.

For a given θ_i and θ, Eq. (16) can be rewritten as a function of frequency

$$U^{\text{array}}(f) = \frac{1}{N} \sum_{n=0}^{N-1} e^{+j2\pi nf/\Delta f_p} \tag{19}$$

where $\Delta f_p = c/n_g \Delta L$ is the period of the frequency response (FSR) and n_g is the group index. The frequency response has the same functional form as the spatial response given by Eq. (18), and the half width is given by $\Delta f_{1/2} = \Delta f_p/N$. The half width corresponds to the resolution as defined by Rayleigh's criterion, which states that the peak transmission of one wavelength must be located at the null of the second wavelength for the two to be resolved. The resolving power is typically expressed as $\lambda/\Delta\lambda_{1/2} = mN$. The periodic frequency response is a result of the path length difference ΔL, while the periodic angular response is a result of the discrete aperture. Note that the angular dispersion is also expressed by the ratio of the angular to wavelength period as follows:

$$D_\alpha = \frac{\Delta\theta_p}{\Delta\lambda_p} = \frac{c}{\lambda^2} \frac{\Delta\theta_p}{\Delta f_p} = \frac{\Delta L}{d\lambda \cos\theta_m}$$

To demonstrate that the envelope factor is a slowly varying function, consider a one-dimensional slit defined by $u(x) = \prod(x/a)$, where $a < d$. It has a far field square-magnitude response given by

$$|U^{\text{env}}(s)|^2 = \left(\frac{n_s}{\lambda L}\right)^2 \left|\frac{\sin(\pi s a)}{\pi s}\right|^2$$

The half width of the envelope factor is $\Delta s_{\frac{1}{2}} = 1/a$. Taking the ratio of the envelope factor half width to the array factor half width shows that the envelope factor is at least N times wider. The envelope factor is also effectively wavelength independent compared to the array factor. The diffraction pattern maximum of the envelope factor occurs at $\theta_i = \theta_m$. This condition is also true for a reflecting facet, where it defines the specular reflection angle. According to Eq. (12), $\theta_i = \theta_m$ is only satisfied for $m = 0$, which is not interesting since an angularly dispersive filter requires $|m| > 0$. However, the specular reflection angle is defined with respect to the normal to the facet, not the grating. Thus, the reflection facets can be tilted with respect to the grating and shift the envelope maximum to be centered about any diffraction order. The reflection grating shown in Figure 4-17c has a tilted, or blazed, angle for this purpose.

Diffraction gratings are a fundamental component of optical spectrum analyzers with 0.1 nm resolutions in the near-infrared wavelength region (1000–1700 nm). For even higher resolutions, scanning Fabry–Perot cavities are used, which are discussed in Chapter 5. Diffraction gratings are typically combined with other optical components to collimate the incident wave and focus the diffracted waves. To avoid the need for collimating optics, both the dispersion and focusing functions can be performed with a concave grating. This approach is advantageous in planar waveguide implementations. Particular arrangements of the source, detector and grating are required for focusing the diffracted light while minimizing aberrations. One important configuration is the Rowland circle [9]. The concave grating forms an arc with a radius of curvature R. The source and detector are placed along a circle of radius R/2, called the Rowland circle, which intersects the grating arc at its midpoint as shown in Figure 4-18. For a source on the Rowland circle, the diffracted waves focus at points on the Rowland circle. The proof is obtained by showing that the path length difference between a point source on the circle and any two adjacent points on the grating (solid circles in Figure 4-18) is a constant equal to $d \sin\theta_i$ where d is the grating pitch [10]. Etched concave gratings in planar waveguide slabs have been demonstrated both with the Rowland configuration [11] and other arrangements to obtain a straight focal line [12], which is convenient for attaching a linear fiber or detector array. The slab confines the incident and diffracted waves in the vertical direction. A cross-talk loss, which is limited by imperfections in the reflective facets, of 25 dB has been reported for an etched grating device [13].

A chirped grating can also be used to angularly disperse light and focus it [14]. The principle of operation is depicted in Figure 4-19. The grating is tilted to radiate light over a broad spectrum out of the singlemode waveguide [15]. Let the local grating period in the waveguide be $\delta(z) = \Lambda(z)/\cos\theta_T$ where θ_T is the tilt angle with

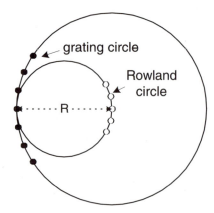

FIGURE 4-18   Rowland circle configuration for a concave grating.

respect to the waveguide axis and $\Lambda(z)$ is the grating period at normal incidence as shown in Figure 4-19b. The angle of observation θ_o depends on the wavelength and the local period $\delta(z)$ according to $n_e\delta(z)[1 + \cos\theta_o] = \lambda$. A negative chirp focuses the light onto a detector array since the longer period diffracts light at a steeper angle than the shorter period. A spectrum analyzer using a tilted and chirped UV-induced grating has been demonstrated in fiber [14] and planar waveguide [16] implementations. A resolution of 0.12 nm and a bandwidth of 14 nm have been achieved [14].

4.4.2   Waveguide Grating Routers

Now, we turn our attention to the WGR, which uses an array of waveguides to perform the function of a diffraction grating. A waveguide grating was first proposed by Smit [17], and a device with nanometer resolution in the long wavelength window was reported first by Takahashi [18]. Integration of the waveguide grating with slab couplers to form an NXN device was demonstrated by Dragone [19]. In this configuration, a multiplexer/demultiplexer that is easily scaled to a large number of ports is achieved with narrow passbands, low loss, and good cross-talk suppression. Other names for the WGR include arrayed waveguide grating (AWG) and a phased array (PHASAR). The wavelength routing capability of the WGR will become evident in the following discussion. The WGR functionality can also be achieved by replacing the slab couplers with MMIs. MMI-based WGRs and switches are discussed at the end of the section.

A WGR schematic is shown in Figure 4-20. It consists of an input waveguide array, two slab couplers interconnected by an array of waveguides (the grating array), and an output waveguide array. The length between adjacent waveguides in the grating array varies by a constant, ΔL. The device may be operated as a $1 \times N$ device having a single input and multiple outputs, $K \times 1$, or a $K \times N$ device having multiple inputs and outputs. The principle of operation is explained using the free space op-

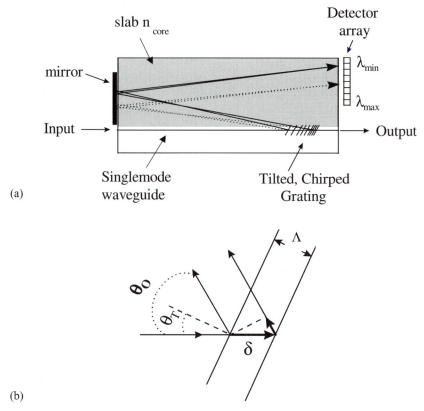

(a)

(b)

FIGURE 4-19   A spectrum analyzer using a chirped, tilted UV-induced Bragg grating in a singlemode waveguide that diffracts into a slab region (a) and the interference between two adjacent reflectors (b).

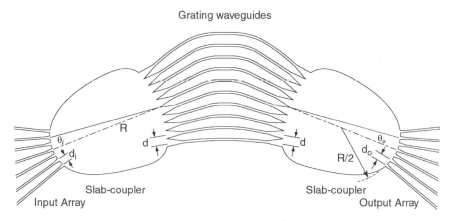

FIGURE 4-20   A WGR schematic showing the slab couplers and the input, grating and output waveguide arrays.

tics analogy shown in Figure 4-21. The diffracted light from the input fiber is Fourier transformed by the first slab coupler (first lens). The waveguide grating array samples the far field like a multi-slit aperture. Then, a position dependent delay is added by the different lengths of the grating arms (the prism) that makes the angle of propagation into the second slab coupler (second lens) wavelength dependent. Subsequently, the location of the focal point along the output array is wavelength dependent. The sampling performed by the waveguide grating array results in multiple diffraction orders at the receiver array. The mth, $m + 1$ and $m - 1$ diffracted orders for a single wavelength are depicted along the output plane in Figure 4-21. For the slab coupler, the imaging may be accomplished using a Rowland configuration. Mutual coupling between waveguides causes the phase center to be displaced relative to the end of the array, so the radius of curvature is offset from its geometrical location to compensate for this phase [20].

Constructive interference occurs at wavelengths that are an integer multiple of the optical path length difference between adjacent grating waveguides. The major contributor is the fixed path length difference, $n_e \Delta L$, where n_e is the waveguide effective index. As a result of the Rowland configuration, propagation through each slab waveguide contributes $n_s d \sin\theta$ to the optical path length difference where d is the grating pitch, θ is the input or output waveguide angle as defined in Figure 4-20, and n_s is the slab effective index. The condition for constructive interference is

$$n_e \Delta L + n_s d[\sin\theta_i + \sin\theta_o] = m\lambda \tag{20}$$

where m is the grating order, and $\theta_i \approx kd_i/R$ and $\theta_o \approx nd_o/R$ are the angles for the kth input and nth output waveguides, respectively. The pitch of the input and output waveguide arrays is designated by d_i and d_o. For the purposes of this discussion, let

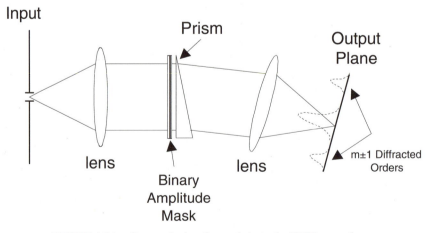

FIGURE 4-21 Geometrical optics analogy to the WGR operation.

Eq. (20) be satisfied at the design center wavelength λ_c for the central input ($k = 0$) and output ($n = 0$) ports, so $m\lambda_c = mc/f_c = n_e\Delta L$. The FSR, Δf_p, is determined by evaluating Eq. (20) at the mth and ($m + 1$)st-order with the frequency dependence of the refractive indices included. The resulting expression for the FSR depends on the input and output port as follows [21]:

$$\Delta f_p = \cfrac{c}{\left[\left(n_e + f_c\cfrac{dn_e}{df}\right)\Delta L + \left(n_s + f_c\cfrac{dn_s}{df}\right)d(\sin\theta_i + \sin\theta_o)\right]} \tag{21}$$

where the approximation $\Delta f_p \ll f_c$ was made. When $d(\sin\theta_i + \sin\theta_o) \ll \Delta L$, Eq. (21) simplifies to $\Delta f_p = c/n_g\Delta L$ where $n_g \equiv n_e + f_c(dn_e/df) = n_e - \lambda_c(dn_e/d\lambda)$ is the group index.

Angular dispersion results from the wavelength dependent, linear phase variation across the output of the grating array. When Fourier transformed, the phase variation causes the position of the focal point to shift with wavelength. The dispersion is defined as the derivative of θ_o with respect to λ and is found by differentiating Eq. (20).

$$\frac{d\theta_o}{d\lambda} = \frac{n_g\Delta L}{n_s d\lambda} = \frac{m}{n_s d}\frac{n_g}{n_e} \tag{22}$$

The dispersion increases with increasing grating order and decreasing grating pitch. At the output array (the z'-plane in Figure 4-22), the linear dispersion is equal to the product of R, the Rowland circle diameter, and the angular dispersion.

$$D_l = \frac{dz'}{df} = R\frac{d\theta_o}{d\lambda}\frac{d\lambda}{df} = -\frac{\lambda_c^2}{c}\frac{Rm}{n_s d}\frac{n_g}{n_e} \approx -\frac{R\Delta L}{f_c d} \tag{23}$$

where the approximations $m\lambda \approx n_g\Delta L$ and $n_g \approx n_s$ were made and the grating arm lengths decrease with increasing m. Three geometrical parameters set the dispersion: the focal length R, the grating order that is proportional to ΔL, and the grating pitch d. Within one FSR, the sign of the dispersion requires that longer wavelengths focus at larger values of z'. The output array pitch divided by the dispersion sets the channel spacing.

$$\Delta f_{ch} = \frac{d_o}{|D_l|} = \frac{d_o df_c}{R\Delta L}\frac{n_s}{n_g} \tag{24}$$

The basic principles of operation for the WGR are similar to the diffraction grating. Both filters are periodic in frequency and have a spatial response that is the product of a periodic array factor and an envelope function. We shall now derive expressions for the frequency response of the WGR to provide insight into how various parameters affect it. A more precise analysis requires a combination of electromagnetic modeling with the following treatment. As an aside, one can simulate the whole de-

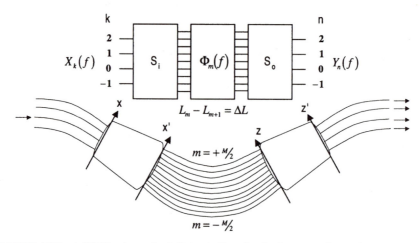

FIGURE 4-22 A WGR schematic defining indices for the input, grating and output arrays, input and output planes for the slab couplers, and the transfer matrix notation.

vice using electromagnetic models such as beam propagation techniques; however, this approach is time consuming since one simulation is needed for each frequency. A more efficient approach is to break the problem into pieces, which are easily described by transfer matrices [22]. The frequency dependence is dominated by the grating array, while the slab couplers are essentially wavelength independent. The input (S_i) and output (S_o) slab transfer matrices are cascaded with the grating array frequency response (Φ) as shown schematically in Figure 4-22 and mathematically as follows:

$$[Y_n(f)] = [S_o(n, m)] [\Phi(m, f)] [S_i(m, k)] [X_k(f)] \qquad (25)$$
$$N \times 1 \qquad N \times M \quad M \times M \quad M \times K \quad K \times 1$$

The dimension of each matrix is indicated along the bottom row for a WGR with K inputs, M grating array waveguides, and N outputs. A detailed electromagnetic model, that accounts for coupling between the waveguides at the input and output of each array, is performed at the design center wavelength to obtain the transfer matrix for each slab coupler. The grating array response is then treated as a diagonal matrix. Our goal is to find an expression for the frequency response for input k and output n in terms of the transfer matrix elements as follows:

$$H_{nk}(f) = \frac{Y_n(f)}{X_k(f)} = \sum_m S_o(n, m)\Phi(m, f)S_i(m, k) \qquad (26)$$

We begin by finding simplified expressions for the slab transfer matrices, assuming no waveguide coupling. Let the kth input, mth grating, and nth output waveguide array fields be designated as follows:

$$u_i(x - kd_i) \supset e^{+j\frac{2\pi n_s}{\lambda R}x'kd_i}U_i(x') \tag{27}$$

$$u_g(z - md) \supset e^{+j\frac{2\pi n_s}{\lambda R}z'md}U_g(z')$$

$$u_o(z' - nd_o)$$

where the fields are normalized so that $\int u^2(x)dx = 1$. Also shown are the far field notations needed for the following analysis, where $\supset$ denotes the Fourier transform and scaling operations of Eq. (22) in Chapter 3. Now, the slab coupler matrix elements are expressed as

$$S_i(m, k) = \int_{A_\infty} u_g(x' - md)e^{+j\frac{2\pi n_s}{\lambda R}x'kd_i}U_i(x')dx' \approx \xi_g e^{+j\frac{2\pi n_s}{\lambda R}mdkd_i}U_i(md) \tag{28}$$

$$S_o(n, m) = \int_{A_\infty} u_o(z' - nd_o)e^{+j\frac{2\pi n_s}{\lambda R}z'md}U_g(z')dz' \approx \xi_o e^{+j\frac{2\pi n_s}{\lambda R}mdnd_o}U_g(nd_o)$$

where the integral is over the array extent A_∞ and $\xi = \int u(x)dx$. The approximate expressions on the right are obtained by assuming that the near field term is much more localized than the far field term, i.e. $u_g(x') \approx \delta(x')$ and $u_o(z') \approx \delta(z')$. The grating array response is a delay given by

$$\Phi(m, f) = e^{+j2\pi m\frac{f}{\Delta f_p}} \tag{29}$$

where $\Delta f_p = c/n_g\Delta L$. The WGR's frequency response is obtained by substituting Eqs. (28) and (29) into Eq. (26). The sum of the phase terms is

$$2\pi m\left[\frac{f}{\Delta f_p} + \frac{n_s d}{\lambda R}(kd_i + nd_o)\right] \tag{30}$$

where the quadratic phase terms in $U_i(x')$ and $U_g(z')$ from the Fraunhofer approximation were assumed to be negligible. A spatial period is defined by $\Delta z_p \equiv \lambda R/n_s d$. By substituting $\Delta z_p/\Delta f_p = D$ into Eq. (30), the frequency response is

$$H_{nk}(f) \approx \xi_g\xi_o U_g(nd_o)\sum_m U_i(md)e^{+j\frac{2\pi m}{\Delta f_p}\left[f + \frac{(kd_i + nd_o)}{D}\right]} \tag{31}$$

$$\approx \xi_g\xi_o U_g(nd_o)u_a\left(f + \frac{(kd_i + nd_o)}{D}\right)$$

The frequency response is the product of an envelope factor U_g times an array factor u_a. The frequency dependence is determined by the array factor, $u_a(f) = \sum_m U_i(md)e^{+j2\pi m(f/\Delta f_p)}$, whose coefficients are samples of the input waveguide's far field. A shift in the output or input waveguide results in a shift in the frequency response. A change in the input waveguide shifts the frequency response by $\Delta f_{shift} = d_i/D$. A shift in the output waveguide defines the channel spacing, $\Delta f_{ch} = d_o/D$. If $d_i = d_o$, then the frequency response $H_{n,k}(f)$ is also obtained for input waveguide $k + p$ and output waveguide $n - p$ where p is an integer, i.e. $H_{n,k}(f) = H_{n-p,k+p}(f)$. This wavelength routing behavior is depicted in Figure 4-23 for a 4 × 4 WGR.

We now consider the impact of finite apertures, $u_g(x') \neq \delta(x')$ and $u_o(z') \neq \delta(z')$, in Eq. (28). For S_i, we define a modified $U_i(md)$ as follows:

$$U_i'(md) = \int_{A_\infty} u_g(x' - md)U_i(x')dx' \tag{32}$$

For S_o, the integral expression in Eq. (28) is substituted into Eq. (26) to account for the finite extent of $u_o(z')$. So,

$$H_{nk}(f) = U_g(nd_o)\int_{A_\infty} u_o(z' - nd_o)u_a(fD + kd_i + z')dz' \tag{33}$$

$$= U_g(nd_o)\int_{A_\infty} u_o(z')u_a(fD + kd_i + nd_o + z')dz'$$

where an exchange of the integration variable was made from the top to the bottom expression. The new array factor is defined as a function of z' as follows:

$$u_a(z') = \sum_m U_i'(md)e^{+j2\pi m\frac{z'}{\Delta z_p}} \tag{34}$$

The effect of a finite extent $u_g(x')$ is to modify the array factor coefficients; whereas, the effect of a finite extent $u_o(z')$ is to make the frequency response a convolution of $u_o(z')$ and $u_a(-z')$ evaluated at $z' = Df + kd_i + nd_o$. As the number of grating waveguides increases, $u_a(z') \rightarrow u_i(-z')$ since $u_i(x) \supset U_i(x') \supset u_i(-x)$. For $\Delta L = 0$, the WGR forms a 1:1 imaging system, and $\theta_i = -\theta_o$. As shown in Chapter 3 for MA filters, narrowband responses require a large number of stages, and therefore, a large M for the WGR. The frequency response can be modified by changing the input and output fields or by introducing phase and amplitude filtering in the grating arms.

Four sources of excess loss impact design considerations for the WGR. First, losses can result from non-adiabatic transitions between the waveguide arrays and

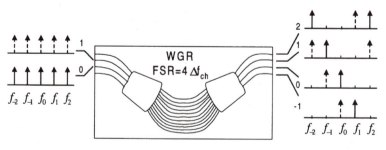

FIGURE 4-23   Frequency routing behavior of a 4 × 4 WGR where $f_n = f_c + n\Delta f_{ch}$.

slab regions. Second, the far field pattern of the input guide must be covered by the grating waveguides; otherwise, power in the tails of the far field that are not coupled to a grating waveguide will be lost. In practice, a large number of grating waveguides are used, making this loss negligible. Third, loss arises from power that is diffracted into the adjacent orders at the output array. Finally, mode mismatch loss occurs when $u_o(z')$ and $u_a(z')$ are not identical.

Using Eqs. (33)–(34) and a Gaussian approximation for the waveguide modes, simple formulas are obtained for the frequency response, passband width, the center channel loss, and the loss uniformity. Following [23], let the input, grating and output waveguides be uncoupled and the waveguide modes be defined by a normalized Gaussian distribution.

$$u(x) = \frac{1}{w_0 \sqrt{\pi}} \exp\{-(x/w_o)^2\} \tag{35}$$

where $\int u(x)dx = 1$ and w_0 is the mode field radius. The far field is

$$|U(x')| = \frac{n_s}{\lambda R} \exp\left\{-\left(\frac{x'}{W}\right)^2\right\} \tag{36}$$

where $W = \lambda R/n_s \pi w_o$. Substituting $u_a(z') \approx u_i(-z')$ into Eq. (33) and neglecting the envelope factor, the frequency response is approximated by

$$H_{00}(\delta f) \approx \frac{\int_{-\infty}^{\infty} u_o(z)u_i\left(-D\delta f - z\right)dz}{\sqrt{\int_{-\infty}^{\infty} |u_i(z)|^2 dz \int_{-\infty}^{\infty} |u_o(z)|^2 dz}} \tag{37}$$

where $n = k = 0$. The normalization is shown explicitly in Eq. (37) to emphasize that it is a mode overlap calculation. A peak transmission of 100% is obtained from Eq. (37) only if the input and output modes are identical. Note that $\delta f = f - f_c$ is used instead of f since we used $u_i(z)$ in the approximation which is not periodic like $u_a(z)$. Assuming that the input and output array waveguides have identical, Gaussian modes, the convolution is a Gaussian function equal to

$$H_{00}(\delta f) \approx \exp\left\{-\frac{1}{2}\left(\frac{z}{w_o}\right)^2\right\}\bigg|_{z=D\delta f} \tag{38}$$

The square magnitude response is

$$|H_{00}(\delta f)|^2 \approx \exp\left\{-\left(\frac{\delta f}{w_o}D\right)^2\right\} \tag{39}$$

The passband width Δf_L, defined as the full width at L dB down from the maximum, is [23]

$$\frac{\Delta f_L}{\Delta f_{ch}} \approx 0.96 \frac{w_o}{d_o} \sqrt{L} \qquad (40)$$

Equation (40) shows that the passband width can be increased by increasing the mode size to output array pitch ratio; however, an increase in cross-talk will result as discussed later.

The envelope factor allows the loss uniformity among the output channels and the diffraction loss to be determined. Evaluating the square magnitude of Eq. (36) relative to $z' = 0$, the loss for the output at $z' = nd_o$ relative to the central output port is approximated by [23]

$$L_u \approx -10 \log\left[\exp\left(-2\left\{\frac{z'}{W}\right\}^2\right) \right] \approx 8.7 \left(\frac{z'}{\Delta z_p} \frac{\pi w_o}{d}\right)^2 \qquad (41)$$

A simple argument can be made for the loss uniformity. The N channels must lie within one spatial period at the output plane, so the maximum deviation from the center guide is $\Delta z_p/2$. Figure 4-24 shows the envelope factor along the output plane.

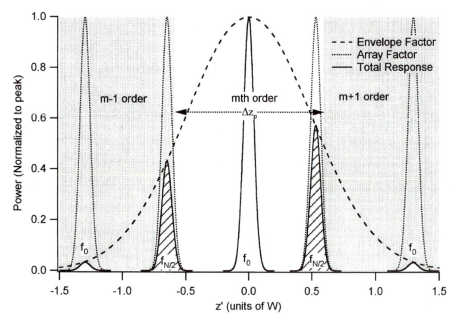

FIGURE 4-24   The far field diffraction pattern envelope, which impacts channel-to-channel loss uniformity, and array factor for two frequencies, f_0 and $f_{N/2}$.

Multiple diffraction orders are shown for an edge channel and the central output channel with frequencies $f_{N/2}$ and f_0, respectively. For the edge channel, the amplitudes of the m and $m - 1$ orders are similar. Since power is conserved, each order carries approximately half the power, so the loss uniformity is roughly 3 dB between the center and edge channels. A beam propagation simulation of an output slab coupler is shown in Figure 4-25. The diffraction orders for the frequency corresponding to the central output port are clearly shown. The diffraction loss of the central output port is approximated by calculating the power in the $m \pm 1$ orders, which occur at $\pm \Delta z_p$.

Several factors influence cross-talk. By increasing the number of grating waveguides, a narrower array factor can be achieved. Then, power is more localized at the output array and less couples into adjacent outputs. The sharper $u_a(z')$ is made, the larger the cross-talk loss will be. If the grating waveguides do not cover the far field of the input waveguide sufficiently and the field is truncated, sidelobes will increase in the far field of the grating array and increase cross-talk levels. A ratio of $Md/W \approx 4.2$ provides a cross-talk loss greater than 40 dB [23]. The output array is equally important in designing for low cross-talk. To reduce cross-talk, the output waveguides must be separated farther apart or more strongly guiding waveguides can be used. Since the Gaussian approximation deviates from the actual mode significantly in the tails of the distribution, the cross-talk must be calculated with the actual waveguide mode fields. With this approach, a waveguide with a normalized frequency of $V = 3$ requires a ratio of $d_o/w_o \approx 2.2$ for a cross-talk loss of 40 dB; while a ratio of $d_o/w_o \approx 3$ is needed for 30 dB cross-talk loss if $V = 2$ [23]. So, the grating array and output array can be designed to provide low cross-talk. In practice,

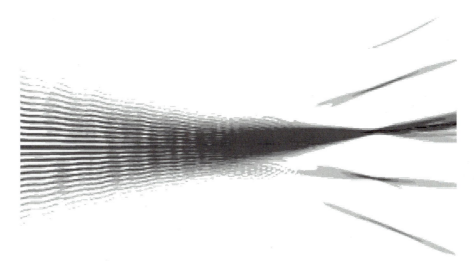

FIGURE 4-25 Beam propagation method simulation for a WGR output slab coupler showing the diffraction orders for one frequency with most of its power coupled to the central output port [obtained using Prometheus by BBV Software, The Netherlands (bbv@bbv.nl)].

the limiting source for cross-talk is phase errors in the grating arms introduced during fabrication, which may be due to local variations in index, core height or thickness, or mask discretization. Longer path length differences, ΔL, cause the device to be more susceptible to fabrication variations as discussed in Chapter 5; consequently, devices with narrow channel spacings have larger cross-talk. Test methods to determine the phase errors and post-fabrication techniques to correct the errors are discussed in Chapter 7.

A basic design procedure for a WGR is outlined assuming Gaussian mode profiles and no coupling between waveguides in the input, grating or output arrays. The design requirements are the channel spacing Δf_{ch}, passband width Δf_L, cross-talk, loss and loss uniformity, and the number of output channels N. The ratio d_o/w_o is chosen to satisfy the cross-talk requirement and the passband width. Typically, the cross-talk is most important, and other methods are used to widen the passband, as discussed below. The mode field radius is determined by the core-to-cladding index of refraction difference and the core size for singlemode operation. Loss and loss uniformity depend on the ratio d/w_o. A loss uniformity of <3 dB requires $N\Delta f_{ch} <$ FSR. By substituting $z'_{max} = (N/2)d_o$ into Eq. (41), we see that the loss uniformity requirement dictates the spatial periodicity d. Then, the channel spacing requirement is met by choosing R in Eq. (24). As an example of some design values, reported parameters for a 100 GHz, 16 channel multiplexer are $R = 9.381$ mm, FSR = 12.8 nm (1600 GHz), $d = d_o = 25$ μm, $m = 118$, and $w_0 = 4.5$ μm (power half width at $1/e^2$) [23]. A small-order design for 1.3/1.55 μm multiplexing was also demonstrated, with a passband loss of 2 dB [24]. By simultaneously changing the FSR and the output waveguide spacing, the passband width can be varied while maintaining a constant channel spacing [25]. This behavior can be verified by substituting FSR = $c/n_g\Delta L$ into the dispersion relation [Eq. (24)].

WGRs realized in doped-silica planar waveguides have excellent multi-channel filter characteristics. Measurements on a 1×48 WGR designed for a Gaussian passband are shown in Figure 4-26 [26]. The passband loss is 3 to 4 dB and the cross-talk loss is greater than 26 dB for adjacent channels. For WDM systems, it is desirable to have a wider, flatter passband than is provided by the Gaussian response. Several techniques that broaden the input or output waveguide modes to widen the passband include: the use of multi-mode output waveguides, which must terminate in a receiver [27], Y-branches [28], MMI couplers [29], and a parabolic waveguide horn [30]. Mode mismatch results in an excess loss of approximately 2.5 to 3 dB. Figure 4-27 shows measurements on a 1×36 WGR designed with a flattened passband [26]. The passband loss is 7 to 8.5 dB and the adjacent channel cross-talk loss is greater than 27 dB. Another technique to flatten the passbands is to make the grating array transfer function approximate a sinc function by introducing phase shifts and loss. Then, the array factor approaches a rectangular response. Once again, mode mismatch introduces excess loss, in addition to any losses incurred in the grating array to approximate the sinc function [31,32]. Experimental results show an excess loss close to 2.5 dB [33]. An approach, which avoids mode mismatch excess loss, is to introduce a spatial dispersion on the input, output, or combination thereof, which cancels the linear dispersion of the grating over the

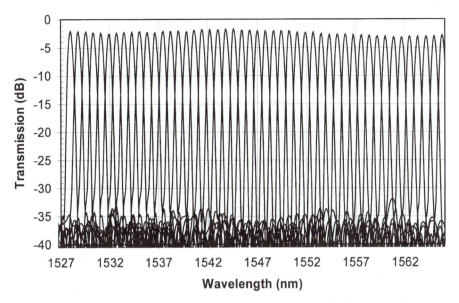

FIGURE 4-26 Experimental results on a 1 × 48 WGR with 100 GHz channel spacing and a Gaussian passband design [26].

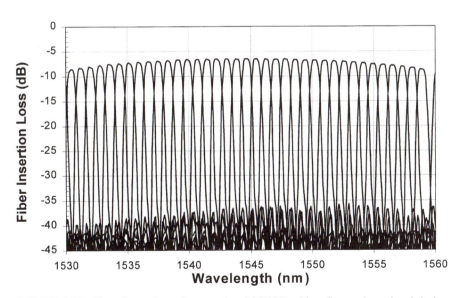

FIGURE 4-27 Experimental results on a 1 × 36 WGR with a flattened passband design [26].

passband [34, 35]. An MZI on the input with a FSR equal to the channel spacing is one example. By moving the input field in the opposite direction as the grating angular dispersion, the frequency response at the passband center is kept almost stationary over the passband width. This behavior can be understood by letting $kd_i \rightarrow kd_i + x(f)$ in Eq. (33). For example, let $x(f) = -\Delta z_p \sin(2\pi f/\Delta f_p)/2\pi$. Over the frequency range where $dx/df \approx -D$, the frequency response is a maximum, i.e. equal to the response at the passband center. A wide, flat passband results while avoiding mode mismatch loss. A practical concern is the alignment of the MZI and WGR center wavelengths.

The temporal dispersion characteristics of the WGR are important for use as a demultiplexer in high bit rate systems. For symmetrical waveguide modes, the frequency response is an even function according to Eq. (31). When inverse Fourier transformed to the time domain, the impulse response is real and even. As discussed in Chapter 3, this symmetry implies that the impulse response has linear phase and therefore zero dispersion. Ideally, WGRs are linear-phase filters and therefore attractive for high bit rate filtering applications. A group delay measurement of a flattened passband WGR is shown in Figure 4-28 [36]. There is no dispersion across the passband. Only in the transition and stop bands, where the zeros are close to the unit circle, does loss begin to introduce a departure from linear phase.

Phase errors introduced during fabrication change the center wavelength of the output waveguides relative to their nominal design values. A technique to overcome

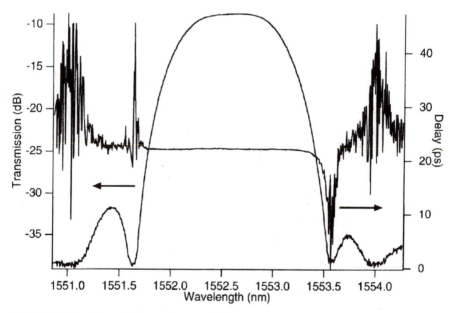

FIGURE 4-28   Transmission and group delay measurement for one output of a flattened passband WGR [36].

this problem by design has been demonstrated [28,37]. Extra input and output waveguides are included in the design, and different pitches are used for the input and output waveguides. From the expression for $H_{nk}(f)$, we see that a frequency shift relative to the channel spacing of $\Delta f_{shift}/\Delta f_{ch} = d_i/d_o$ is achieved by shifting the input one position. For example, a ratio of $d_i/d_o = 9/10$ allows the center wavelength to be compensated by $1/10$ of a channel spacing while keeping the channel width and spacing constant. The center wavelengths of the output channels are then obtained with much tighter tolerances by choosing the best input waveguide. The WGR architecture has been realized in many material systems including silica, InGaAsP, silicon-on-insulator, and polymers. The WGR has found numerous applications, beyond de/multiplexing, which include add/drop [38], cross-connects [39,40], channel equalization [41], and tunable lasers [42]. Chirped grating array waveguides, as opposed to uniformly varying in length, have been employed to reduce transmission in the adjacent FSRs for multifrequency lasers [43] and to provide a 2 × N cross-connect functionality where N channels can be interchanged between two fibers [44].

A WGR can also be made by substituting MMI couplers in place of the slab couplers [45]. MMI couplers offer uniform splitting ratios that are insensitive to fabrication variations, but they are limited to a small number of inputs and outputs compared to slab couplers. Unlike slab coupler WGRs, the grating arm lengths do not increase in a uniform fashion between adjacent waveguides when MMIs are used. Besse et al. [46] derived a numbering scheme for the grating arms so that the arm lengths can be viewed as portions of a prism, i.e. adjacent numbered arms vary by ΔL. The arm lengths can then be easily determined for any size ($N \times N$) device, and the resulting frequency response is equivalent to a cascade of MZIs with geometrically increasing ΔLs as discussed for the $\log_2 N$ cascade architecture in Section 4.2. A drawback is that the waveguides may cross-over each other to achieve the necessary lengths. To avoid waveguide crossings, a set of optimized lengths have been reported for $N = 3$–10 [47]. For demultiplexer and router applications, the cross-talk performance is limited by the smaller number of grating array waveguides compared to slab coupler designs. One approach to reducing the cross-talk is to use a nonuniform splitter, which can suppress the sidelobes by 32 dB compared to 13 dB for the uniform splitting case [47]. Very compact devices can be made with MMI couplers in high-Δ waveguides. For example, a 4-channel device was made with dimensions of 2.8mm × 100 μm [48]. The previously mentioned methods to flatten the passband are not applicable for MMI-based WGRs, since they would change the MMI splitting ratios. As discussed in Chapter 6, flat passbands can be achieved by combining ARMA filters and MMIs.

The WGR, whether implemented with slab or MMI couplers, is fundamentally a multi-branch interferometer. Just as the input to an MZI can be switched between the output ports by changing the relative phase between the interferometer arms, the WGR can function as a 1 × N switch. A multi-port interferometric switch composed of two 4 × 4 MMIs with uniform splitting ratios is shown in Figure 4-29 [49]. Each grating arm is nominally the same length, i.e. $\Delta L = 0$, except for the small path length differences introduced by the phase shifters ϕ. For a given input port m, the relative phases between the grating arms are set (by an appropriate choice of ϕ) to

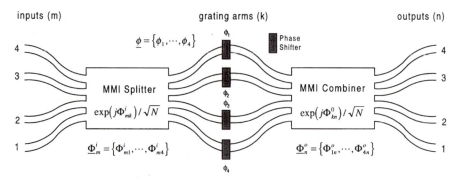

FIGURE 4-29 An MMI switch showing phase shifters in the interferometer arms.

route the signal to a desired output port n. The relative phases for the MMIs (Φ^i_{mk} and Φ^o_{kn}) are easily calculated from the phases given in Chapter 2 for a single MMI. Considering only the combiner for a moment, let $\underline{\Phi}^o_n$ represent the relative output phases for an input on the nth port. From the discussion in Chapter 3 on reciprocal devices, we know that if the relative phases going into the combiner from the grating arms equal $-\underline{\Phi}^o_n$, then the output will be on the nth port as desired. The choice of input ports determines the relative phases into the grating arms, denoted by $\underline{\Phi}^i_m$. The desired switching condition is achieved when the phase in the grating arms satisfies $\underline{\phi} = -\underline{\Phi}^o_n - \underline{\Phi}^i_m$. If $\Delta L \neq 0$, then the switch is wavelength dependent, and the switching condition is calculated for the design center wavelength. The remaining channels at $\lambda_c \pm n\Delta\lambda_{ch}$ are routed to the other output ports according to the routing behavior discussed in [46–47].

4.5 LATTICE FILTERS

A multi-stage filter can be realized by concatenating MZIs in a lattice architecture as shown in Figure 4-30 [50]. The major advantage of this architecture is that a very low loss passband can be achieved [51] and only $N + 1$ couplers are needed for an N-stage filter. Each stage has a unit delay; so, the transmission is defined by a Fourier series. As a result, MA lattice filters are also referred to as Fourier filters [52]. Applications include bandpass filters [53], gain equalization [52], and dispersion compensation [51]. Two approaches have been used to design MA lattice fil-

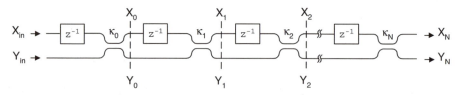

FIGURE 4-30 The MA lattice filter schematic.

ters. First, the filter synthesis problem can be solved by finding the coupling ratios that give a best fit to the desired function using a nonlinear optimization algorithm. Alternatively, an approach using recursion relations to translate between the filter coefficients and coupling ratios, derived by Jinguji and Kawachi [51], may be used. The latter approach is taken in the following discussion.

4.5.1 The Z-Transform Description and Synthesis Algorithm

Figure 4-30 shows the couplers and unit delays for the MA lattice architecture. The transfer matrix is the product of the propagation matrix and coupler matrix. For a lossless filter and a path length difference between the arms of ΔL for each stage, the transfer matrix for the nth stage is

$$\begin{bmatrix} X_n(z) \\ Y_n(z) \end{bmatrix} = \Phi_n(z) \begin{bmatrix} X_{n-1}(z) \\ Y_{n-1}(z) \end{bmatrix} \tag{42}$$

$$\Phi_n(z) = \begin{bmatrix} c_n & -js_n \\ -js_n & c_n \end{bmatrix} \begin{bmatrix} e^{-j\phi_n}z^{-1} & 0 \\ 0 & 1 \end{bmatrix} = \begin{bmatrix} c_n e^{-j\phi_n}z^{-1} & -js_n \\ -js_n e^{-j\phi_n}z^{-1} & c_n \end{bmatrix} \tag{43}$$

where the directional coupler has a power coupling ratio κ_n. The through-port coupling is $c_n = \cos\theta_n = \sqrt{1 - \kappa_n}$, and the cross-port coupling is $-js_n = -j\sin\theta_n = -j\sqrt{\kappa_n}$. The ϕ_n term represents a phase shift, whereby the optical path length deviates from its nominal value. For an Nth-order filter, the inputs and outputs are related as follows:

$$\begin{bmatrix} X_n(z) \\ Y_n(z) \end{bmatrix} = \Phi_N \cdots \Phi_1 \Phi_0 \begin{bmatrix} X_{in}(z) \\ Y_{in}(z) \end{bmatrix} = \begin{bmatrix} z^{-1}A_N(z) & -jB_N^R(z) \\ -jz^{-1}B_N(z) & A_N^R(z) \end{bmatrix} \begin{bmatrix} X_{in}(z) \\ Y_{in}(z) \end{bmatrix} \tag{44}$$

where $A_N(z)$ and $B_N(z)$ are Nth-order polynomials in z^{-1} and $\phi_0 = 0$. The input X_{in} in Figure 4-30 has an added delay that allows Φ_0 to be written in the form of Eq. (43) instead of requiring a special definition. This extra delay is removed when calculating the frequency response. The reverse polynomials, denoted by $A_N^R(z)$ and $B_N^R(z)$, are obtained from $A_N(z)$ and $B_N(z)$ by reversing the order of the coefficients, but keeping the phase terms in their original order. The reverse polynomials are defined in terms of the forward polynomials as follows:

$$A_N^R(z) = z^{-N}e^{-j\phi_{tot}^N}A_N^*(z^{*-1}) \tag{45}$$

$$B_N^R(z) = z^{-N}e^{-j\phi_{tot}^N}B_N^*(z^{*-1}) \tag{46}$$

where $\phi_{tot}^N = \sum_{n=1}^{N}\phi_n$. Loss is included in the Z-transform description by substituting γz^{-1} for z^{-1}. The incorporation of loss into Eqs. (45)–(46) changes the relationship between the roots of the forward and reverse polynomial. Consider a polynomial with a single zero at $z = \gamma r e^{-j\phi}$.

$$A_1(z) = 1 - re^{-j\phi} z^{-1}\gamma \tag{47}$$

Its reverse polynomial is given by

$$A_1^R(z) = -r + e^{j\phi} z^{-1}\gamma \tag{48}$$

The zero of the reverse polynomial occurs at $z = e^{j\phi}\gamma/r$. The two roots are reflected about a circle of radius equal to γ instead of the unit circle. Since any higher order polynomial can be written as the product of single stages, it is true in general that the forward and reverse polynomials are reflected about a circle of radius γ. The first few orders of forward and reverse polynomials in terms of the coupling coefficients, loss and phase terms are shown below.

$$A_0 = c_0$$

$$A_1 = c_1 c_0 e^{-j\phi_1} \gamma z^{-1} - s_0 s_1$$

$$A_2 = c_0 c_1 c_2 e^{-j(\phi_1+\phi_2)} \gamma^2 z^{-2} - s_1(s_2 c_0 e^{-j\phi_1} + c_2 s_0 e^{-j\phi_2})\gamma z^{-1} - s_2 c_1 s_0$$

$$A_0^R = c_0$$

$$A_1^R = -s_1 s_0 e^{-j\phi_1} \gamma z^{-1} + c_0 c_1$$

$$A_2^R = -s_2 c_1 s_0 e^{-j(\phi_1+\phi_2)} \gamma^2 z^{-2} - s_1(c_2 s_0 e^{-j\phi_1} + s_2 c_0 e^{-j\phi_2})\gamma z^{-1} + c_0 c_1 c_2$$

$$B_0 = s_0$$

$$B_1 = s_1 c_0 e^{-j\phi_1} \gamma z^{-1} + c_1 s_0$$

$$B_2 = s_2 c_1 c_0 e^{-j(\phi_1+\phi_2)} \gamma^2 z^{-2} + s_1(c_0 c_2 e^{-j\phi_1} - s_2 s_0 e^{-j\phi_2})\gamma z^{-1} + c_2 c_1 s_0$$

$$B_0^R = s_0$$

$$B_1^R = c_1 s_0 e^{-j\phi_1} \gamma z^{-1} + s_1 c_0$$

$$B_2^R = c_2 c_1 s_0 e^{-j(\phi_1+\phi_2)} \gamma^2 z^{-2} + s_1(-s_0 s_2 e^{-j\phi_1} + c_2 c_0 e^{-j\phi_2})\gamma z^{-1} + s_2 c_1 c_0$$

The transfer functions are given by the $A(z)$ and $B(z)$ polynomials and their reverses. For the remainder of the derivation, let $\gamma = 1$ to simplify the notation. The Nth-order polynomials are defined in terms of the next lower order polynomial as follows:

$$\begin{bmatrix} z^{-1}A_N(z) & -jB_N^R(z) \\ -jz^{-1}B_N(z) & A_N^R(z) \end{bmatrix} = \begin{bmatrix} z^{-1}e^{-j\phi_N}c_N & -js_N \\ -jz^{-1}e^{-j\phi_N}s_N & c_N \end{bmatrix} \begin{bmatrix} z^{-1}A_{N-1}(z) & -jB_{N-1}^R(z) \\ -jz^{-1}B_{N-1}(z) & A_{N-1}^R(z) \end{bmatrix} \tag{49}$$

From Eq. (49), the following step-up recursion relations are defined.

$$A_N(z) = c_N e^{-j\phi_N} z^{-1} A_{N-1}(z) - s_N B_{N-1}(z) \tag{50}$$

$$B_N(z) = s_N e^{-j\phi_N} z^{-1} A_{N-1}(z) + c_N B_{N-1}(z) \tag{51}$$

The step-up recursion relations allow the transfer functions to be easily calculated from knowledge of the coupling ratios and phases. Equations (50) and (51) can also be expressed in the time domain using the polynomial coefficients.

$$\underline{a}_N = c_N e^{-j\phi_N}\begin{bmatrix} 0 \\ \underline{a}_{N-1} \end{bmatrix} - s_N \begin{bmatrix} \underline{b}_{N-1} \\ 0 \end{bmatrix} \tag{52}$$

$$\underline{b}_N = s_N e^{-j\phi_N}\begin{bmatrix} 0 \\ \underline{a}_{N-1} \end{bmatrix} + c_N \begin{bmatrix} \underline{b}_{N-1} \\ 0 \end{bmatrix} \tag{53}$$

where

$$A_N(z) = a_{0,N} + a_{1,N}z^{-1} + \cdots + a_{N,N}z^{-N} \tag{54}$$

$$\underline{a}_N(z) = [a_{0,N}, a_{1,N}, \cdots, a_{N,N}]^T$$

$$B_N(z) = b_{0,N} + b_{1,N}z^{-1} + \cdots + b_{N,N}z^{-N}$$

$$\underline{b}_N(z) = [b_{0,N}, b_{1,N}, \cdots, b_{N,N}]^T$$

Note that $a_{0,0} = c_0$ and $b_{0,0} = s_0$. Expressions for the first and last coefficients for each polynomial are

$$a_{0,N} = \begin{cases} -s_1 s_0 & \text{for } N = 1 \\ -s_N s_0 \prod\limits_{n=1}^{N-1} c_n & \text{for } N > 1 \end{cases} \tag{55}$$

$$a_{N,N} = e^{-j\phi_{tot}^N} \prod_{n=1}^{N} c_n \tag{56}$$

$$b_{0,N} = s_0 \prod_{n=1}^{N} c_n \tag{57}$$

$$b_{N,N} = e^{-j\phi_{tot}^N} s_N \prod_{n=0}^{N-1} c_n \tag{58}$$

for $N \geq 1$. Note that $a_{0,N}$ is real and negative while $b_{0,N}$ is real and positive. The remaining two coefficients are complex in general because of the phase terms. For filter design, we need to find the coupling ratios and phase terms given one of the forward or reverse polynomials as the desired function. Using Eqs. (55)–(58), we obtain

$$-\frac{a_{0,N}}{b_{0,N}} = \frac{b_{N,N}}{a_{N,N}} = \frac{s_N}{c_N} = \tan\theta_n \tag{59}$$

If either the first or last coefficients of A_N and B_N are known, then we can find the Nth coupling ratio from Eq. (59). The coupling ratio can be written as follows:

$$\kappa_N = \frac{\left|\dfrac{b_{N,N}}{a_{N,N}}\right|^2}{1 + \left|\dfrac{b_{N,N}}{a_{N,N}}\right|^2} = \frac{\left|\dfrac{a_{0,N}}{b_{0,N}}\right|^2}{1 + \left|\dfrac{a_{0,N}}{b_{0,N}}\right|^2} \tag{60}$$

By multiplying Eq. (49) by $\Phi_n^{-1}(z)$, the following step-down recursion relations are defined.

$$A_{N-1}(z) = ze^{+j\phi_N}\{c_N A_N(z) + s_N B_N(z)\} \tag{61}$$

$$B_{N-1}(z) = -s_N A_N(z) + c_N B_N(z) \tag{62}$$

Equations (55)–(58) guarantee that the order is reduced in Eqs. (61) and (62). Equations (61) and (62) are the fundamental equations for the synthesis algorithm, where the coupling ratios and phase terms are determined from knowledge of $A_N(z)$ and $B_N(z)$. Note that both κ_N and ϕ_N must be known to apply the step-down recursions. The coupling ratio is already known, and the phase ϕ_N is found in two steps. First, Eq. (62) is solved, which only requires knowledge of A_N, B_N and κ_N. Then, Eq. (58) is used to define the total phase for the reduced order filter as follows:

$$\phi_{tot}^{N-1} = -\arg(b_{N-1,N-1}) \tag{63}$$

Next, the bracketed term in (61) is given the special designation $\widetilde{A}_{N-1}(z)$. According to Eq. (56), the argument of its last coefficient must satisfy

$$\arg(\widetilde{a}_{N-1,N-1}) + \phi_N = -\arg(a_{N-1,N-1}) = \phi_{tot}^{N-1} \tag{64}$$

Substituting Eq. (63) into Eq. (64) gives the phase for the Nth stage.

$$\phi_N = -\arg(\widetilde{a}_{N-1,N-1}) - \arg(b_{N-1,N-1}) \tag{65}$$

Typically, only one polynomial is specified by the design, say $A_N(z)$. We need to find a valid $B_N(z)$ in order to apply the step-down recursion relations. The necessary relationship is given by equating the determinant of the transmission matrix defined in Eq. (44) to the product of the determinants of the individual transmission matrices, $\det(\Phi_n) = e^{-j\phi_n}z^{-1}$. The following expression results:

$$B_N(z)B_N^R(z) = -A_N(z)A_N^R(z) + z^{-N}e^{-j\phi_{tot}^N} \tag{66}$$

Given $A_N(z)$, the roots of the above equation are calculated. These roots occur in pairs about the circle of radius γ, one root from each pair belongs to $B_N(z)$ and the other to

$B_N^R(z)$. The choice of which roots are assigned to the forward and which to the reverse polynomial is known as spectral factorization. For example, one choice is to assign the minimum phase roots to $B_N(z)$ and the maximum-phase roots to $B_N^R(z)$.

The roots of $B_N(z)$ only specify it to within a multiplicative constant, α. Let $B_N(z) = \alpha\Pi_{n=1}^N(1 - z_n z^{-1}) = \alpha B_N'(z)$. The product of the first and last coefficients of $A_N(z)$ and $B_N(z)$ are equal according to Eqs. (55)–(58). The scaling factor is then defined as follows:

$$\alpha = \sqrt{\frac{-a_{0,N}a_{N,N}}{b_{0,N}'b_{N,N}'}} \tag{67}$$

The steps in designing an optical lattice filter from the specification of an Nth-order polynomial are summarized in Figure 4-31. The input may be either $A_N(z)$ or $B_N(z)$. The design can be tailored for a specific loss, γ, if it is known *a priori*. A prelimi-

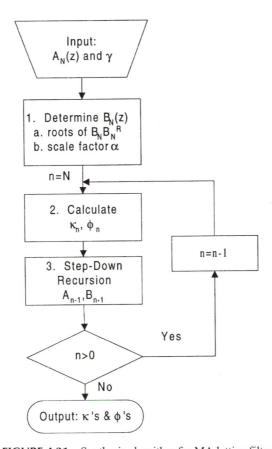

FIGURE 4-31   Synthesis algorithm for MA lattice filters.

nary check on the input polynomial magnitude response is useful to determine if its maximum is ≤ 1. Otherwise, the input polynomial must be scaled accordingly for a passive filter. The next step is to determine $B_N(z)$ by spectral factorization. The desired roots are chosen and a scale factor calculated. Then, the algorithm enters a loop where the nth coupling ratio and phase term are found. The step-down recursions are used to find A_{n-1} and B_{n-1} for the next iteration. The output consists of all the coupling ratios and phase terms.

The polynomials that define the MA lattice filter satisfy power conservation. In the frequency domain, Eq. (66) is

$$|A_N(\omega)|^2 + |B_N(\omega)|^2 = 1 \tag{68}$$

since $X^*(z^{*-1})$ in the Z domain is equal to $X^*(\omega)$ in the frequency domain. For a given input, the two outputs are power complementary.

Next, we investigate the impact of symmetry on the polynomials. A symmetrical filter is defined as one where the couplers and phase terms satisfy $\kappa_n = \kappa_{N-n}$ for $n = 0, \ldots, N$ and $\phi_n = \phi_{N+1-n}$ for $n = 1, \ldots, N$. An N-stage lattice filter is shown in Figure 4-32a with the input and output highlighted for the $B_N(z)$ response. The filter stages are reversed in order in Figure 4-32b. Since the filter is reciprocal, the highlighted path is equal to the $B_N^R(z)$ response. For a symmetrical filter, the responses in Figure 4-32 are indistinguishable, so $B_N(z) = B_N^R(z)$. Therefore, a symmetrical filter has a linear-phase $B_N(z)$.

An example of lattice filter synthesis is now presented. Let the desired response be $H_2(z) = 0.25[z^{-2} - 2 \cos(\phi_d)z^{-1} + 1]$. The roots are on the unit circle, so $H_2(z)$ is a linear-phase polynomial. The case for $\phi_d = 0$ is considered first. The polynomial $H_2(z)$ can be realized with either $A_2(z)$ or $B_2(z)$. The design steps and solutions for each case are summarized in Table 4-2. Starting with the design for $A_2(z)$, $H_2(z)$ is

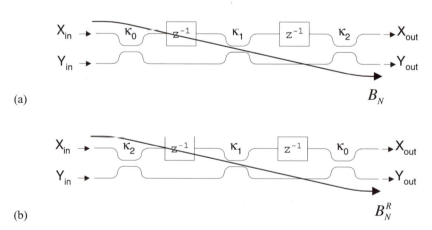

(a)

(b)

FIGURE 4-32   A schematic of a two-stage lattice filter (a) and its flipped version (b) showing that if the filter is symmetric (i.e. $\kappa_0 = \kappa_2$), then the cross-port transmission responses are identical since (a) and (b) are indistinguishable in such a case.

TABLE 4-2   Solutions for a Second-Order MA Lattice Filter with $\phi_d = 0$

Design Step	Soln	$\phi_d = 0$ degrees	Design Step	$\phi_d = 0$ degrees
$A_2(z)$ $A_2(z)A_2^R(z) - z^{-N}$		$-0.250 + 0.500z^{-1} - 0.250z^{-2}$ $0.0625 - 0.250z^{-3} - 0.625z^{-2}$ $-0.250z^{-1} + 0.0625z^{-4}$	$B_2(z)$ $B_2(z)B_2^R(z) + z^{-N}$	$0.250 - 0.500z^{-1} + 0.250z^{-2}$ $0.0625 - 0.250z^{-1} + 1.375z^{-2}$ $-0.250z^{-3} + 0.0625z^{-4}$
Roots of $B_2(z)B_2^R(z)$	1	0.1716, −1	Roots of $A_2(z)A_2^R(z)$	0.1716, −1
	2	5.8284, −1		5.8284, −1
	3	0.1716, −1		0.1716, −1
	4	5.8284, −1		5.8284, −1
$B_2(z)$	1	$0.6036 + 0.5000z^{-1} - 0.1036z^{-2}$	$A_2(z)$	$-0.6036 - 0.5000z^{-1} + 0.1036z^{-2}$
	2	$0.1036 - 0.5000z^{-1} - 0.6036z^{-2}$		$-0.1036 + 0.5000z^{-1} + 0.6036z^{-2}$
	3	$0.6036 + 0.5000z^{-1} - 0.1036z^{-2}$		$-0.6036 - 0.5000z^{-1} + 0.1036z^{-2}$
	4	$0.1036 - 0.5000z^{-1} - 0.6036z^{-2}$		$-0.1036 + 0.5000z^{-1} + 0.6036z^{-2}$
Coupling Ratio Solutions	1	0.8536, 0.5000, 0.1464	Coupling Ratio Solutions	0.8536, 0.5000, 0.8536
	2	0.1464, 0.5000, 0.8536		0.1464, 0.5000, 0.1464
	3	0.8536, 0.5000, 0.1464		0.8536, 0.5000, 0.8536
	4	0.1464, 0.5000, 0.8536		0.1464, 0.5000, 0.1464
Phase Solutions (radians)	1	0.000, −3.1416	Phase Solutions (radians)	0, 0
	2	3.1416, 0		3.1416, 3.1416
	3	0.000, −3.1416		0, 0
	4	3.1416, 0		3.1416, 3.1416

multiplied by -1 so that the first term in $A_2(z)$ is negative in agreement with Eq. (55). This multiplication does not change the magnitude or group delay response. Next, the roots of $B_2(z)B_2^R(z)$ are determined using Eq. (66). A root from each reciprocal pair is chosen to define $B_2(z)$. In general there are 2^N solutions for $B_N(z)$. Since one root pair from $B_2(z)B_2^R(z)$ is on the unit circle, $B_2(z)$ has only two distinct solutions in this example. Note that $B_2(z)$ is not symmetrical and that the coupling ratios and phases are not symmetrical either. The two solutions for the coupling ratios and phase terms are the reverse of each other. The design for $B_2(z) = H_2(z)$ is shown in the last two columns of Table 4-2. Now, $B_2(z)$ is symmetrical, and the resulting coupling ratios and phase terms are symmetrical. There are two distinct solutions.

The design of $A_2(z) = -H_2(z)$ for $\phi_d = \pi/6$ is summarized in Table 4-3. The polynomial $H_2(z)$ is a high pass filter with its maximum transmission at $z = -1$ ($\omega = \pi$). For $\phi_d \neq 0$, the maximum transmission is less than one. A gain term can be included so that $|A_2(-1)| = |\Gamma|H_2(-1)| = 1$. The last two columns in Table 4-3 summarize the design for $\Gamma = 1$ and $\Gamma = 1.0718$, respectively. For $\Gamma = 1$, there are four different solutions although, only two are unique since the other two can be obtained by reversing the order of the coupling ratios and phase terms from the first solution. For $\Gamma = 1.0718$, one zero for $B_2(z)$ is on the unit circle at $z = -1$ as expected by power conservation since $A_2(-1) = 1$. The coupling ratios for $\Gamma = 1.0718$ are larger than those for $\Gamma = 1$. A graph of $A_2(z)$ for $\phi_d = 0$ and $\phi_d = \pi/6$ with $\Gamma = 1.0718$ is shown in Figure 4-33. The passband has 100% transmission in both cases in contrast to the cascade filter architecture.

A large number of identical periods can easily be calculated using Chebyshev's identity. Given a unimodular matrix (a matrix whose determinant is one)

$$\Phi = \begin{bmatrix} A & C \\ B & D \end{bmatrix}$$

to be raised to the Nth power, the solution is [54]

$$\Phi^N = \begin{bmatrix} \dfrac{A \sin(Nx) - \sin[(N-1)x]}{\sin(x)} & \dfrac{B \sin(Nx)}{\sin(x)} \\[2em] \dfrac{C \sin(Nx)}{\sin(x)} & \dfrac{D \sin(Nx) - \sin[(N-1)x]}{\sin(x)} \end{bmatrix} \tag{69}$$

where $\cos(x) = (A + D)/2$. To calculate the transfer matrix for an Nth-order lattice filter with identical stages, we make Eq. (43) unimodular by removing the average phase term, so

$$\Phi_n(\omega) = \begin{bmatrix} c_n\zeta_n & -js_n\zeta_n^* \\ -js_n\zeta_n & c_n\zeta_n^* \end{bmatrix}$$

where $\zeta_n = e^{-j\frac{1}{2}(\omega + \phi_n)}$. Then, $x = c_n \cos[\frac{1}{2}(\omega + \phi_n)]$ and the transfer functions are obtained by substituting the coefficients of Φ_n and x into Eq. (69). A simple expres-

TABLE 4-3 Solutions for a Second-Order MA Lattice Filter with $\phi_d = \pi/6$

Design Step	Soln	$\phi_d = 30°$	$\phi_d = 0°$ w/Gain $= 1.0718$
$A_2(z)$		$-0.250 + 0.433z^{-1} - 0.250z^{-2}$	$-0.2679 + 0.4641z^{-1} - 0.2679z^{-2}$
$A_2(z)A_2^R(z) - z^{-N}$		$0.0625 - 0.2165z^{-1} - 0.6875z^{-2}$	$-0.0718 + 0.2487z^{-1} + 0.6410z^{-2}$
		$-0.2165z^{-3} + 0.0625z^{-4}$	$+0.2487z^{-3} - 0.07128z^{-4}$
Roots of	1	$0.1801, -0.5993$	$0.1896, 1.0000 \angle -3.1385$
$B_2(z)B_2^R(z)$	2	$5.5519, -0.5993$	$5.2745, 1.0000 \angle -3.1385$
	3	$0.1801, -1.6687$	$0.1896, 1.0000 \angle -3.1385$
	4	$5.5519, -1.6687$	$5.2745, 1.0000 \angle -3.1385$
$B_2(z)$	1	$0.7609 + 0.3189z^{-1} - 0.0821z^{-2}$	$+0.6154 + 0.4987z^{-1} - 0.1167z^{-2}$
	2	$0.1371 - 0.6788z^{-1} - 0.4560z^{-2}$	$+0.1167 - 0.4987z^{-1} - 0.6154z^{-2}$
	3	$0.4560 + 0.6788z^{-1} - 0.1371z^{-3}$	$+0.6154 + 0.4987z^{-1} - 0.1167z^{-2}$
	4	$0.0821 - 0.3189z^{-1} - 0.7609z^{-2}$	$+0.1167 - 0.4987z^{-1} - 0.6154z^{-2}$
Coupling Ratio	1	$0.9026, 0.2892, 0.0974$	$0.8406, 0.4641, 0.1594$
Solutions	2	$0.2311, 0.6483, 0.7689$	$0.1594, 0.4641, 0.8406$
κ_n	3	$0.7689, 0.6483, 0.2311$	$0.8406, 0.4641, 0.1594$
		$0.0974, 0.2892, 0.9026$	$0.1594, 0.4641, 0.8406$
Phase Solutions	1	$0.000, 3.1416$	$0.0012, -3.1428$
ϕ_n	2	$3.1416, 0.0000$	$3.1404, 0.0012$
(radians)	3	$0.000, 3.1416$	$0.0012, -3.1428$
	4	$3.1416, 0.0000$	$3.1404, 0.0012$

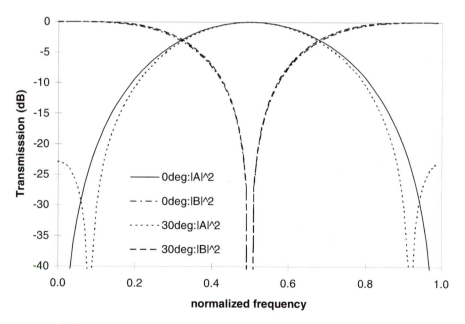

FIGURE 4-33 Magnitude response for a second-order MA lattice filter.

sion for the cross-port response is found by setting $\omega + \phi_n = 2\pi m$ where m is an integer. Then, $\cos(x) = \pm c_n = \pm \cos\theta_n$ and

$$|t_x|^2 = \left| \frac{Y_{out}}{X_{in}} \right| = \left| \frac{X_{out}}{Y_{in}} \right|^2 = \sin^2(N\theta_n) \tag{70}$$

Full coupling to the cross-port is achieved for $N\theta_n = (2m + 1)\pi/2$. This result is equivalent to the result obtained in Section 4.6 for a coupled-mode filter with uniform coupling.

4.5.1.1   *Gain Equalization Filter*   As discussed in Chapter 1, gain equalization filters (GEF) compensate for the wavelength dependent gain of optical amplifiers. An erbium-doped fiber amplifier gain characteristic is used to design a MA optical lattice GEF. The desired response is the negative of the gain spectrum, in dB, shown in Figure 4-34. The wavelength dependence of the gain is not periodic. In addition, the endpoints may not have the same gain in general. For this example, we shall approximate the desired response with a periodic filter over the range of 1540 to 1560 nm. Two approaches for defining a periodic desired response are: (1) let the region of

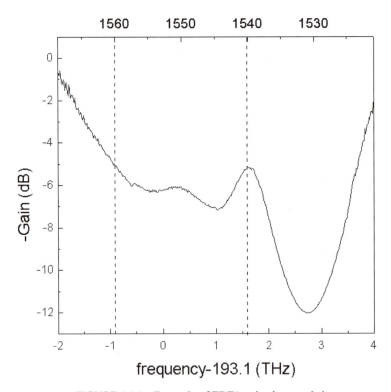

FIGURE 4-34   Example of EDFA gain characteristic.

interest represent half the FSR and define the other half to be symmetric about $\nu = 0$, or (2) make the FSR larger than the region of interest and optimize the response only over the region of interest. These approaches are illustrated in Figure 4-35. When the first approach is used, it yields a linear-phase filter. The second approach yields a mixed-phase filter. A 12th-order linear-phase filter was designed using a least squares fit. Its roots are located symmetrically about $\nu = 0$. The roots between $\nu = 0$ and $\nu = 0.5$ were chosen as the starting values for the optimization of a sixth-order

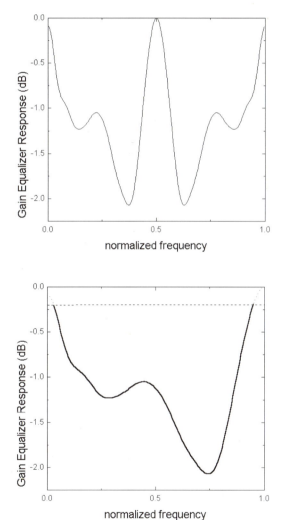

(a)

(b)

FIGURE 4-35 Approaches for defining the desired periodic response for a GEF: (a) a linear-phase response symmetric about v = 0 and (b) a mixed-phase response where the endpoints of the FSR are not defined.

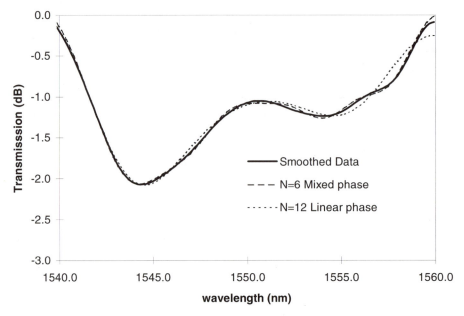

FIGURE 4-36   A sixth-order, mixed-phase MA lattice GEF compared to the smoothed data and a 12th-order linear-phase MA response.

mixed-phase filter. The responses are shown in Figure 4-36. The mixed-phase design has a maximum deviation of 0.05 dB from the smoothed input data. The corresponding polynomials and parameters for the sixth-order filter, assuming a minimum-phase solution for $B_6(z)$, are given in Table 4-4. Because dispersion can be an issue for high bit rate systems, the group delay and dispersion of the mixed-phase filter were calculated. The period is 20 nm which corresponds to a unit delay of $T = 0.28$ ps and a path length difference of 83 μm for $n_g = 1.45$. The normalized group delay calculated from $A_6(z)$ was multiplied by T to yield the response shown in Figure 4-37. The dispersion was calculated by taking the finite difference of the group delay with respect to wavelength. The dispersion is less than 0.15 ps/nm. For such a small unit de-

TABLE 4-4   Design Parameters for a Sixth-Order MA Lattice GEF

$|a_n|$	$\arg(a_n)$	$|b_n|$	$\arg(b_n)$	κ_n	ϕ_n
0.0468	0.000	0.4059	1.5711	0.9776	−1.7188
0.0561	−1.2783	0.1856	−1.4000	0.1146	3.1745
0.3387	1.7858	0.1460	−1.8622	0.2137	−4.8206
0.7331	0.7444	0.0487	−2.2361	0.6133	−0.7999
0.3075	2.4134	0.0259	−2.2721	0.3510	2.6154
0.1431	−0.9489	0.0102	−1.9695	0.0225	−0.9308
0.0615	0.6613	0.0071	−0.9094	0.0131	

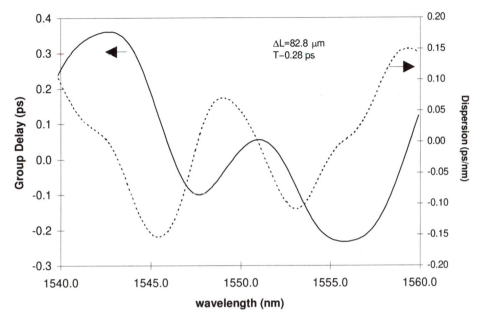

FIGURE 4-37 Group delay and dispersion for the sixth-order MA lattice GEF.

lay, the resulting group delay and dispersion are practically negligible; so, a linear-phase design is not critical in this case, and the filter order can be significantly reduced by employing a mixed-phase design.

4.5.1.2 Dispersion Compensation Filter An important filter application for WDM systems is dispersion compensation. Let the ideal response be a constant dispersion and unity magnitude response across a desired bandwidth, $\nu_{min} \leq \nu \leq \nu_{max}$, i.e.

$$|H(\nu)| = 1 \tag{71}$$

$$D(\nu) = \frac{d\tau_n}{d\nu} = D_d \tag{72}$$

where τ_n is the normalized group delay and D_d is a constant equal to the desired normalized dispersion. A linear group delay response can be approximated using the non-linear phase response of the zeros in a MA filter. The magnitude and phase response are related by the zeros in a MA filter; consequently, they cannot be specified independently.

For this example, the ideal dispersion is defined as $D(\nu_1 \pm 0.1) = -D_d$ and $D(\nu_2 \pm 0.1) = +D_d$ where $\nu_1 = 1/4$ and $\nu_2 = 3/4$. One design method is to use frequency sampling. The following approximate frequency response is assumed [55]:

$$|H(\nu)| = 1 \qquad (73)$$

$$\Phi(\nu) = -\frac{D_d}{(2\pi)^2} \sin(2\pi\nu)$$

The group delay is

$$\tau_n = -\frac{d\Phi(\nu)}{d\nu} = \frac{D_d}{(2\pi)} \cos(2\pi\nu)$$

which approximates a linear group delay with a negative slope near $\nu = 1/4$ and a positive slope near $\nu = 3/4$. The frequency response $H(\nu) = e^{j\Phi(\nu)}$ is sampled, an inverse FFT is performed, and the resulting filter coefficients are truncated for the desired order. As an example, let $D_d = 12$ and the filter order be $N = 8$. The resulting zeros, coupling ratios and phase terms are shown in Table 4-5. Half of the zeros are minimum-phase and half are maximum-phase. The coupling ratio values cover a wide range with some values being very close to zero. A more linear delay over each passband can be achieved by optimizing the zero locations with respect to an error criterion. A tradeoff between the magnitude error and group delay or dispersion error is accomplished by adjusting ξ in the following error definitions:

$$E_\tau = \sum_{i=1}^{P} \{\xi[|H(\omega_i)| - 1]^2 + (1 - \xi)[\tau(\omega_i) - \tau_d(\omega_i)]|^2\} \qquad (74)$$

$$E_D = \sum_{i=1}^{P} \{\xi[|H(\omega_i)| - 1]^2 + (1 - \xi)[D(\omega_i) - D_d(\omega_i)]|^2\}$$

where $0 \le \xi \le 1$ and the ω_i's are the sampled frequencies in each passband. Three cases are demonstrated: the inverse FFT method (IFFT) discussed above and two optimized responses using $\xi = 0.5$. The filters are assumed to be lossless. The first

TABLE 4-5   Design Values for an Eighth-Order Dispersion Compensator Using the Inverse Fast Fourier Transform Approach. The phase is Given in Radians.

$|z_n|$	$\arg(z_n)$	κ_n	ϕ_n
2.673	1.392	0.941	3.141
2.673	−1.392	0.470	0.000
2.205	0.438	0.627	−0.001
2.205	−0.438	0.005	3.143
0.374	1.750	0.623	0.001
0.453	2.704	0.005	−3.142
0.453	−2.704	0.627	3.142
0.374	−1.750	0.470	−3.142
		0.059	

optimization minimizes the magnitude and group delay errors over the two pass-bands. In the second optimization, the magnitude and dispersion errors are minimized. The magnitude responses are shown in Figure 4-38. There is minimum ripple for the IFFT case with loss ≤ 0.3 dB across the FSR. For the group delay optimized response, the loss is slightly larger with a maximum of 0.5 dB. For the dispersion optimized response, the loss is much larger, having a maximum loss of 5.2 dB across the FSR. The group delay responses are plotted in Figure 4-39. Since the desired response was chosen to have a constant negative dispersion for half the FSR and a constant positive dispersion for the other half, the ideal group delay is V-shaped. The differences between the group delay responses are small. The dispersion responses are shown in Figure 4-40. It is clear that constant dispersion is best approximated by the second optimization process, but the tradeoff is higher losses across the FSR. By increasing the filter order, the desired dispersion can be more closely approximated.

The group delay and dispersion are proportional to T and T^2, respectively. It is desired to make T as large as possible; however, the tradeoff is that the passband width and FSR scale as $1/T$. As an example, lets choose a passband width of 20 GHz over which we want to compensate a signal. The above designs then require FSR = 100GHz, which translates to a unit delay of $T = 100$ ps and a differential path length of 2.07 mm for $n_g = 1.45$. The dispersion is then given by $12T/\Delta\lambda$, where $\Delta\lambda$ is the passband width in nm. The result is a dispersion of ± 750 ps/nm.

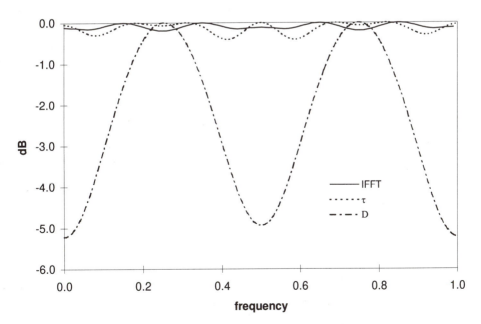

FIGURE 4-38 Magnitude response of an eighth-order MA dispersion compensator comparing different designs.

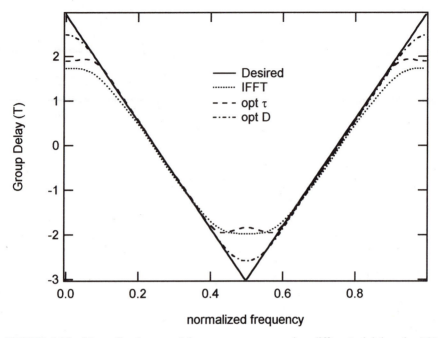

FIGURE 4-39   Normalized group delay responses comparing different eighth-order MA dispersion compensator designs.

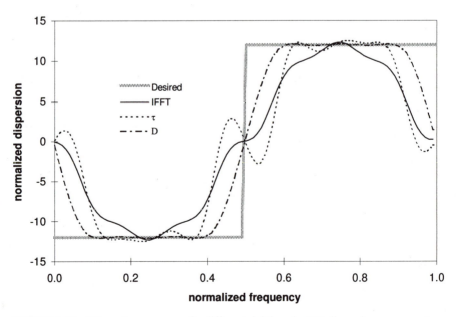

FIGURE 4-40   Dispersion responses for different eighth-order MA dispersion compensators.

Standard singlemode fiber (SMF) has a dispersion of approximately +17 ps/nm-km at 1550 nm; therefore, the filters in this example can compensate for 44 km of SMF with tradeoffs between the passband loss and dispersion ripple across the passband. By setting the FSR equal to the channel spacing in a WDM system, each filter can compensate multiple channels. For the same filter design, the impact of increasing the unit delay by a factor of two is to decrease the passband width and the FSR by a factor of 2. Simultaneously, the peak dispersion increases by a factor of 4, so the filter could compensate 176 km of SMF over a passband of 10 GHz by doubling the unit delay.

Planar waveguide MA dispersion compensators were first demonsrated by Takiguchi et al., [56] using 12 stages on a 52×71 mm^2 chip as shown in Figure 4-41. This particular architecture uses a combination of two different relative delays. The group delay for 20 km of SMF over a 22 GHz range centered at 1550 nm is shown in Figure 4-42a. The delay after transmission through the fiber followed by the compensating filter is significantly flatter as shown in Figure 4-42b. The waveguides had a $\Delta = 1.5\%$, a minimum bend radius of $R = 2$ mm, and a total path length of 50 cm. The passband loss was 8 dB. In another experiment, optical dispersion was compensated for 2.5 Gb/s transmission over a 40 km length of standard SMF using an equalizer with 5 stages and a total insertion loss of 3.5 dB [57]. A figure of merit for dispersion compensating devices is defined as the ratio of the dispersion to the loss. The maximum dispersion for this device was +836 ps/nm yielding a figure of merit of 239, which is competitive with dispersion compensating fiber. A tunable version has also been demonstrated that uses symmetric MZIs with heaters for each coupler in the design [58]. An 8-stage tunable filter capable of compensating 50 km of standard SMF was demonstrated. The dispersion was tunable over the range of −681 to +786 ps/nm as shown in Figure 4-43. The bandwidth for dispersion compensation was 16.3 GHz. The power transmittance varies depending on the group delay, but is <1 dB over the passband as shown in Figure 4-44.

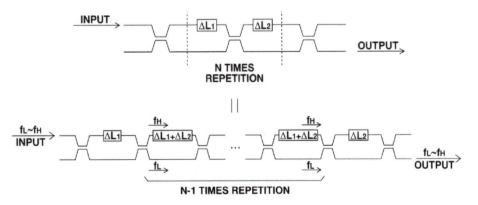

FIGURE 4-41 Circuit design schematic for an experimental dispersion compensator [56]. Reprinted by permission. Copyright © 1994 IEEE.

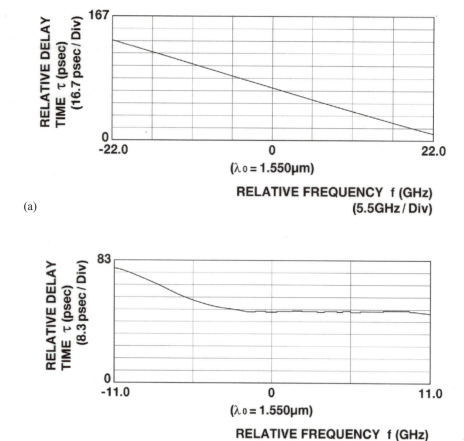

(a)

(b)

FIGURE 4-42   Relative delay for (a.) transmission through 20 km of 1.3 μm dispersion zero fiber and (b.) after compensation by a 12th-order filter [56]. Reprinted by permission. Copyright © 1994 IEEE.

4.5.2   Generalized Lattice Filters

Several generalizations to the lattice filter were introduced by Li and co-workers [52,59], which impact the number of required stages and the filter's robustness to fabrication variations. First, the delays in each stage may be integer multiples of the unit delay. Second, the relative delay may be either in the top or bottom arm of each stage. Finally, non-integer multiples of the unit delay for each stage are used when optimizing a design, particularly when the coupler wavelength dependence is important. The transfer functions for the general architecture are described. Then, design examples and experimental results for a wideband multiplexer and a gain equalizer are presented.

A generalized lattice filter consists of N stages, and each stage has a relative de-

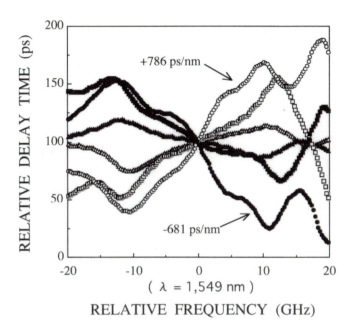

FIGURE 4-43 Relative group delay for a tunable dispersion compensator [58]. Reprinted by permission.

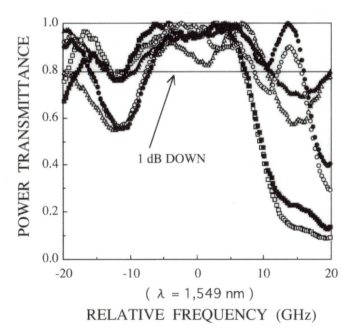

FIGURE 4-44 Power transmittance for a tunable dispersion compensator [58]. Reprinted by permission.

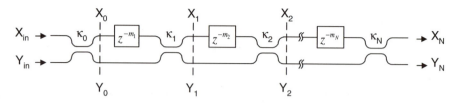

FIGURE 4-45   Generalized lattice filter schematic with Z-transform notation.

lay which is m_n times the unit delay as shown in Figure 4-45. A positive delay indicates that the top waveguide is longer, and a negative delay indicates that the bottom waveguide is longer. For now, we shall consider integer values for the m_n's. The sequence of relative delays is denoted by $m_1/m_2/\cdots/m_N$. A specific example is given in Figure 4-46 for a 3 stage filter with relative delays of $1/-2/4$. A single stage filter has an order equal to $|m_1|$. The order of an N-stage filter is $P = \sum_{n=1}^{N}|m_n|$, so the filter order is no longer equal to the number of stages. The transfer function for any input to output can be written by adding the contributions from all the optical paths. The number of paths for an N-stage filter is $Q = 2^N$. Using the same definitions as Eq. (42), the polynomials are given by

$$A_N^P(z) = [\alpha_0 + \alpha_1 z^{-m_1} + \cdots + \alpha_Q z^{-(m_1 + \cdots + m_N)}]z^{m_{\mathrm{neg}}} \tag{75}$$

$$B_N^P(z) = [\beta_0 + \beta_1 z^{-m_1} + \cdots + \beta_Q z^{-(m_1 + \cdots + m_N)}]z^{m_{\mathrm{neg}}} \tag{76}$$

where α_n and β_n are the transmission coefficients for each path, m_{neg} is the sum of the negative relative delays, the superscript P denotes the polynomial order, and the subscript N denotes the output of the Nth stage. Multiplication by m_{neg} makes the Z-transform causal. The polynomial coefficients $a_{p,P}$ and $b_{p,P}$ are the sum of all paths with the same length.

$$A_N^P(z) = a_{0,P} + a_{1,P}z^{-1} + \cdots + a_{P,P}z^{-P} \tag{77}$$

$$B_N^P(z) = b_{0,P} + b_{1,P}z^{-1} + \cdots + b_{P,P}z^{-P} \tag{78}$$

Some coefficients may be zero, depending on the particular delay sequence. For example, consider a second-order filter with relative delays of $2/-4$, so $m_{\mathrm{neg}} = -4$. The

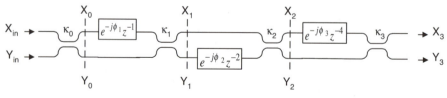

FIGURE 4-46   A generalized lattice filter schematic for a three-stage filter with delays of $1/-2/4$.

delays associated with each path in Eqs. (75)–(76) are $[0, m_1, m_2, m_1 + m_2] - m_{neg} = [0,2,-4,-2] + 4 = [4,6,0,2]$. This delay sequence produces an even filter function. Relative delays of $\pm 1/\pm 2/\pm 4/\cdots/\pm 2^{N-1}$ give a non-zero value for every coefficient.

An advantage of generalized lattice filters is the ability to increase the number of coefficients in the approximating polynomial without increasing the number of filter stages proportionally. An N-stage filter can have $P + 1$ nonzero coefficients; however, it has only $2N + 1$ design parameters, consisting of $N + 1$ couplers and N delays. When the number of polynomial coefficients exceeds the number of design variables, then we don't have design control over every coefficient. An arbitrary polynomial of order P cannot be realized in general. In such a case, multiple delay sequences are considered, and the resulting coefficients for each are optimized to best fit the desired polynomial.

The generalized lattice filter can also be described by a transfer matrix. The transfer matrix for the nth stage where $m_n > 0$ is given by

$$\begin{bmatrix} X_n \\ Y_n \end{bmatrix} = \begin{bmatrix} z^{-m_n}e^{-j\phi_n}c_n & -js_n \\ -jz^{-m_n}e^{-j\phi_n}s_n & c_n \end{bmatrix} \begin{bmatrix} X_{n-1} \\ Y_{n-1} \end{bmatrix} \tag{79}$$

for $n = 0, \ldots, N$. For $m_n < 0$,

$$\begin{bmatrix} X_n \\ Y_n \end{bmatrix} = \begin{bmatrix} c_n & -jz^{-m_n}e^{-j\phi_n}s_n \\ -js_n & z^{-m_n}e^{-j\phi_n}c_n \end{bmatrix} \begin{bmatrix} X_{n-1} \\ Y_{n-1} \end{bmatrix} = z^{m_n}e^{-j\phi_n}\begin{bmatrix} z^{-m_n}e^{j\phi_n}c_n & -js_n \\ -jz^{-m_n}e^{j\phi_n}s_n & c_n \end{bmatrix}\begin{bmatrix} X_{n-1} \\ Y_{n-1} \end{bmatrix} \tag{80}$$

By rewriting the matrix as in the last equality, the form is identical to the case for $m_n > 0$ and to the lattice filter matrix in Eq. (43). For $n = 0$, $m_0 = 0$ and $\phi_0 = 0$. The transfer matrix is causal for both $m_n > 0$ and $m_n < 0$. The transfer matrix description allows the recursion relations developed previously to be applied to the generalized lattice filter.

4.5.2.1 Wide Bandpass Filter

Wide bandpass filters find applications in combining wavelength bands, for example, a pump band and a signal band. In this example, the 1300 and 1550 nm windows are combined. A wideband generalized lattice filter design is explored starting with 3 stages having delays of $1/\pm 2/4$. The coupling ratios are assumed to be independent of wavelength, but their sensitivity to fabrication variations is explored. For this example, we assume that the coupling length varies by 5% around its nominal value in a systematic manner, i.e. either all of the coupling ratios will shift to higher or lower values. The frequency response for the nominal design and those with variations are shown in Figure 4-47 for a delay sequence of $1/2/4$. The responses for a delay sequence of $1/-2/4$ are shown in Figure 4-48. Note that the short frequency stopband continues to have high rejection for the $1/-2/4$; whereas, both stopbands are severely degraded for the $1/2/4$ sequence. In general, sequences with shorter coupling ratios are less sensitive to fabrication variations as demonstrated by this example.

For a wide bandpass filter covering a hundred or more nanometers, the wavelength dependence of the couplers must be considered in the design. As an example,

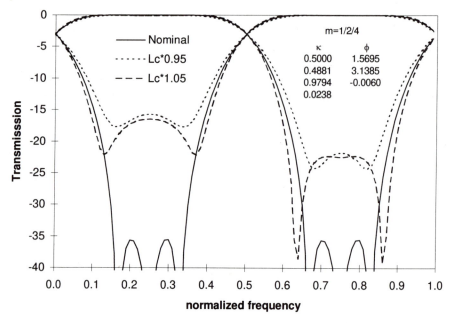

FIGURE 4-47   Sensitivity to coupler tolerances for a broadband filter with m = 1/2/4.

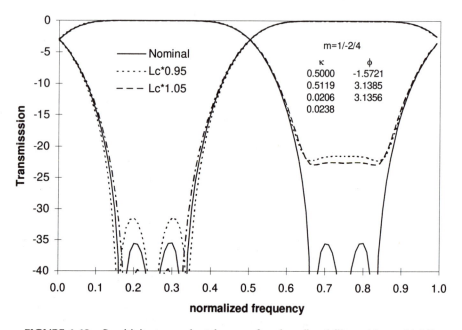

FIGURE 4-48   Sensitivity to coupler tolerances for a broadband filter with m = 1/–2/4.

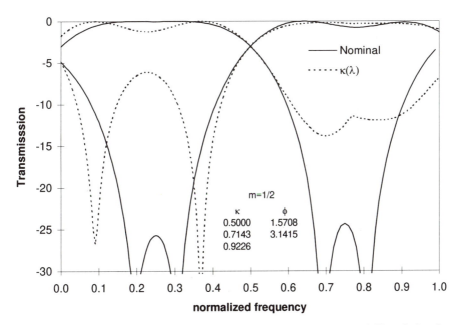

FIGURE 4-49 Impact of coupler wavelength dependence on a broadband filter design for m = 1/2.

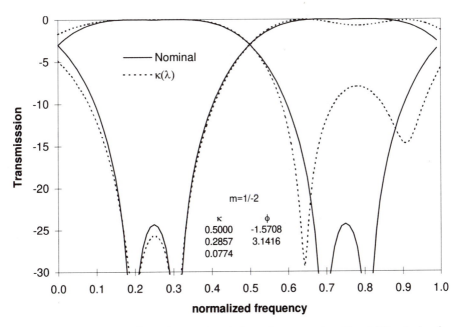

FIGURE 4-50 Impact of coupler wavelength dependence on a broadband filter design for m = 1/–2.

a coupler design having a full coupling length of 500 μm at 1550 nm and 1000 μm at 1300 nm was assumed. The impact of coupler wavelength dependence on a 2-stage design is shown in Figure 4-49 for delays of 1/2 and in Figure 4-50 for delays of 1/–2. The results are similar to the fabrication sensitivity study. The shorter coupler design is less sensitive to the wavelength dependence. To improve the design further, non-integer values are allowed for the relative delays, and both the relative delay sequence and coupling ratios are optimized. The filter is then quasi-periodic, since the integer relationship between consecutive stages has been removed. An experimental demonstration is shown in Figure 4-51 [60]. Three sections, consisting of two stages each, were used to realize a response having a 50 dB rejection over a 100 nm wide stopband.

4.5.2.2 Gain Equalization Filter   A six-stage, generalized lattice GEF was fabricated by Li et al. [61]. The shape of the square magnitude response was opti-

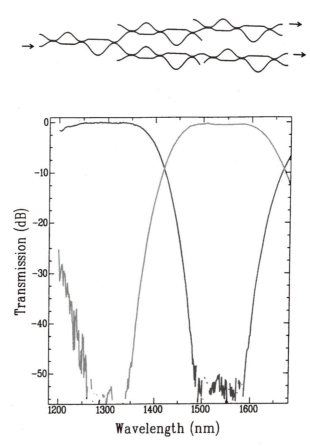

FIGURE 4-51   Experimental results for a three-section wideband Fourier filter with two stages in each section [60]. Reprinted by permission.

mized for a 2000 km system containing 40 km spans and two-stage erbium doped fiber amplifiers in each span. The results are shown in Figure 4-52. The desired equalization function is closely approximated over a 30 nm bandwidth, which is significantly larger than the 20 nm bandwidth approximated by the sixth-order lattice filter in Section 4.5.1.1. The minimum loss for the fabricated filter was 1.5 dB. The delay sequence was $1/-2/-2/-2/-4/-4$ with relative path length differences of 13.56, -27.09, -26.93, -26.80, -53.16, and -53.70 μm. The coupler lengths were 1002, 861, 571, 1112, 1367, 842 and 1180 μm. The waveguides were 5×5 μm with $\Delta = 0.63\%$ and center-to-center waveguide separations of 9.25 to 9.5 μm. No heaters were used to tune the phase after fabrication for either the wideband or GEF generalized lattice filters, demonstrating the robustness of the designs to fabrication variations.

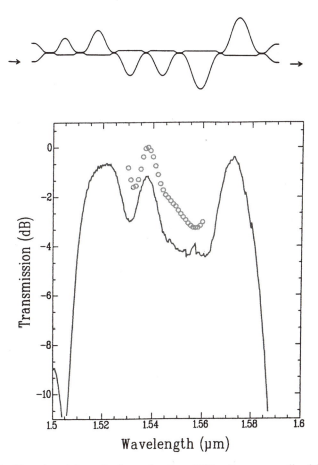

FIGURE 4-52 Experimental results for a six-stage GEF using a generalized lattice filter [61]. Reprinted by permission.

4.6 COUPLED-MODE FILTERS

A different approach for realizing optical MA filters is to split the signal between orthogonal modes, whether polarization or spatial modes, instead of between distinct paths. The differential delay is proportional to the difference in effective indices of the modes. Birefringent filters are discussed first, and their similarity to other MA filters is highlighted. Filters with a periodic variation in the refractive index are then introduced. Coupled-mode theory for two forward propagating modes is presented for comparison to the Z-transform techniques.

To describe the characteristics of birefringent filters, we begin by describing the transfer matrices for propagation through birefringent plates and polarizers. There are two input and two output fields that are orthogonally polarized, denoted by $E_x^{in,out}$ and $E_y^{in,out}$. A polarized signal along the x-direction is represented by

$$\begin{bmatrix} E_x \\ E_y \end{bmatrix} = \begin{bmatrix} 1 \\ 0 \end{bmatrix}$$

and a polarizer oriented along the x-direction has a transfer matrix

$$\Phi^{pol} = \begin{bmatrix} 1 & 0 \\ 0 & 0 \end{bmatrix}$$

where

$$\begin{bmatrix} E_x^{out} \\ E_y^{out} \end{bmatrix} = \Phi^{pol} \begin{bmatrix} E_x^{in} \\ E_y^{in} \end{bmatrix}$$

The transfer matrices are called Jones matrices. When a polarized signal is launched along the 45° axis of a birefringent plate, the polarization of the output signal depends on the plate thickness. The relative phase difference between light propagating along each principal axis is $\Delta\phi = 2\pi(n_e - n_o)L/\lambda$, where n_e and n_o denote the refractive index of the extraordinary and ordinary rays. The delay matrix is given by

$$\Phi(f) = \begin{bmatrix} e^{-j\pi f/\Delta f} & 0 \\ 0 & e^{+j\pi f/\Delta f} \end{bmatrix} \text{ or } \Phi(z) = \sqrt{z}\begin{bmatrix} z^{-1} & 0 \\ 0 & 1 \end{bmatrix} \tag{81}$$

where $\Delta n = n_e - n_o$ and $\Delta f = c/\Delta nL$. The input signal is projected onto each principal axis of the crystal using the following rotation matrix:

$$R(\psi) = \begin{bmatrix} \cos\psi & \sin\psi \\ -\sin\psi & \cos\psi \end{bmatrix} \tag{82}$$

where ψ is the angle of the plate relative to a chosen coordinate axis. The polarization out of the plate must be projected onto the coordinate axis, which is accomplished using $R(-\psi)$. The transfer matrix for one plate is given by

$$\begin{bmatrix} E_x^{out} \\ E_y^{out} \end{bmatrix} = R(-\psi)\Phi(f)R(+\psi)\begin{bmatrix} E_x^{in} \\ E_y^{in} \end{bmatrix} \tag{83}$$

From Eq. (81), we know that the frequency response is periodic with period Δf, so the FSR is inversely proportional to the birefringence. Using Eq. (83), the transmission through a birefringent plate placed between two parallel polarizers oriented in the x-direction is given by $|E_x^{out}/E_y^{out}|^2 = \cos^2(\Delta\phi/2)$, assuming an input polarization along the x-direction. The filter response depends on the input state of polarization, which is a problem for optical fiber communication systems where the polarization state is not preserved along the fiber. A Lyot-Ohman filter is formed by cascading birefringent plates with thicknesses that vary in a geometric progression given by [L, 2L, 4L, ...], and separating them by parallel polarizers. The transmission response is given by [54]

$$\left| \cos\left(\frac{\Delta\phi}{2}\right)\cos\left(2\frac{\Delta\phi}{2}\right)\cdots\cos\left(2^{N-1}\frac{\Delta\phi}{2}\right) \right|^2 = \left| \frac{\sin\left(2^N\dfrac{\Delta\phi}{2}\right)}{2^N \sin\left(\dfrac{\Delta\phi}{2}\right)} \right|^2 \tag{84}$$

assuming an x-polarized input. Equation (84) is equivalent to the response of a uniformly illuminated diffraction grating and to the $\log_2 N$ cascade architecture discussed in Section 2.2. Another birefringent filter, which requires only two polarizers and is composed of plates with uniform thicknesses, is a Solc filter. The polarizers are at the input and output. The angle of each birefringent plate is varied to produce different filter coefficients. The Solc filter is analogous to the waveguide lattice filter. The rotation matrix performs the function of the coupler matrix. Birefringent filters have been proposed [62] and demonstrated for dispersion compensation applications [63]. In the latter case, six TiO_2 crystals with $n_e = 2.709$ and $n_o = 2.451$ were used to produce a dispersion of 150 ps/nm over a 48.5 GHz bandwidth.

While coupling at discrete points has been discussed so far, we now focus on coupled-mode filters characterized by a continuous, weaker coupling. Coupled-mode theory is typically used to analyze such filters. The coupled-mode solutions are outlined for two co-propagating waves. The derivation is essentially equivalent to the one for the directional coupler in Chapter 2, except that the index perturbation is periodic. Let the index perturbation be described by [64]:

$$\delta n(x, y, z) = \delta n_{dc}(x, y, z) + \delta n_{ac}(x, y, z) \cos\left[\frac{2\pi}{\Lambda}z + \phi(z)\right] \tag{85}$$

where δn_{dc} and δn_{ac} are the dc and ac index change that vary slowly compared to the grating period, Λ. An index change with a raised Gaussian profile is depicted in Figure 4-53. Any chirp of the grating period is included via the term $\phi(z)$. Chirp is discussed further in Chapter 5 for the development of coupled-mode solutions for counter-propagating modes. Let the transverse electric field be described by

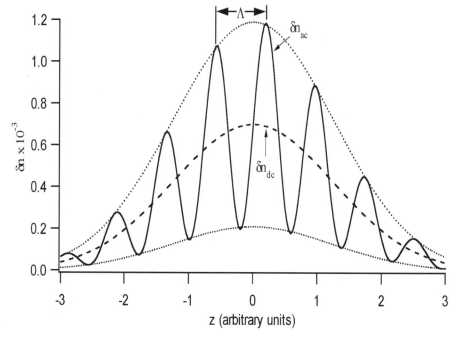

FIGURE 4-53   A refractive index profile for co-propagating mode coupling.

$$\vec{E}(x, y, z, t) = [\vec{e}_a(x, y)A(z)e^{-j\beta_a z} + \vec{e}_b(x, y)B(z)e^{-j\beta_b z}]e^{+j\omega t} \qquad (86)$$

where $A(z)$ and $B(z)$ are the amplitudes of the two coupled waves, z is the propagation distance along the waveguide, and $\vec{e}_{a,b}(x, y)$ are the transverse mode profiles of the unperturbed waveguide(s). For coupling to occur, the difference in propagation constants must be matched by the grating spatial frequency as follows: $\beta_a - \beta_b = 2\pi\Delta n_e/\lambda_c = 2\pi/\Lambda$, where Δn_e is the difference in effective indices of modes a and b at the center wavelength, given by $\lambda_c = \Delta n_e \Lambda$. The coupled-mode equations are derived by substituting Eq. (86) into the Helmholtz equation in Chapter 2. The term with $k_0^2 n_0(x, y)^2$ is replaced by k_0^2 times $(n_0 + \delta n)^2 \approx n_0^2 + 2n_0\delta n$, where $n_0(x, y)$ is the unperturbed refractive index distribution and $\delta n(x, y, z)$, given by Eq. (85), is the index perturbation. The amplitudes of the two modes are assumed to vary slowly, so the second derivatives of $A(z)$ and $B(z)$ are neglected. The synchronous (phase-velocity matched) terms contribute to coupling, so the next step is to group terms with similar exponents. A wavelength-dependent variable δ is defined for convenience; it describes the deviation of the wavelength from λ_c [64].

$$\delta \equiv \frac{1}{2}\left(\beta_a - \beta_b - \frac{2\pi}{\Lambda}\right) = \pi\Delta n_e\left(\frac{1}{\lambda} - \frac{1}{\lambda_c}\right) \qquad (87)$$

Two equations result with exponential terms of $\exp(+j\delta)$ and $\exp(-j\delta)$. One equation is multiplied by $\vec{e}_a^*(x, y)$ and the other by $\vec{e}_b^*(x, y)$. Then, they are integrated over x and y. Three coupling coefficients are defined after the integration [64].

$$\sigma_{aa} \equiv k_0 \delta \bar{n}_{dc} \int\int_A |\vec{e}_a(x, y)|^2 \, dxdy \tag{88}$$

$$\sigma_{bb} \equiv k_0 \delta \bar{n}_{dc} \int\int_A |\vec{e}_b(x, y)|^2 \, dxdy \tag{89}$$

$$\kappa_{ab} = \kappa_{ba}^* \equiv \frac{k_0 \delta \bar{n}_{ac}}{2} \int\int_A \vec{e}_a(x, y) \cdot \vec{e}_b^*(x, y) dxdy \tag{90}$$

where $\delta \bar{n}_{dc}$ and $\delta \bar{n}_{ac}$ represent a uniform index change across an area A. A normalization of $\int\int_{A_\infty} |\vec{e}_{a,b}(x, y)|^2 \, dxdy = 1$ was chosen for $\vec{e}_a$ and $\vec{e}_b$. Equations (88) and (89) describe coupling arising from the dc portion of the index perturbation, while Eq. (90) describes coupling from the ac portion. To simplify the resulting equations, the mode amplitudes are recast in terms of $R(z) \equiv A(z)e^{+j(\sigma_{aa}+\sigma_{bb})z/2}e^{-j(\delta z - \phi/2)}$ and $S(z) \equiv B(z)e^{+j(\sigma_{aa}+\sigma_{bb})z/2}e^{+j(\delta z - \phi/2)}$. The coupled-mode equations are then given by [64]

$$\frac{dR}{dz} = -j\hat{\sigma}R(z) - j\kappa^*S(z) \tag{91}$$

$$\frac{dS}{dz} = +j\hat{\sigma}S(z) - j\kappa R(z) \tag{92}$$

where $\kappa = \kappa_{ab}$ and

$$\hat{\sigma} \equiv \delta + \frac{\sigma_{aa} - \sigma_{bb}}{2} - \frac{1}{2}\frac{d\phi}{dz}$$

Analytic solutions can be found for a uniform grating where $\kappa(z)$, $\phi(z)$, $\sigma_{aa}(z)$ and $\sigma_{bb}(z)$ are constants. As an example, consider the coupler in Figure 4-54. The

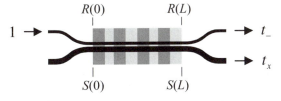

FIGURE 4-54   Grating-assisted mode coupling for a coupler with very different waveguide widths. The index perturbation δn is indicated by the shaded regions.

waveguides have very different widths, so the nominal coupling is quite small. By introducing a periodic index variation whose period matches the difference in propagation constants of the two waveguides, coupling is assisted. The bar-state t_- and cross-state t_x amplitude transmission solutions for a uniform grating of length L are given by

$$t_- = \left.\frac{R(L)}{R(0)}\right|_{S(0)=0} = \cos(\gamma L) - j\frac{\hat{\sigma}}{\gamma}\sin(\gamma L) \tag{93}$$

$$t_x = \left.\frac{S(L)}{R(0)}\right|_{S(0)=0} = -j\frac{\kappa}{\gamma}\sin(\gamma L) \tag{94}$$

where $\gamma = \sqrt{\kappa^2 + \hat{\sigma}^2}$ and the boundary conditions were assumed to be $R(0) = 1$ and $S(0) = 0$. The maximum cross-coupling occurs at $\hat{\sigma} = 0$, corresponding to a wavelength of

$$\lambda_{\max} = \frac{\lambda_c}{1 - (\sigma_{aa} - \sigma_{bb})\Lambda/2\pi}$$

and is given by $|t_x(\lambda_{\max})|^2 = \sin^2(\kappa L)$. This result is equivalent to Eq. (70) for an N-stage uniform lattice filter with $N\theta_n \quad \kappa L$. The bandwidth, defined by the first zeros on either side of $\lambda_{\max}$ for the cross-state transmission of a uniform grating, is given by [64]

$$\frac{\Delta\lambda}{\lambda} = \frac{2\lambda}{\Delta n_g L}\sqrt{1 - \left(\frac{\kappa L}{\pi}\right)^2} \tag{95}$$

where $\Delta\lambda$ is the bandwidth and $\kappa L \le \pi$. For a weak grating, $\Delta\lambda/\lambda = 2/N$ where N is the number of periods. The center wavelength can be tuned over a significant range by inducing small changes in Δn since $\delta\lambda_c/\lambda_c = \delta\Delta n/\Delta n$ for a constant Λ. For bandpass filters, we want $|t_x(\lambda_{\max})|^2 = 1$, so $\kappa L = \pi/2$ gives the shortest filter and minimizes the sidelobes, which increase with κL [64].

To analyze a nonuniform grating, one can solve the coupled-mode equations directly, or segment the grating into approximately uniform sections and multiply the solutions of each section. In the latter case, the solution for a uniform grating is conveniently written in a transfer matrix form for easy concatenation. The transfer matrix using the coupled-mode solutions [64] is given as follows:

$$\begin{bmatrix} R_{n+1} \\ S_{n+1} \end{bmatrix} = \begin{bmatrix} F_{11}^n & F_{12}^n \\ F_{21}^n & F_{22}^n \end{bmatrix}\begin{bmatrix} R_n \\ S_n \end{bmatrix} \tag{96}$$

where

$$F_{11}^n = \cos(\gamma_n\,\Delta z) - j\frac{\hat{\sigma}_n}{\gamma_n}\sin(\gamma_n\,\Delta z) \tag{97}$$

$$F_{12}^n = -j\frac{\kappa_n}{\gamma_n}\sin(\gamma_n \, \Delta z) \qquad (98)$$

$$F_{21}^n = -j\frac{\kappa_n}{\gamma_n}\sin(\gamma_n \, \Delta z) \qquad (99)$$

$$F_{22}^n = \cos(\gamma_n \, \Delta z) + j\frac{\hat{\sigma}_n}{\gamma_n}\sin(\gamma_n \, \Delta z) \qquad (100)$$

The product of transfer matrices, $F = F^N F^{N-1} \cdots F^1$, yields the overall response. The grating's amplitude cross-state and bar-state transmission are given by $t_x = F_{21}$ and $t_- = F_{11}$. For large bandwidths, the waveguide modal dispersion must be included.

Acousto-optic filters use surface acoustic waves (SAWs) in x-cut y-propagating lithium niobate, which has good acousto-optic and piezoelectric figures of merit, to create a periodic index grating by the photoelastic effect that couples the two polarizations of the fundamental mode. The acoustic wave is generated using interdigitated transducers (IDT). For a birefringence of $\Delta n = 0.08$ at 1550 nm, the required period of the index change is $\Lambda = 20$ μm for phase matching [65]. The acoustic period depends on the SAW velocity (V_s) and frequency (f_s) as $\Lambda_a = V_s/f_s$ where $V_s = 3.7$ km/s for LiNbO$_3$. For $\Delta n \Lambda_a = 1550$ nm, $f_s = 175$ MHz. The passband width is limited by the number of periods, $\Delta\lambda_{FWHM}/\lambda = 0.8\lambda/n_g L$ where L is the device length. The switching time is proportional to the acoustic transit time, given by $\tau = L/V_s$. For $L = 20$ mm and $\lambda = 1550$ nm, $\Delta\lambda = 1.5$ nm, $\tau = 6$ ms, and the RF power for switching is 10 mW [65]. Since the index grating is induced by a traveling acoustic wave, the phase match condition includes a Doppler shift equal to f_s. By changing the frequency f_s of the drive, the period Λ changes and the center wavelength of the filter shifts. To reduce the sidelobes in the filter response, the induced index change $\delta n(z)$ is tapered in a manner equivalent to using different window functions to reduce sidelobes in digital MA filters. This tapering is referred to as apodization. The transfer matrix approach defined by Eqs. (96)-(100), was used to model the transmission for a uniform grating versus one with a Gaussian profile truncated to a length equal to twice the FWHM. The grating length was chosen to be 2 cm, or 1000 periods long. The coupling coefficients were chosen so that $\sum_{n=1}^N \kappa_n L_n = \pi/2$ where N was the number of uniform segments for the Gaussian profile. The cross-state transmission is shown in Figure 4-55 and the bar-state transmission in Figure 4-56. Although the sidelobes are greatly reduced for the Gaussian profile, the $\Delta\lambda_{FWHM}$ is larger (2 nm compared to 1.2 nm for the uniform profile).

Long period gratings (LPG) [66] couple light from the fundamental mode to one of the cladding modes of an optical filter, thereby producing a wavelength dependent loss for the fundamental mode. An important application is for realizing gain equalization filters [67]. A periodic index variation is induced in a fiber, typically by UV exposure of a fiber with a Ge-doped core. The period is inversely proportional to the difference in the effective indices of the fundamental and a particular

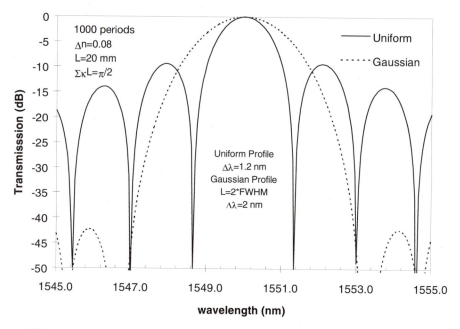

FIGURE 4-55   Cross-state transmission response for a coupled-mode MA filter with a uniform coupling strength profile compared to one with a Gaussian profile.

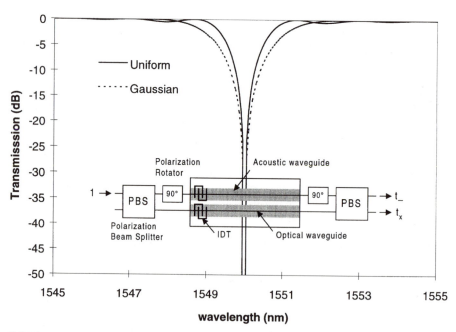

FIGURE 4-56   Bar-state transmission for the coupled-mode MA filter. An acousto-optic filter schematic with polarization diversity is shown in the inset.

cladding mode, $\Lambda_p = \lambda_c/(n_f - n_{cl}^P)$ where n_f is the effective index of the fundamental mode and n_{cl}^P is the effective index of the pth cladding mode. Typical periods for dispersion shifted fiber range from 200 to 600 μm. The out-of-band losses for the fundamental mode are small, typically < 0.2 dB [66]. The choice of cladding modes, the grating length, and the apodization profile for the coupling strength are used to tailor the filter shape. Gratings with different periods are concatenated so that coupling to multiple cladding modes may be used to approximate a desired filter response. A gain response with <1 dB ripple was achieved for an erbium doped fiber amplifier with 22 dB gain over a 40 nm bandwidth using long period grating filters

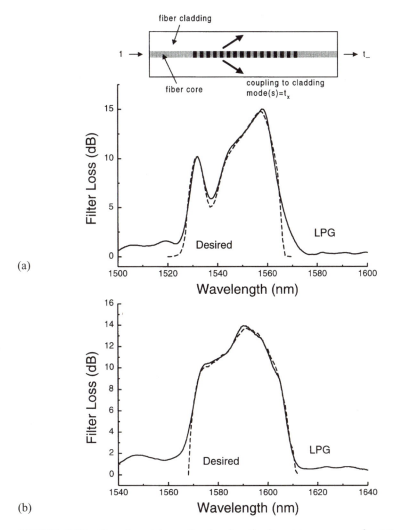

(a)

(b)

FIGURE 4-57   Experimental results showing the bar-state response for LPG filters designed to equalize the amplifier gain over two 40 nm bandwidths [69]. The principle of operation is illustrated in the inset.

[68]. By splitting the incoming signal into two bands, then amplifying and recombining them, an amplifier with an effective bandwidth >80 nm was demonstrated. Six long period grating (LPG) filters were used to equalize the gain over each band. The desired and measured spectra for the long period grating filters are shown in Figure 4-57 [69].

REFERENCES

1. N. Takato, T. Kominato, A. Sugita, K. Jinguji, H. Toba, and M. Kawachi, "Silica-Based Integrated Mach–Zehnder Multi/Demultiplexer Family with Channel Spacing of 0.01–250 nm," IEEE J. Selected Areas Commun., vol. 8, no. 6, pp. 1120–1127, 1990.

2. K. Jinguji, N. Takato, A. Sugita, and M. Kawachi, "Mach–Zehnder interferometer type optical waveguide coupler with wavelength-flattened coupling ratio," Electron. Lett., vol. 16, no. 17, pp. 1326–1327, 1990.

3. N. Takato, K. Jinguji, M. Yasu, H. Toba, and M. Kawachi, "Silica-Based Single-Mode Waveguides on Silicon and Their Application to Guided-Wave Optical Interferometers," J. Lightw. Technol., vol. 6, no. 6, pp. 1003–1010, 1988.

4. B. Verbeek, C. Henry, N. Olsson, K. Orlowsky, R. Kazarinov, and B. Johnson, "Integrated Four-Channel Mach–Zehnder Multi/Demultiplexer Fabricated with Phosphorous Doped SiO_2 Waveguides on Si," J. Lightw. Technol., vol. 6, no. 6, pp. 1011–1015, 1988.

5. S. Suzuki, Y. Inoue, and T. Kominato, "High-Density Integrated 1 × 16 FDM Multi/Demultiplexer," IEEE Lasers and Electro-Optics Society (LEOS) 1994 7th Annual Meeting. Boston, MA, October 31 - November 3, 1994, pp. 263–264.

6. K. Sasayama, M. Okuno, and K. Habara, "Coherent optical transversal filter using silica-based single-mode waveguides," Electron. Lett., vol. 25, no. 22, pp. 1508–1509, 1989.

7. K. Sasayama, M. Okuno, and K. Habara, "Coherent Optical Transversal Filter Using Silica-Based Waveguides for High-Speed Signal Processing," J. Lightw. Technol., vol. 9, no. 10, pp. 1225–1230, 1991.

8. K. Sasayama, M. Okuno, and K. Habara, "Photonic FDM Multichannel Selector Using Coherent Optical Transversal Filter," J. Lightw. Technol., vol. 12, no. 4, pp. 664–669, 1994.

9. H. Rowland, "On concave gratings for optical purposes," Amer. J. Sci., vol. 3, no. 26, pp. 87–98, 1883. Also, H. Rowland, Phil. Mag., 13, p. 467, 1882.

10. M. Klein and T. Furtak, Optics. New York: John Wiley & Sons, 1986.

11. J. Soole, A. Scherer, H. Leblanc, N. Andreadakis, R. Bhat, and M. Koza, "Monolithic InP-based grating spectrometer for wavelength-division multiplexed systems at 1.5 μm," Electron. Lett., vol. 27, no. 2, pp. 132–134, 1991.

12. P. Clemens, G. Heise, R. Marz, H. Michel, A. Reichelt, and H. Schneider, "Flat-field spectrograph in SiO_2/Si," IEEE Photon. Technol. Lett., vol. 4, pp. 886–887, 1992.

13. E. Koteles, J. He, B. Lamontagne, A. Delage, L. Erickson, G. Champion, and M. Davies, "Recent Advances in InP-Based Waveguide Grating Demultiplexers," in 1998 OSA Technical Digest Series Vol. 2. Optical Fiber Communication Conference. San Jose, CA, February 22–27, 1998, pp. 82–83.

14. J. Wagener, T. Strasser, J. Pedrazzani, J. DeMarco, and D. DiGiovanni, "Fiber grating op-

tical spectrum analyzer tap," European Conf. on Optical Communications (ECOC), 1997, pp. 65–68, PD V.5.

15. T. Erdogan and J. Sipe, "Tilted Fiber Phase Gratings," *J. Opt. Soc. Am. A,* vol. 13, no. 2, pp. 296, 1996.

16. C. Madsen, J. Wagener, T. Strasser, M. Milbrodt, E. Laskowski, and J. DeMarco, "Planar Waveguide Grating Optical Spectrum Analyzer," in Optical Society of America Technical Digest Vol. 4, Integrated Photonics Research Conf. Victoria, Canada, March 29-April 3, 1998, pp. 99–101.

17. M. Smit, "New Focusing and Dispersive Planar Component Based on an Optical Phased Array," *Electron. Lett.,* vol. 24, no. 7, pp. 385–386, 1988.

18. H. Takahashi, S. Suzuki, K. Kato, and I. Nishi, "Arrayed-Waveguide Grating for Wavelength Division Multi/Demultiplexer With Nanometre Resolution," *Electron. Lett.,* vol. 26, no. 2, pp. 87–88, 1990.

19. C. Dragone, "An $N \times N$ Optical Multiplexer Using A Planar Arrangement of Two Star Couplers," *IEEE Photonics Technol. Lett.,* vol. 3, no. 9, pp. 812–815, 1991.

20. C. Dragone, C. Henry, I. Kaminow, and R. Kistler, "Efficient multichannel integrated optics star coupler on silicon," *IEEE Photon. Technol. Lett.,* vol. 1, no. 8, pp. 241–243, 1989.

21. H. Takahashi, K. Oda, H. Toba, and Y. Inoue, "Transmission characteristics of arrayed waveguide $N \times N$ wavelength multiplexer," *J. Lightw. Technol.,* vol. 13, no. 3, pp. 447–455, 1995.

22. M. Smit, "Integrated Optics in Silicon-based Aluminum Oxide," PhD Dissertation. The Netherlands: Delft University of Technology, 1991.

23. M. Smit and C. Van Dam, "PHASAR-Based WDM-Devices: Principles, Design and Applications," *IEEE J. Selected Topics Quant. Electron.,* vol. 2, no. 2, pp. 236–250, 1996.

24. R. Adar, C. Henry, C. Dragone, R. Kistler, and M. Milbrodt, "Broad-Band Array Multiplexers Made with Silica Waveguides on Silicon," *J. Lightw. Technol.,* vol. 11, no. 2, pp. 212–219, 1993.

25. Y. Li and C. Henry, "Silicon optical bench waveguide technology," in Optical Fiber Telecommunications IIIB, I. Kaminow and T. Koch, Eds. San Diego: Academic Press, 1997, pp. 319–376.

26. Y.P. Li and L.G. Cohen, "Planar Waveguide DWDMs for Telecommunications: Design and Tradeoffs," 13th Annual National Fiber Optic Engineers Conference (NFOEC), San Diego, CA, September 22–24, 1997.

27. M. Amersfoot, C. de Boer, F. van Ham, M. Smit, P. Demeester, J. van der Tol, and A. Kuntze, "Phased-array wavelength demultiplexer with flattened wavelength response," *Electron. Lett.,* vol. 30, no. 4, pp. 300–302, 1994.

28. C. Dragone, "Frequency routing device having a wide and substantially flat passband," U.S. Patent, no. 5,412,744, Issued May 1995.

29. M. Amersfoot, J. Soole, H. Leblanc, N. Andreadakis, A. Rajhel, and C. Caneau, "Passband broadening of integrated arrayed waveguide filters using multimode interference couplers," *Electron. Lett.,* vol. 32, no. 5, pp. 449–451, 1996.

30. K. Okamoto and A. Sugita, "Flat Spectral Response Arrayed-Waveguide Grating Multiplexer with Parabolic Waveguide Horns," *Electron. Lett.,* vol. 32, no. 18, pp. 1661–1663, 1996.

31. C. Dragone, "Frequency routing device having a spatially filtered optical grating for providing an increased passband width," U.S. Patent, no. 5,467,418, Nov. 14, 1995.

32. K. Okamoto and H. Yamada, "Arrayed-waveguide grating multiplexer with flat spectral response," *Opt. Lett.,* vol. 20, no. 1, pp. 43–45, 1995.

33. C. Dragone, T. Strasser, G. Bogert, L. Stulz, and P. Chou, "Waveguide grating router with maximally flat passband produced by spatial filtering," *Electron. Lett.,* vol. 33, pp. 1312–1314, 1997.

34. C. Dragone, "Frequency routing device having a width and substantially flat passband," U.S. Patent, no. 5,488,680, Jan. 31, 1996.

35. G. Thompson, R. Epworth, C. Rogers, S. Day, and S. Ojha, "An original low-loss and pass-band flattened SiO$_2$ on Si planar wavelength demultiplexer," in Vol. 2 1998 OSA Technical Digest Series (Optical Society of America, Washington, DC, 1998), Optical Fiber Communication Conference1998, p. 77.

36. B. Eggleton, G. Lenz, N. Litchinitser, D. Patterson, and R. Slusher, "Implications of fiber grating dispersion for WDM communication systems," *IEEE Photon. Technol. Lett.,* vol. 9, no. 10, pp. 1403–1405, 1997.

37. H. Uetsuka, K. Akiba, H. Okano, Y. Kurosawa, and K. Morosawa, "Novel 1 × N guided-wave multi/demultiplexer for FDM," Vol. 8 of 1995 Technical Digest Series (Optical Society of America, Washington, DC), Optical Fiber Communication Conference1995.

38. Y. Tachikawa, Y. Inoue, M. Ishii, and T. Nozawa, "Arrayed-Waveguide Grating Multiplexer with Loop-Back Optical Paths and Its Applications," *J. Lightw. Technol.,* vol. 14, no. 6, pp. 977–984, 1996.

39. H. Li, C. Lee, W. Lin, and S. Didde, "8-wavelength Photonic Integrated 2 × 2 WDM Cross-connect Switch Using 2xn Phased-array Wave-guide Grating (pawg) Multi/demultiplexers," *Electron. Lett.,* vol. 33, no. 7, pp. 592–593, 1997.

40. H. Li, C. Lee, S. Zhong, Y. Chen, M. Dagenais, and D. Stone, "Multiwavelength integrated 2 × 2 optical cross-connect switch and lambda-partitioner with 2xN phased-array waveguide gratingin self-loopback configuration," in Vol. 2 1998 OSA Technical Digest Series (Optical Society of America, Washington, DC, 1998), Optical Fiber Communication Conference1998, pp. 79–80.

41. M. Zirngibl, C. Joyner, and B. Glance, "Digitally tunable channel dropping filter/equalizer based on waveguide grating router and optical amplifier integration," *IEEE Photon. Technol. Lett.,* vol. 6, no. 4, pp. 513–515, 1994.

42. M. Zirngibl, C. Joyner, L. Stulz, U. Koren, M. Chien, M. Young, and B. Miller, "Digitally tunable laser based on the integration of a waveguide grating multiplexer and an optical amplifier," *IEEE Photon. Technol. Lett.,* vol. 6, no. 4, pp. 516–518, 1994.

43. C. Doerr, M. Zirngibl, and C. Joyner, "Chirping of the waveguide grating router for free-spectral-range mode selection in the multifrequency laser," *IEEE Photon. Technol. Lett.,* vol. 4, pp. 500–502, 1996.

44. C. Doerr, C. Joyner, L. Stulz, and R. Monnard, "Wavelength-division multiplexing cross connect in InP," *IEEE Photon. Technol. Lett.,* vol. 10, no. 1, pp. 117–119, 1998.

45. L. Lierstuen and A. Sudbo, "8-channel wavelength division multiplexer based on multimode interference couplers," *IEEE Photon. Technol. Lett.,* vol. 7, no. 9, pp. 1034–1036, 1995.

46. P. Besse, M. Bachmann, C. Nadler, and H. Melchior, "The integrated prism interpretation of multileg Mach–Zehnder interferometers based on multimode interference couplers," *Optical Quantum Electron.,* vol. 27, pp. 909–920, 1995.

47. M. Paiam and R. MacDonald, "Design of phased-array wavelength division multiplexers using multimode interference couplers," *Appl. Opt.,* vol. 36, no. 21, pp. 5097–5108, 1997.

48. C. Van Dam, M. Amersfoot, G. ten Kate, F. van Ham, M. Smit, P. Besse, M. Bachmann, and H. Melchior, "Novel InP-based phased-array wavelength demultiplexer using a generalized MMI-MZI configuration," Proceedings of the 7th European Conference on Integrated Optics (ECIO). Delft, The Netherlands, 1995, pp. 275–278.

49. P. Besse, M. Bachmann, and H. Melchior, "Phase relations in multi-mode interference couplers and their application to generalized integrated Mach–Zehnder optical switches," 6th European Conference on Integrated Optics, 1993, Vol. 2, pp. 22–23.

50. B. Moslehi, J. Goodman, M. Tur, and H. Shaw, "Fiber-Optic Lattice Signal Processing," *Proc. IEEE,* vol. 72, no. 7, pp. 909–930, 1984.

51. K. Jinguji and M. Kawachi, "Synthesis of Coherent Two-Port Lattice-Form Optical Delay-Line Circuit," *J. Lightw. Technol.,* vol. 13, pp. 72–82, 1995.

52. Y. Li and C. Henry, "Silica-based Optical Integrated Circuits," *IEE Proc.-Optoelectron.,* vol. 143, no. 5, pp. 263–280, 1996.

53. M. Kuznetsov, "Cascaded Coupler Mach–Zehnder Channel Dropping Filters for Wavelength-Division-Multiplexed Optical Systems," *J. Lightw. Technol.,* vol. 12, no. 2, pp. 226–230, 1994.

54. P. Yeh, Optical Waves in Layered Media. New York: John Wiley & Sons, 1988.

55. K. Takiguchi, K. Jinguji, K. Okamoto, and Y. Ohmori, "Variable Group-Delay Dispersion Equalizer Using Lattice-Form Programmable Optical Filter on Planar Lightwave Circuit," *IEEE J. Selected Areas Commun.,* vol. 2, no. 2, pp. 270–276, 1996.

56. K. Takiguchi, K. Okamoto, S. Suzuki, and Y. Ohmori, "Planar Lightwave Circuit Optical Dispersion Equalizer," *IEEE Photonics Technol. Lett.,* vol. 6, no. 1, pp. 86–88, 1994.

57. K. Takiguchi, K. Okamoto, and K. Moriwaki, "Dispersion Compensation Using a Planar Lightwave Circuit Optical Equalizer," *IEEE Photon. Technol. Lett.,* vol. 6, no. 4, pp. 561–564, 1994.

58. K. Takiguchi, K. Jinguji, K. Okamoto, and Y. Ohmori, "Dispersion Compensation Using a Variable Group-Delay Dispersion Equalizer," *Electron. Lett.,* vol. 31, no. 25, pp. 2192–2194, 1995.

59. C. Henry, E. Laskowski, Y. Li, C. Mak, and H. Yaffe, "Monolithic optical waveguide filters based on Fourier expansion," United States Patent, vol. 5,596,661, Jan. 21, 1997.

60. Y. Li, C. Henry, E. Laskowski, H. Yaffe, and R. Sweatt, "Monolithic optical waveguide 1.31/1.55 μm WDM with −50 dB cross-talk over 100 nm bandwidth," *Electron. Lett.,* vol. 31, no. 24, pp. 2100–2101, 1995.

61. Y. Li, C. Henry, E. Laskowski, C. Mak, and H. Yaffe, "Waveguide EDFA Gain Equalisation Filter," *Electron. Lett.,* vol. 31, no. 23, pp. 2005–2006, 1995.

62. T. Ozeki, "Optical Equalizers," *Opt. Lett.,* vol. 17, no. 5, pp. 375–377, 1992.

63. M. Sharma, H. Ibe, and T. Ozeki, "Optical Circuits for Equalizing Group Delay Dispersion of Optical Fibers," *J. Lightwave Technol.,* vol. 12, no. 10, pp. 1759–1765, 1994.

64. T. Erdogan, "Fiber Grating Spectra," *J. Lightwave Technol.,* vol. 15, no. 8, pp. 1277–1294, 1997.

65. D. Smith, R. Chakravarthy, Z. Bao, J. J. Baran, J.L., A. d'Alessandro, D. Fritz, S. Huang, X. Zou, S. Hwang, A. Willner, and K. Li, "Evolution of the Acousto-optic wavelength routing switch," *J. Lightw. Technol.,* vol. 14, no. 6, pp. 1005–1019, 1996.

66. A. Vengsarkar, P. Lemaire, J. Judkins, V. Bhatia, T. Erdogan, and J. Sipe, "Long Period

Fiber Gratings as Band Rejection Filters," *J. Lightwave Technol.,* vol. 14, no. 1, pp. 58–65, 1996.

67. A. Vengsarkar, J. Pedrazzani, J. Judkins, P. Lemaire, N. Bergano, and C. Davidson, "Long Period Fiber Grating Based Gain Equalizers," *Opt. Lett.,* vol. 21, no. 5, pp. 336–338, 1996.

68. P. Wysocki, J. Judkins, R. Espindola, M. Andrejco, A. Vengsarkar, and K. Walker, "Erbium-doped fiber amplifier flattened beyond 40 nm using long-period grating," Optical Fiber Communication Conference. Dallas, TX: OSA, February 16–21, 1997, PD2.

69. A. Srivastava, Y. Sun, J. Sulhoff, C. Wolf, M. Zirngibl, R. Monnard, A. Chraplyvy, A. Abramov, R. Espindola, T. Strasser, J. Pedrazzani, A. Vengsarkar, J. Zyskind, J. Zhou, D. Ferrand, P. Wysocki, J. Judkins, and Y. Li, "1 Tb/s transmission of 100 WDM 10 Gb/s channels over 400 km of TrueWave Fiber," in OSA Technical Digest Series, Vol. 2, Optical Fiber Communications Conference. San Jose, CA, 1998, p. PD10.

PROBLEMS

1. Design a 4-stage lattice filter to compensate a constant dispersion of +500 ps/nm. Each stage should have a delay that is an integer multiple of the unit delay. The passband width (FWHM) should be at least 20 GHz and the peak-to-peak passband dispersion ripple should be <50 ps/nm. Show the magnitude, group delay and dispersion responses.

2. Calculate the resolution, angular dispersion and FSR (angular and temporal) at $\lambda_c = 1550$ nm for a diffraction grating, assuming 600 grooves/mm, a blaze angle of $\alpha = 30°$, $\theta_i = -\theta_m = \alpha$ (known as the Littrow configuration), and a beam width of 10 mm.

3. Design a WGR with a Y-branch receiver waveguide to create a flat passband having a full width at 1 dB down from the peak transmission equal to 40% of the channel spacing. Use a center wavelength of 1550 nm, a FSR = 1600 GHz and a channel spacing of 100 GHz. Calculate the passband loss and loss uniformity.

4. Design a 2-stage 1×9 demultiplexer using 3×3 MMI couplers with uniform splitting ratios. Give the delays and phases required for each arm. Assume a channel spacing of 100 GHz.

CHAPTER FIVE

Multi-Stage AR Architectures

Optical filters with feedback paths and at least one all-pole response are discussed in this chapter. Filters consisting of ring resonators, thin film dielectric stacks, and Bragg gratings are included in this category. In many cases, these filters have both AR and ARMA responses; however, the pole and zero locations of the ARMA response are coupled, which prohibits the realization of a general ARMA response. Filters capable of realizing general ARMA responses are discussed in Chapter 6. A major advantage for AR and ARMA filters is their steeper rolloff and larger stopband rejection characteristics compared to MA filters with the same number of stages. Although a more ideal magnitude response can be achieved, one drawback is that a dispersionless AR or ARMA filter cannot be realized.

Feedback paths may be realized by looping the delay line back on itself to form a ring or by reflecting the signal between two partial reflectors as in a Fabry–Perot cavity. Filters with ring-resonator feedback paths are discussed first, including both cascade and lattice structures. Using the Z-transform description and synthesis algorithm developed for the lattice structure, design examples are presented for bandpass, gain equalization and dispersion compensating filters. The maximum FSR for ring-based filters is limited by the minimum bend radius. The use of Vernier operation to effectively extend the FSR is discussed, along with the limitations that it introduces. Reflective AR filters are addressed next. Unlike the ring-based designs, reflective-cavity filters do not suffer from FSR concerns. The reflectors may be discrete or distributed. Thin film filters consist of discrete reflections originating at the interfaces between different dielectrics and are easily modeled using 2×2 transfer matrices. On the other hand, Bragg gratings are conveniently modeled as distributed reflectors using coupled-mode theory. Both Z-transform and coupled-mode analyses are presented for reflective filters, showing the contribution of each approach to the understanding, design, and analysis of AR optical filters.

5.1   RING CASCADE FILTER

As discussed for MA filters, cascading single-stage AR filters as shown in Figure 5-1 is a straightforward way to realize higher order responses. Only one output of each all-pole filter is connected to the next stage. Filter synthesis is simple because each stage is independent, so the coupling ratios are set according to the desired pole value for each stage. Each stage has an unused input and output port that can serve as post-fabrication test ports.

The transfer function is given in Eq. (1) for an Nth-order, AR cascade filter.

$$H_N(z) = \prod_{n=1}^{N} \left\{ \frac{-\kappa_n \sqrt{\gamma} e^{-j\phi_n} z^{-1}}{1 - c_n^2 \gamma e^{-j\phi_n} z^{-1}} \right\} = \frac{\Gamma_N z^{-N/2}}{A_N(z)} \tag{1}$$

It is the product of single-stage transfer functions as discussed in Chapter 3. As before, the unit delay is $z^{-1} = e^{-j\Omega T}$ where $T = n_g L/c = 1/FSR$ and L is the nominal ring perimeter, which is assumed to be the same for all of the stages. The ring loss, $-20\log_{10}\gamma = \alpha_R L$, is assumed to be identical for each stage. All sources of loss including propagation, bend and transition losses are included in the average loss per unit length, α_R. Phase terms ϕ_n are included for each stage. For filter synthesis, each stage is used to realize one pole of $A_N(z)$. Without limiting the realizable pole locations, we assume that the coupling ratios within each stage are equal. The pole for the nth stage is located at $p_n = c_n^2 \gamma e^{-j\phi_n}$, where $c_n = \sqrt{1 - \kappa_n}$.

The filter operates by injecting light into the filter at X_0 and receiving it at X_N for an odd order filter, as shown in Figure 5-1, and receiving at Y_N for an even order filter. To test an individual stage after fabrication, light can be injected at Y_0 and received at Y_1 to test stage 1. Similarly, the input and output are $X_1 \rightarrow X_2$ for stage 2, and $Y_2 \rightarrow Y_3$ for stage 3.

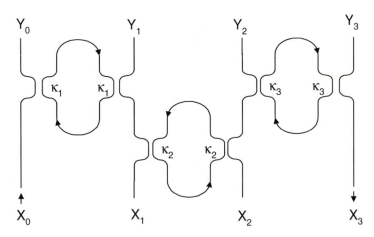

FIGURE 5-1   AR ring cascade filter architecture.

For filter synthesis, we start with a desired response and find an AR function that best approximates it. Then, the poles of the AR function are used to set the coupling coefficients and phase of each stage. As an example, let the desired filter be a fourth-order Chebyshev Type I ARMA filter with a cutoff of 0.1 × FSR and a pass-band ripple of 0.0043 dB, which equates to a stopband rejection of 30 dB for the power complementary response. The cascade architecture does not have a power complementary response; however, the lattice architecture discussed in Section 5.2 does. All the zeros of the Chebyshev Type I response are located at $z = -1$, so the stopband does not have equiripple. Type I more closely resembles an achievable AR response than the Type II, which has an equiripple stopband. The roots for the AR response were determined by starting with the roots of the Type II response and op-timizing their values to minimize the following error criterion:

$$\varepsilon = \sum_i \left[|H_D(\nu_i)| - |H_{AR}^N(\nu_i)| \right]^2 \tag{2}$$

where H_D indicates the desired response, H_{AR}^N is the Nth-order fit, and the sampled frequencies are indicated by ν_i. The results are shown in Figure 5-2. The polynomial coefficients for the Chebyshev design are shown along with the power complemen-tary responses. Note that the AR response decreases smoothly away from the pass-

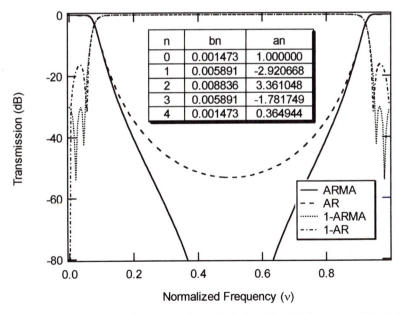

n	bn	an
0	0.001473	1.000000
1	0.005891	-2.920668
2	0.008836	3.361048
3	0.005891	-1.781749
4	0.001473	0.364944

FIGURE 5-2   Comparison of $N = 4$ ARMA Chebyshev Type 1 filter to an $N = 4$ AR re-sponse. Also shown are the power complementary responses and the coefficients for the ARMA filter.

band but is never zero because there are no zeros to make sharp nulls. When the AR response is implemented using a cascade architecture, the peak transmission is <−20 dB as shown in Figure 5-3. The results are shown for both a lossless filter and one with 0.5 dB per feedback path length. Table 5-1 shows the resulting coupling ratios and phase terms for each design. Since an optical amplifier may be required to compensate the loss, another design approach is considered. We now optimize the AR response with respect to the shape and absolute magnitude. The maximum transmission result, denoted by T_{max}, is also shown in Figure 5-3. The peak transmission is now unity for a lossless filter; however, the passband width is significantly smaller. The resulting filter response does not approximate the shape of the desired function very well. This example demonstrates that there is a tradeoff between obtaining a best fit to a desired function and achieving maximum passband transmission for the cascade filter.

This tradeoff is investigated further using the transfer function for a single stage. The ratio of maximum to minimum transmission is given by the following expression:

$$\frac{T_{max}}{T_{min}} = \frac{|H(z)|^2_{z=1}}{|H(z)|^2_{z=-1}} = \frac{(1 + c_n^2 \gamma)^2}{(1 - c_n^2 \gamma)^2} \tag{3}$$

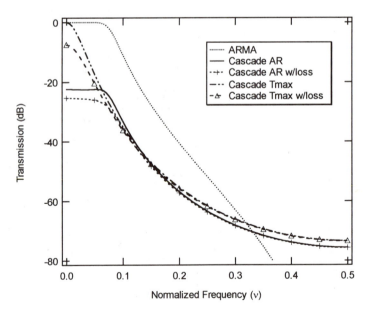

FIGURE 5-3 Comparison of the cascade implementation for the optimized AR response and of a response optimized for peak passband transmission. Both a lossless case and a case with 0.5 dB per feedback path length are shown.

TABLE 5-1   Coupling Ratios and Phases (rads) for Each Stage of the AR Cascade Filter Optimized Designs

Optimized Filter Shape			
N	κ	κ_{loss}	ϕ
1	0.129	0.077	0.445
2	0.129	0.077	−0.445
3	0.309	0.268	0.150
4	0.309	0.268	−0.150
Optimized Transmission ($T_{max} = 1$)			
N	κ	κ_{loss}	ϕ
1	0.192	0.144	0.001
2	0.192	0.144	−0.001
3	0.194	0.146	0.000
4	0.304	0.262	0.000

For a sharp filter response and large rejection, the pole approaches the unit circle, or γ in the lossy case, which occurs as κ_n goes to zero. The maximum transmission for a single stage is given by the following expression:

$$T_{max} = |H(z)|^2_{z=1} = \frac{\kappa_n^2 \gamma}{(1 - c_n^2 \gamma)^2} \tag{4}$$

The limit of Eq. (4) goes to zero for a lossy filter as κ_n goes to zero. This tradeoff between peak transmission and pole magnitude for a lossy cascade filter is independent of the number of stages. When the poles are staggered about the unit circle to provide a wider passband as shown in Figure 5-3, the peak transmission is substantially reduced just as noted in Chapter 4 for the cascade MA filter. The tradeoff between loss and passband width was demonstrated experimentally using a three-stage cascade ring filter and Ge-doped silica waveguides [1]. The feedback path length was 28 mm, $\Delta = 0.75\%$, and the ring loss was 0.25 dB/cm. The phases for each stage were set for two cases: (1) a maximum passband transmission response, and (2) a flat passband response. The results are shown for a single polarization in Figure 5-4. The difference in maximum transmission is 4.5 dB and the difference in the rejection, defined as the ratio of maximum to minimum transmission, is 5 dB between the two responses.

5.2   RING LATTICE FILTER

The MA and AR lattice architectures were first described using Z-transforms for circuits consisting of fiber delay lines and discrete couplers [2,3]. The AR lattice fil-

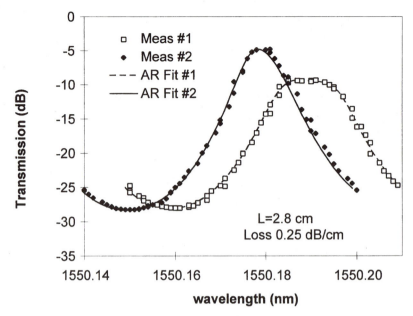

FIGURE 5-4 Measured data and the AR fit for an N=3 cascade ring filter where the phases were adjusted to optimize the passband width or the peak passband transmission.

ter architecture using ring resonators is shown in Figure 5-5a. It consists of N rings with nominally equal perimeters connected by couplers with coupling ratios $\{\kappa_0, \ldots, \kappa_N\}$. This layout avoids waveguide crossings, which lead to undesirable coupling and excess loss. A Z-transform description is presented first, followed by the derivation of a filter synthesis algorithm. The sensitivity of the filter response to fabrication variations is explored next. Filter design examples are then given for bandpass, gain equalization, and dispersion compensation applications.

5.2.1 The Z-Transform Description

We now develop a Z-transform description for the multi-stage AR ring lattice architecture shown in Figure 5-5a [4]. The architecture has two inputs, E_{i1} and E_{i2}, and two outputs, E_{o1} and E_{o2}. The goal is to obtain a description of the transfer functions for each input and output combination.

For the Z-transform description, intermediate signals propagating in the forward $\{T_0, \ldots, T_N\}$ and reverse $\{R_0, \ldots, R_N\}$ directions are labeled in Figure 5-5a. The frequency response is derived starting with the transmission matrix for a single stage. A single stage consists of half a delay followed by a coupler as shown in Figure 5-5b. The phase ϕ_n is split equally between the upper and lower half delays as indicated in the Z-transform schematic of Figure 5-5b. To simplify the notation, let ζ include the phase, unit delay and loss, i.e. $\zeta_n = \gamma e^{-j\phi_n} z^{-1}$. The coupling for the

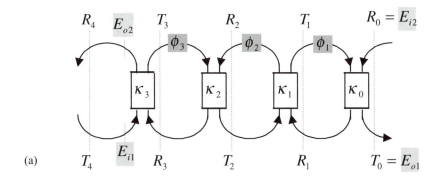

(a)

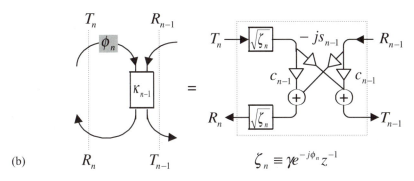

(b)

$$\zeta_n \equiv \gamma e^{-j\phi_n} z^{-1}$$

FIGURE 5-5 AR lattice filter architecture: (a) multi-stage waveguide schematic, and (b) single-stage waveguide schematic and its Z-transform representation.

through-port is denoted by $c_n = \sqrt{1 - \kappa_n}$ and for the cross-port by $-js_n = -j\sqrt{\kappa_n}$ for each directional coupler. The outputs for a single stage are written in terms of the inputs as

$$T_{n-1} = -js_{n-1}\sqrt{\zeta_n}T_n + c_{n-1}R_{n-1} \tag{5}$$

$$R_n = \sqrt{\zeta_n}[c_{n-1}\sqrt{\zeta_n}T_n - js_{n-1}R_{n-1}]$$

The matrix form of Eq. (5) defines the scattering matrix.

$$\begin{bmatrix} R_n \\ T_{n-1} \end{bmatrix} = S_n \begin{bmatrix} T_n \\ R_{n-1} \end{bmatrix} \text{ where } S_n = \begin{bmatrix} c_{n-1}\zeta_n & -js_{n-1}\sqrt{\zeta_n} \\ -js_{n-1}\sqrt{\zeta_n} & c_{n-1} \end{bmatrix} \tag{6}$$

For a lossless filter, S_n is unitary with a determinant equal to $e^{-j(\omega + \phi_n)}$. Note that $s_{12} = s_{21}$ as expected for a reciprocal device. For filter design, the transfer matrix is required, which expresses R_n and T_n in terms of R_{n-1} and T_{n-1}. After some algebraic manipulations of Eq. (5), the transfer matrix for the nth stage is obtained.

$$\Phi_n = \frac{1}{-js_n\sqrt{\zeta_{n+1}}}\begin{bmatrix} 1 & -c_n \\ c_n\zeta_{n+1} & -\zeta_{n+1} \end{bmatrix} \tag{7}$$

$$\text{where } \begin{bmatrix} T_{n+1}(z) \\ R_{n+1}(z) \end{bmatrix} = \Phi_n \begin{bmatrix} T_n(z) \\ R_n(z) \end{bmatrix} \tag{8}$$

for $0 \le n \le N$. Note that the determinant of Φ_n is 1. The relationship between the input and output stages is expressed by concatenating Φ_n's as follows:

$$\begin{bmatrix} T_{n+1}(z) \\ R_{n+1}(z) \end{bmatrix} = \Phi_N \cdots \Phi_0 \begin{bmatrix} T_0(z) \\ R_0(z) \end{bmatrix} \tag{9}$$

Let Φ_{N0} represent the product of matrices $\Phi_N\Phi_{N-1}\cdots\Phi_0$, then the individual terms in the Nth-order transfer matrix are expressed in terms of two polynomials as follows:

$$\Phi_{N0} = \frac{1}{\displaystyle\prod_{n=0}^{N}(-js_n\sqrt{\zeta_{n+1}})}\begin{bmatrix} A_N(z) & B_N^R(z) \\ \zeta_{N+1}B_N(z) & \zeta_{N+1}A_N^R(z) \end{bmatrix} \tag{10}$$

where $A_N(z)$ and $B_N(z)$ are Nth-order polynomials in z^{-1}. For lossless filters, the reverse polynomials are defined in terms of the forward polynomials as follows:

$$A_N^R(z) = (-1)^{N-1}z^{-N}e^{-j\phi_{tot}}A_N^*(z^{*-1}) \tag{11}$$

$$B_N^R(z) = (-1)^{N-1}z^{-N}e^{-j\phi_{tot}}B_N^*(z^{*-1}) \tag{12}$$

where $\phi_{tot} = \sum_{n=1}^{N}\phi_n$. For lossy filters, z^{-1} is replaced by γz^{-1} in Eqs. (11) and (12). The roots of the reverse polynomials are mirror images about the circle of radius γ to the roots of their forward polynomial counterparts.

Assuming no input on E_{i2}, the transfer functions for E_{i1} to each output are expressed as follows:

$$H_{11}(z) = \frac{E_{o1}}{E_{i1}} = \frac{T_0(z)}{\sqrt{\zeta_{N+1}}T_{N+1}(z)} = \frac{(-j)^{N+1}\sqrt{\sigma_N}e^{-j\phi_{tot}}\gamma^N z^{-N}}{A_N(z)} \tag{13}$$

$$H_{21}(z) = \frac{E_{o2}}{E_{i1}} = \frac{R_{N+1}(z)}{\zeta_{N+1}T_{N+1}(z)} = \frac{B_N(z)}{A_N(z)} \tag{14}$$

where $\sigma_N = \prod_{n=0}^{N}\kappa_N$. Note that $H_{11}(z)$ is an AR filter. Since $H_{21}(z)$ has both poles and zeros, it is an ARMA filter. Similarly, assuming no input at E_{i1}, the remaining transfer functions are

$$H_{22}(z) = \frac{E_{o2}}{E_{i2}} = \frac{R_{N+1}(z)}{\sqrt{\zeta_{N+1}}R_0(z)} = H_{11}(z) \tag{15}$$

$$H_{12}(z) = \frac{E_{o1}}{E_{i2}} = \frac{T_0(z)}{R_0(z)} = \frac{-B_N^R(z)}{A_N(z)} \tag{16}$$

The roots of $A_N(z)$ must lie within the unit circle for a stable filter. The location of the poles (and zeros for the ARMA response) are a nonlinear function of the coupling ratios as demonstrated by the A_N and B_N polynomials of zeroth-, first-, and second-order listed below.

$$A_0 = 1$$

$$A_1 = 1 - c_0 c_1 e^{-j\phi_1} \gamma z^{-1}$$

$$A_2 = 1 - c_1(c_1 e^{-j\phi_1} + c_2 e^{-j\phi_2})\gamma z^{-1} + c_0 c_2 e^{-j(\phi_1+\phi_2)}\gamma^2 z^{-2}$$

$$A_0^R = -1$$

$$A_1^R = -c_0 c_1 + e^{-j\phi_1} \gamma z^{-1}$$

$$A_2^R = -c_0 c_2 + c_1(c_2 e^{-j\phi_1} + c_0 e^{-j\phi_2})\gamma z^{-1} - e^{-j(\phi_1+\phi_2)}\gamma^2 z^{-2}$$

$$B_0 = c_0$$

$$B_1 = c_1 - c_0 e^{-j\phi_1} \gamma z^{-1}$$

$$B_2 = c_2 - c_1(c_0 c_2 e^{-j\phi_1} + e^{-j\phi_2})\gamma z^{-1} + c_0 e^{-j(\phi_1+\phi_2)}\gamma^2 z^{-2}$$

$$B_0^R = -c_0$$

$$B_1^R = -c_0 + c_1 e^{-j\phi_1} \gamma z^{-1}$$

$$B_2^R = -c_0 + c_1(e^{-j\phi_1} + c_0 c_2 e^{-j\phi_2})\gamma z^{-1} - c_2 e^{-j(\phi_1+\phi_2)}\gamma^2 z^{-2}$$

Loss is shown explicitly for the forward and reverse polynomials to clarify that only the coupler transmission coefficients are reversed and not the polynomial coefficients, which include the loss and phase. If there are no phase errors, then the coefficients are real valued. The polynomial coefficients depend only on the c_n terms, unlike the MA lattice filter where both the c_n and s_n terms appear.

Any variations in loss from one stage to another can be included as a separate, exponentially decreasing term by replacing the phase term with $e^{-(j\phi_n+\Delta\alpha_n L)}$. The middle coefficient from $A_2(z)$ and $A_2^R(z)$ with nonuniform loss are shown below.

$$a_{1,2} = c_1(c_0 e^{-\Delta\alpha_1 L - j\phi_1} + c_2 e^{-\Delta\alpha_2 L - j\phi_2})\gamma \tag{17}$$

$$a_{1,2}^R = c_1(c_2 e^{-\Delta\alpha_1 L - j\phi_1} + c_0 e^{-\Delta\alpha_2 L - j\phi_2})\gamma \tag{18}$$

Note that only c_2 and c_0 switch places between Eqs. (17) and (18) while $\Delta\alpha_1$ and $\Delta\alpha_2$ must remain associated with ϕ_1 and ϕ_2. Therefore, the reverse polynomial coefficients cannot be defined in terms of the forward ones without prior knowledge of $\Delta\alpha_1$ and $\Delta\alpha_2$. Unless otherwise noted, the loss is assumed to be uniform, and included in γ so that $\Delta\alpha_n = 0$ for $n = 1, \ldots, N$.

5.2.2 Synthesis Algorithm

For AR lattice filter synthesis, the coupling ratios for specific pole locations must be determined. For a first-order filter, the pole is $p_1 = c_0 c_1 \gamma e^{-j\phi_1}$. The H_{21} zero is $z_1 = (c_0/c_1)\gamma e^{-j\phi_1}$, while the zero for H_{12} is $z_1^R = (c_1/c_0)\gamma e^{-j\phi_1}$. Loss decreases the magnitude of the pole and zeros. Note that the phases of the pole and zeros are the same, so a general ARMA response cannot be synthesized! The magnitude of the pole is smaller than the minimum-phase zero. As the order increases, the dependence of the polynomial roots on the coupling ratios and phases becomes more complicated, making a direct filter synthesis approach difficult. Instead, a synthesis algorithm is derived using recursion relations. First, the step-up recursion relations are examined. From Eqs. (7)–(10), we have

$$\begin{bmatrix} A_n(z) & B_n^R(z) \\ B_n(z) & A_n^R(z) \end{bmatrix} = \begin{bmatrix} 1 & -c_n \\ c_n \zeta_{n+1} & \zeta_{n+1} \end{bmatrix} \begin{bmatrix} A_{n-1}(z) & B_{n-1}^R(z) \\ B_{n-1}(z) & A_{n-1}^R(z) \end{bmatrix} \tag{19}$$

Given a set of coupling ratios and phase terms, the Nth-order polynomials are defined recursively.

$$A_N(z) = A_{N-1}(z) - c_N \zeta_N B_{N-1}(z) \tag{20}$$

$$B_N(z) = c_N A_{N-1}(z) - \zeta_N B_{N-1}(z) \tag{21}$$

Equations (20) and (21) can also be expressed in the time domain, using the polynomial coefficients as follows:

$$\underline{a}_N = \begin{bmatrix} \underline{a}_{N-1} \\ 0 \end{bmatrix} - c_N \gamma e^{-j\phi_N} \begin{bmatrix} 0 \\ \underline{b}_{N-1} \end{bmatrix} \tag{22}$$

$$\underline{b}_N = c_N \begin{bmatrix} \underline{a}_{N-1} \\ 0 \end{bmatrix} - \gamma e^{-j\phi_N} \begin{bmatrix} 0 \\ \underline{b}_{N-1} \end{bmatrix} \tag{23}$$

where

$$A_N(z) = a_{0,N} + a_{1,N} z^{-1} + \cdots + a_{N,N} z^{-N} \tag{24}$$

$$\underline{a}_N = [a_{0,N}, a_{1,N}, \cdots, a_{N,N}]^T$$

$$B_N(z) = b_{0,N} + b_{1,N} z^{-1} + \cdots + b_{N,N} z^{-N}$$

$$\underline{b}_N = [b_{0,N}, b_{1,N}, \cdots, b_{N,N}]^T$$

For the zeroth-order polynomials, $a_{0,0} = 1$ and $b_{0,0} = c_0$. Expressions for the first and last coefficients of each polynomial follow from Eqs. (22) and (23).

$$a_{0,N} = 1 \tag{25}$$

$$a_{N,N} = (-1)^N c_0 c_N e^{-j\phi_{tot}} \gamma^N \qquad (26)$$

$$b_{0,N} = c_N \qquad (27)$$

$$b_{N,N} = (-1)^N c_0 e^{-j\phi_{tot}} \gamma^N \qquad (28)$$

The coupling ratio c_N is obtained from $b_{0,N}$, and the total phase can be obtained from either $a_{N,N}$ or $b_{N,N}$. If γ is known *a priori*, then c_0 can be obtained from $b_{N,N}$. Thus, the polynomial $B_N(z)$ plays a key role in determining the coupling ratios.

Given a desired AR function of order N, $H_D(z) = \Gamma_N/A_N(z)$, a method to determine an equivalent realization using couplers and phase terms in a lattice architecture is needed that can easily be extended to higher orders. The problem is similar to one in digital signal processing relating the tap coefficients of the transversal architecture to the reflection coefficients of the lattice architecture, which is solved using the Levinson recursion algorithm [5] as described in Chapter 3. Analogous recursion relations were first developed for multi-cavity reflective optical lattice filters by Dowling and MacFarlane [6]. We start with the algorithm for the ring lattice filter, and then discuss the similarities for the reflective lattice architecture in Section 5.4.1.

For either the ring or reflective architectures, the algorithm is referred to as the modified Levinson algorithm. It is applicable to both synthesizing and analyzing AR lattice filters. As outlined in Figure 5-6 for the ring architecture, the goal is to

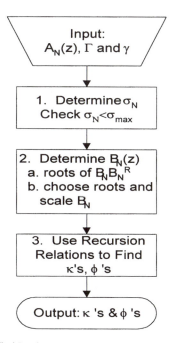

FIGURE 5-6   The modified Levinson algorithm for AR planar waveguide lattice filters.

calculate the κ's and ϕ's to realize an AR lattice filter equivalent of $H_D(z)$. First, a valid $B_N(z)$ is found given $A_N(z)$, then step-down recursion relations are used to reduce the order of $B_N(z)$, finding one coupling ratio and phase term from each order reduction. The inputs are $A_N(z)$, Γ, and γ. The nominal ring loss is required since the pole locations depend on the loss. An underlying assumption is that the $A_N(z)$ polynomial is a valid one, which can be verified by determining that all of its roots lie within the unit circle. For a passive filter, $\Gamma^2 = \sigma_N \gamma^N$ which must be ≤ 1 since σ_N is the product of coupling ratios and $\gamma \leq 1$.

The heart of the algorithm is the definition of the step-down recursion relations. A uniform waveguide loss is assumed, and independent phase terms for each stage are included. By inverting Eq. (19), the following step-down recursion relations are obtained:

$$A_{N-1}(z) = \frac{1}{\kappa_N}[A_N(z) - c_N B_N(z)] \tag{29}$$

$$B_{N-1}(z) = \frac{1}{\kappa_N \zeta_N}[c_N A_N(z) - B_N(z)] \tag{30}$$

The coupling ratio κ_{n-1} is determined from $b_{0,n-1}$ as its order is reduced from $n = N$, ..., 1. However, the following relationship must be met so that the order reduction is satisfied in Eqs. (29) and (30):

$$a_{0,N} a_{N,N} = b_{0,N} b_{N,N} \tag{31}$$

As a check, it is noted that Eqs. (25)–(28) satisfy this relationship. Both $A_N(z)$ and $B_N(z)$, as well as γ and ϕ_N, are needed for the step-down recursion relations.

The first step in synthesizing a desired $A_N(z)$ is to calculate a valid $B_N(z)$. The determinant of Φ_{N0} is equal to the product of the individual determinants of Φ_n for $0 < n < N$, yielding the following relationship:

$$B_N(z)B_N^R(z) = A_N(z)A_N^R(z) + \gamma^N(-1)^N e^{-j\phi_{tot}} \sigma_N z^{-N} \tag{32}$$

Assuming a lossless filter, Eqs. (11) and (12) are substituted into Eq. (32) to yield

$$B_N(z)B_N^*(z^{*-1}) = A_N(z)A_N^*(z^{*-1}) + \sigma_N \tag{33}$$

The gain, σ_N, is needed to solve Eq. (33). If σ_N is known, then $B_N(z)B_N^*(z^{*-1})$ is determined uniquely; otherwise, the physically realizable range of solutions is $0 \leq \sigma_N \leq 1$. Since the maximum transmission cannot exceed unity for a passive filter, the following equation sets an upper bound on σ, denoted by σ_{max}.

$$T_{max} = \frac{\sigma_{max} \gamma^N}{\min\{|A_N(\omega)|^2\}} = 1 \tag{34}$$

Given a value for σ_N, the roots of $B_N(z)B_N^*(z^{*-1})$ are found using Eq. (33). The roots of $B_N(z)B_N^*(z^{*-1})$ occur in reciprocal pairs about the unit circle for a lossless filter. For a lossy filter, the symmetry about the unit circle is broken, but the roots occur in reciprocal pairs about the circle of radius γ. One root from each pair is chosen to define B_N. One choice is to use the minimum-phase roots for the filter synthesis algorithm. These roots define B_N to within a multiplicative factor, denoted by α,

$$B_N(z) = \alpha \prod_{n=1}^{N}(1 - z_n z^{-1}) = \alpha B_N'(z) \tag{35}$$

where z_n designates a root of B_N and $B_N'(z) = 1 + b_{1,N}'z^{-1} + \cdots + b_{N,N}'z^{-N}$. Minimum-phase roots are attractive since the magnitude and group delay are then Hilbert transforms of each other [7]; so given one response, the other can be calculated. Equation (31) allows α to be determined by solving $a_{0,N}a_{N,N} = \alpha^2 b_{0,N}'b_{N,N}'$. Since $a_{0,N} = 1$ and $b_{0,N}' = 1$,

$$\alpha = \sqrt{\frac{a_{N,N}}{b_{N,N}'}} = b_{0,N} = c_N \tag{36}$$

Thus, the coupling ratio $\kappa_N = 1 - c_N^2$ is determined.

To use the step-down recursion relations for B_{N-1}, the phase ϕ_N must be known. The relations in Eqs. (25)–(28) require that the ratio of $a_{N-1,N-1}$ to $b_{N-1,N-1}$ be real and equal to c_{N-1}. This condition provides the following relationship for ϕ_N:

$$\phi_N = \arg[a_{N-1,N-1}] - \arg[\tilde{b}_{N-1,N-1}] \tag{37}$$

where $\tilde{B}_{N-1}(z) = (z/\gamma\kappa_N)[c_N A_N(z) - B_N(z)] = e^{-j\phi_N}B_{N-1}(z)$. In this manner, complete information is obtained for solving the step-down recursion relations. As a check on the output of the algorithm, the following three conditions must be met: (1) the solution produces κ_n's and ϕ_n's which are physically realizable, (2) the sum of the individual phase terms equals ϕ_{tot}, and (3) the product of the coupling ratios equals σ_N.

We choose a second-order filter as an example to demonstrate the synthesis algorithm. Let the magnitude and angle of the desired poles be located at $p_{1,2} = 0.9 \angle \pm\pi/6$ radians, so $A_2(z) - 1 - 1.5588z^{-1} + 0.81z^{-2}$. For an Nth-order AR filter, there are $2N + 1$ inputs (magnitude and phase for N poles and the gain Γ). There are $N + 1$ coupling ratios to be determined and N phase terms. Three values for the gain are considered: (1) $\sigma_2 < \sigma_{max}$, (2) $\sigma_2 = \sigma_{max}$ and (3) $\sigma_2 > \sigma_{max}$. First, a lossless case ($\gamma = 1$) is considered. The different root combinations for $B_2(z)$ are used to calculate the possible solutions, not just the minimum-phase case. For $\sigma_2 = 0.9\sigma_{max}$, different solutions are obtained depending on the choice of roots for $B_2(z)$ as shown in Table 5-2. By reversing the order of coupling ratios and phase terms, the remaining two solutions of the four possible solutions are obtained. The solutions are identical for $\sigma_2 = \sigma_{max} = 9.025 \times 10^{-3}$, with values of $\kappa = \{0.1900, 0.2500, 0.1900\}$ and $\phi = \{0,0\}$.

TABLE 5-2   Solutions for $N = 2$, $\sigma < \sigma_{max}$

κ	ϕ (radians)	roots of $B_2(z)$
0.2437, 0.2509, 0.1325	0, 0	0.9663 $\angle$ ±0.5149
0.1900, 0.2244, 0.1900	+0.1828, −0.1828	0.9663 $\angle$ +0.5149
		1.0349 $\angle$ −0.5149

The polynomial $B_2(z)$ has roots at $1.0 \angle \pm 0.5215$ radians. The reverse polynomial has the identical roots; therefore, all the solutions are identical. For $\sigma_2 = 1.1\sigma_{max}$, the resulting solution is $\kappa = \{0.1900, -0.0617, 0.1900\}$ and $\phi = \{0,0\}$, which is not physically realizable.

Next, a ring loss of 0.5 dB is introduced. When $\sigma_2 = 9.025 \times 10^{-3}$ is used, which is equal to σ_{max} for the lossless case, a valid solution for κ and ϕ is not found. Since T_{max} must be <1 for any ring loss, we conclude that σ_{max} should be calculated for a particular γ by recomputing the coefficients of $A_N(z)$ to remove the ring loss. Then, the condition $T_{max} = 1$ is used to find σ_{max} as indicated in Eq. (38).

$$T_{max} = \frac{\sigma_{max}}{\min\{|A_N^0(\gamma)|^2\}} = 1 \qquad (38)$$

where $A_N^0(\gamma) \equiv \{1, a_{1,N}\gamma^{-1}, a_{2,N}\gamma^{-2}, \cdots, a_{N,N}\gamma^{-N}\}$. Note that σ_{max} is a function of γ. When this method is employed, a value of $\sigma_{max} = 2.078 \times 10^{-3}$ results. The solutions for $\sigma = \sigma_{max}$ are identical with $\kappa = \{0.0912, 0.2480, 0.0912\}$ and $\phi = \{0,0\}$. The roots of $B_2(z)$ are equal to $0.9441 \angle \pm 0.5215$ radians.

The synthesis algorithm is easily tested by starting with a known set of coupling ratios and phases, using the step-up recursion relations to determine $A_N(z)$ and $B_N(z)$, and then applying the modified Levinson algorithm to find the κ_n's and ϕ_n's. Because of the κ_n term in the denominator of the step-down recursion relations, the algorithm is very sensitive to small κ_n, which are typically associated with poles close to the unit circle.

5.2.2.1   *Design Optimization*   For AR lattice filter synthesis, two optimization approaches can be taken: (1) optimize the coupling ratios and phase terms directly, or (2) obtain an optimum AR response first and then calculate the coupling ratios and phase terms. In the first approach, the resulting coupling ratios may not be physically realizable unless a constraint is introduced. Once a stable filter is found in the second approach, physically realizable coupling ratios are given by the modified Levinson algorithm and all 2^N solutions are found. The second approach is described here with the steps outlined in Figure 5-7.

We choose a fourth-order Chebyshev Type II ARMA filter with a cutoff frequency = 0.3*FSR and 30 dB stopband rejection as the desired function, $H_D(z)$. The best fit AR function is determined by minimizing the error defined as the difference be-

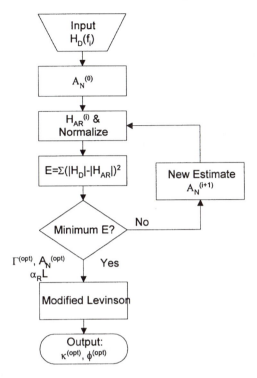

FIGURE 5-7 The AR ring filter synthesis optimization flowchart.

tween H_D and $H_{AR}^N(z) = \Gamma/A_N(z)$ of order $N = 1$, 2 or 3. The gain Γ is chosen for maximum transmission of H_{AR}^N. Figure 5-8 shows the $N = 4$ ARMA response compared to the optimized AR designs of order 1, 2 and 3. Also shown is a best fit $N = 3$ MA design. It is obvious that significantly better rolloff and passband rejection can be achieved with the AR designs over the MA design. Table 5-3 shows the resulting coupling ratios for lossless filters and for filters with 0.5 dB loss/feedback path of order $N = 1$–3. The coupling ratios are symmetric, and the optimal values increase with increasing ring loss. The phase terms ϕ_n are zero in this example. The maximum transmission in the passband and the maximum rejection, defined as the ratio of the maximum to minimum transmission, are also shown. The impact of ring loss is to decrease the peak transmission and the maximum rejection. For lossless filters, 100% passband transmission can be achieved with the lattice architecture in contrast to the cascade architecture! The tradeoff is in the added filter complexity since the stages are coupled instead of independent.

5.2.2.2 Multiple Solutions There are infinitely many coupling ratio and phase term solutions that have the same $A_N(z)$ polynomial, since σ_N gives physically realizable solutions for any value in the range $0 \le \sigma_N \le \sigma_{max}$. Once σ_N is set, there are 2^N solutions for $B_N(z)$. Next, we explore the design space for σ_N;

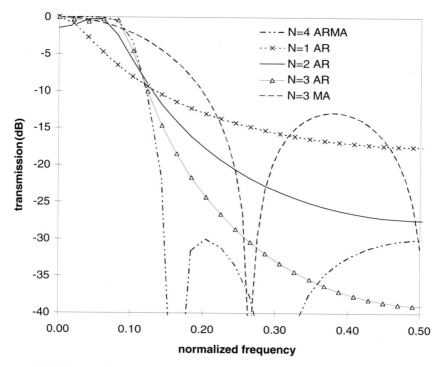

FIGURE 5-8   Comparison of $N = 4$ ARMA to AR and MA optimized designs.

then, we investigate the ARMA magnitude and group delay responses for different choices of $B_N(z)$.

Using $A_3(z)$ defined by $H_{AR}^3(z)$ for the lossless case in Table 5-3, Figure 5-9 demonstrates coupling ratio solutions for different values of σ_3/σ_{max}. All of the solutions have the same $A_3(z)$, but the $B_3(z)$ polynomials are different. The end coupling ratios, κ_0 and κ_3, vary rapidly as $\sigma_3 = \sigma_{max}$ is approached. At $\sigma_3 = \sigma_{max}$, the coupling ratios are symmetric, i.e. $\kappa_0 = \kappa_3$ and $\kappa_1 = \kappa_2$. Figure 5-10 shows that the peak transmission for the AR response increases as the ratio of σ_3/σ_{max} increases, but the shape remains the same. The corresponding ARMA responses are changing since different $B_3(z)$ polynomials result for each value of σ_3. Figure 5-11 shows the change in the roots of $B_3(z)$ as a function of σ_3. The solution that is plotted is the minimum-phase solution. All of the solutions converge at σ_{max}, when all of the roots of $B_3(z)$ are on the unit circle. For a filter with loss, the roots converge to a circle with radius γ. Since the maximum transmission in the passband is typically desired, σ_N can be determined by setting it equal to σ_{max}.

The difference in the ARMA responses between the 2^N solutions is explored using an $N = 2$ example with the solutions shown in Table 5-4. The ring loss is assumed to be 0.25 dB/cm, and the ring perimeter to be 2.7 cm. The notation (x,o) in-

TABLE 5-3   Coupling Ratios for an AR Lattice Filter with and without Loss

No. Stages	Coupling Ratios $\{\kappa_0, \kappa_1, \ldots, \kappa_N\}$	T_{max} (dB)	T_{max}/T_{min} (dB)
1			
No Loss	{0.235, 0.235}	0	17.5
With Loss	{0.246, 0.246}	−1.6	15.5
2			
No Loss	{0.356, 0.145, 0.356}	0	27.3
With Loss	{0.407, 0.151, 0.407}	−1.8	23.9
3			
No Loss	{0.451, 0.138, 0.138, 0.451}	0	39.1
With Loss	{0.548, 0.166, 0.166, 0.548}	−1.9	33.3

dicates that the first root of $B_2(z)$ is minimum-phase and the second root is maximum-phase. The coupling ratios are significantly different for each solution. One solution may be preferable over another if the coupling ratios are smaller, since shorter couplers are less sensitive to fabrication variations. Alternatively, the magnitude or phase response of the ARMA solution may determine the design choice.

In the lossless case, the ARMA magnitude responses are identical. For lossy fil-

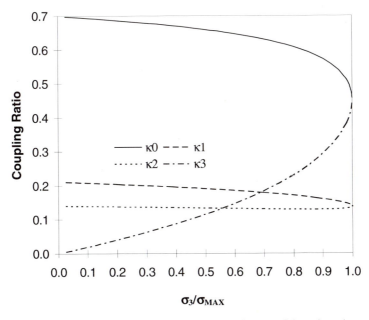

FIGURE 5-9   Levinson recursion solutions as a funtion of the gain σ_3/σ_{max}.

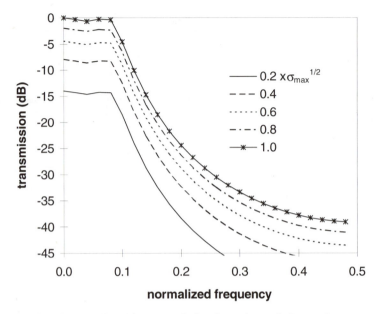

FIGURE 5-10   Resulting transmission for various solutions σ_3/σ_{max}.

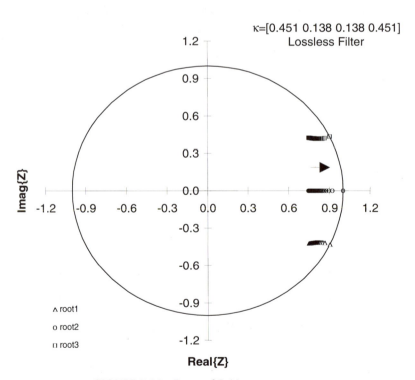

FIGURE 5-11   Roots of $B_3(z)$ versus σ_3.

TABLE 5-4   Design Parameters for a Second-Order Filter to Demonstrate Differences in the ARMA Responses

Solution	$B_2(z)$ roots: $|r| \angle \varphi$ rad	κ's	ϕ's (rad)
1 (x,x)	$0.67 \angle 0.10, 0.42 \angle -0.40$	0.9443, 0.5095, 0.4947	−0.2356, −0.0694
2 (x,o)	$0.67 \angle 0.10, 2.03 \angle -0.40$	0.7320, 0.3633, 0.8950	+0.0270, −0.3320
3 (o,o)	$1.27 \angle 0.10, 2.03 \angle -0.40$	0.4947, 0.5095, 0.9443	−0.0694, −0.2356
4 (o,x)	$1.27 \angle 0.10, 0.42 \angle -0.40$	0.8950, 0.3633, 0.7320	−0.3320, +0.0270

ters, the ARMA magnitude responses are different since the roots are not reciprocal pairs about the unit circle. The ARMA magnitude responses for each solution are plotted in Figure 5-12. The solutions for the (x,o) and (o,x) cases are so close as to be indistinguishable; however, there is a significant difference in the (x,x) and (o,o) solutions. As expected, the maximum-phase solution has a larger stopband rejection since the zeros are closer to the unit circle than the minimum-phase case.

The ARMA group delay characteristics are different for each choice of $B_N(z)$, independent of the loss, since maximum- and minimum-phase zeros produce group delays of opposite signs. Figure 5-13 shows the normalized ARMA group delay responses for each solution. The AR response is also shown for comparison. The actual group delay is obtained by multiplying the normalized group delay by the unit delay. For example, $T = 130$ ps for $n_g = 1.445$ and $L = 2.7$ cm. The group delay for the ARMA passband is more constant for the minimum-phase response, which implies

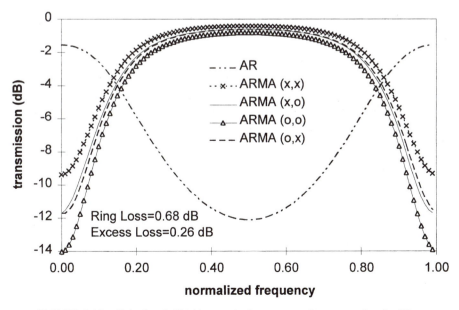

FIGURE 5-12   Calculated ARMA magnitude responses for a second-order filter.

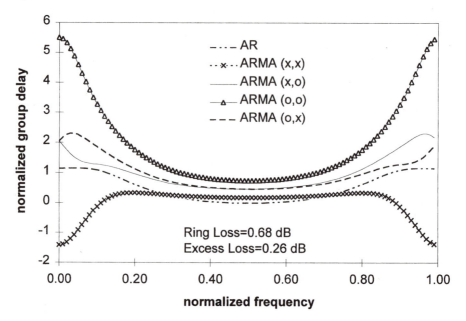

FIGURE 5-13   Calculated ARMA group delay responses for second-order filter.

a smaller dispersion; whereas, it has more of a parabolic response for the maximum-phase case. The group delay response may impact the choice of solutions when dispersion across the ARMA passband must be minimized.

The special case of even symmetry is now investigated. A symmetric filter of odd order, $N = 2M - 1$, is analyzed. The coupling ratios and phases are symmetric about the center-line of the filter, i.e. $\kappa_0 = \kappa_N$, $\kappa_1 = \kappa_{N-1}, \ldots, \kappa_{M-1} = \kappa_M$ and $\phi_1 = \phi_N$, $\phi_2 = \phi_{N-1}, \ldots, \phi_{M-1} = \phi_{M+1}, \ldots, \phi_M$. This symmetry implies that $H_{12}(z) = H_{21}(z)$, so $B_N(z) = -B_N^R(z)$. Thus, $B_N(z)$ is a linear-phase polynomial. The roots of the numerator for the ARMA response are either on the unit circle or occur in conjugate reciprocal pairs about the unit circle for a lossless filter. Therefore, the ARMA and AR group delay responses are equal for a symmetric filter.

The fabrication of a perfectly symmetric filter requires elimination of fabrication variations that introduce asymmetries in the κ's and ϕ's. A folded architecture using a 100% reflector avoids this problem. It has only two ports, so the AR response is reflected from the input port while the ARMA response is transmissive as before. The reflector can be constructed by depositing a high-reflectance thin-film filter on the edge of a planar waveguide chip, inserting a thin-film filter deposited on a thin substrate into a pre-etched trough [8], or constructing a Bragg reflection grating in each waveguide path that crosses the reflector plane. The same transfer matrices are used to describe the fields in the folded architecture as used for the conventional architecture. Figure 5-14 shows a dotted path for the conventional structure and relates the field quantities to the folded structure. The intermediate field quantities T_n

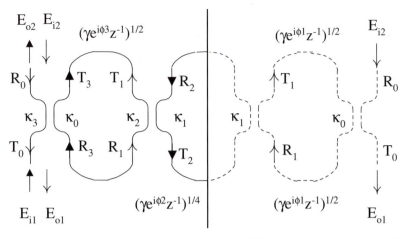

FIGURE 5-14   Relationship of the fields in a folded architecture to those in the general architecture.

and R_n are defined at the coupler output ports on the right side for the conventional architecture, and on the left side for the fields reflected off the mirror and propagating opposite to the conventional architecture. The ARMA output is E_{o2} when E_{i1} is the input. Assuming a 100%, lossless reflector, the relationships between the coupling ratios and phase terms and the polynomial coefficients $A_N(z)$ and $B_N(z)$ are identical to those for the general architecture. Only odd filter orders can be realized using this folded architecture. By using a wavelength dependent reflector, the filter can be tested in a different wavelength band than it operates. The phases and coupling ratios are then determined for each half of the filter by a greatly simplified analysis algorithm, described in Chapter 7. Such a simplification is not afforded by the symmetric MA lattice filter architecture since only the $B(z)$ polynomial is linear-phase; whereas, $A(z)$ is mixed phase in general.

5.2.2.3   *Alternate Design Method Using the Zero Locations*   An alternate design algorithm that uses the zero locations instead of the pole locations was proposed by Orta et al. [9]. Since the maximum transmission in the passband is desired for the AR response, the zeros of the ARMA response for a lossless filter are located on the unit circle by power conservation. The zeros can be obtained from standard ARMA filter designs such as the Butterworth or Chebyshev filters. Once the zero locations are known, $B_N(z)$ is defined to within a multiplicative constant and $A_N(z)$ can be obtained by finding the roots of Eq. (33). Given $B_N(z)$, we need a value for σ_N to determine the roots for $A_N(z)$. A sixth-order, lossless filter was designed [9] using the zero locations for a Chebyshev filter given by

$$z_k = \exp\left(2j \cos^{-1}\left\{ \sin\left(\frac{\Delta\theta_B}{4} \right) \cos\left[\frac{(2k-1)\pi}{2N} \right] \right\} \right) \qquad (39)$$

where $k = 1, \ldots, N$ and $\Delta\theta_B$ is the bandwidth. By choosing $\sigma = 138 \times 10^{-12}$ and $\Delta_B = 0.1\pi$, a 30 dB rejection in the ARMA response is obtained as shown in Figure 5-15. The coupling ratios are shown in the inset. The coupling ratios are small, particularly for the couplers in the center of the filter where 1% coupling is required. A second example illustrates the effect of increasing σ, while using the same zero locations. The spectral responses are shown in Figure 5-16. The rejection is 90 dB across the ARMA stopband, but the rolloff of the passband is much more gradual. For example, the 20 dB width of the previous example is $0.082 \times$ FSR; whereas, it is $0.240 \times$ FSR for the latter case. The coupling ratios are much larger as expected in general for smaller magnitude poles. Smaller σ's correspond to smaller coupling coefficients and therefore sharper responses. In the previous design method, based on determining the desired pole locations, varying σ changed the zero locations and thus the peak transmission, but it did not affect the AR response. In the latter design method, the choice of σ dramatically impacts the filter's AR spectral response but does not change the peak transmission.

5.2.3   Sensitivity to Fabrication Variations

Successful fabrication of AR lattice filters depends on the sensitivity of the important filter characteristics to manufacturing variations. The factors of concern are the ring loss, coupling ratios and effective path lengths for each ring. Each can vary either systematically or randomly. We investigate the sensitivity of the AR lattice filter to fabrication variations using a third-order lattice filter with coupling ratios

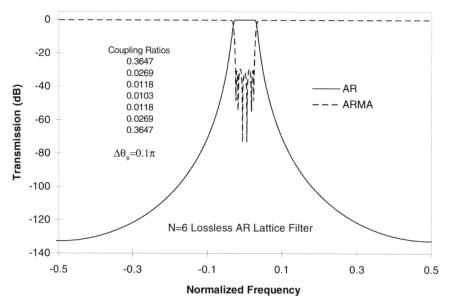

FIGURE 5-15   Sixth order AR bandpass filter (lossless) with 30-dB rejection [9]. Reprinted by permission. Copyright © 1995 IEEE.

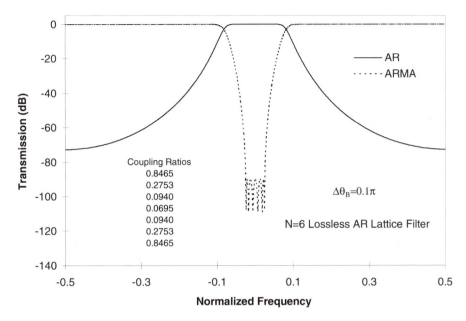

FIGURE 5-16 Sixth-order AR bandpass filter with 90 dB rejection.

$\{0.580, 0.166, 0.166, 0.580\}$ and a loss of 0.5 dB/ring. Random variations are explored using Monte Carlo simulations that add a random variable to each input parameter of interest. The worst cases are determined by the deviation of each individual pole in the AR transfer function from its nominal value. In addition, the norm of the difference between the randomly influenced poles and their nominal values is tracked. The worst case deviations of each individual pole and the largest norm are found over 300 runs.

First, we introduce random variations in the ring loss between stages. The ring loss for each stage is assumed to be uniformly distributed in the range of 0–1.0 dB/ring The worst case deviations for the AR magnitude response are shown in Figure 5-17a. There is very little impact in the resulting passband transmission, passband width or stopband rejection. Figure 5-17b shows the impact of ring loss varying systematically from 0 to 1.0 dB/ring. The peak transmission decreases and the rejection decreases with increasing loss. The passband becomes slightly narrower, and passband ripples that appear in the lossless case are smoothed by the ring loss.

The impact of random coupling ratio variation is evaluated next by adding a Gaussian distributed random variable with zero mean and a standard deviation of $\sigma_\kappa = 3\%$ to each nominal coupling ratio. This level of variation is consistent with typical fabrication variations for silica planar waveguides. Figure 5-18a shows the worst cases. This level of variation can change filter characteristics such as passband width and rejection; however, the general filter shape remains unaltered. Figure 5-18b shows the results for a systematic change in all of the coupling ratios, i.e., $\kappa_n + \Delta\kappa$ for $n = 0, 1, \ldots, N$ and $\Delta\kappa = \{-0.1, -0.05, 0, 0.05, 0.1\}$. This magnitude of

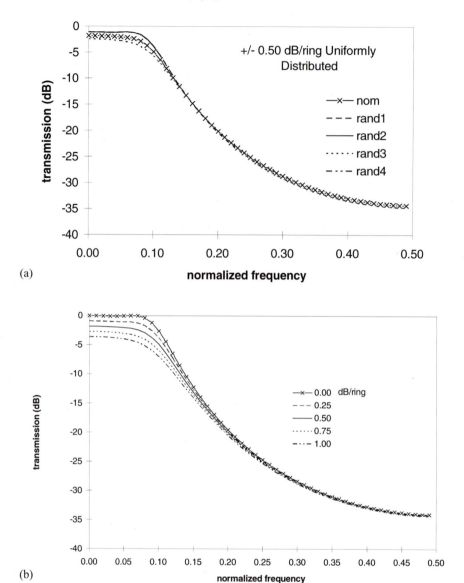

FIGURE 5-17   Impact of random (a) and systematic (b) loss variation on a third-order lattice filter.

systematic change has a dramatic impact on the passband width and the maximum rejection; however, the peak transmission is relatively insensitive.

A systematic variation in the effective path lengths shifts the transfer function with respect to the center wavelength but does not change the shape unless the FSR is significantly affected. In the latter case, there is a compression or expansion of

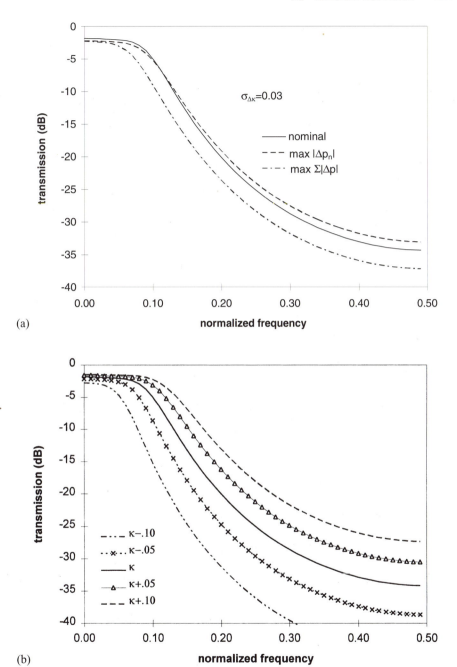

(a)

(b)

FIGURE 5-18   The effect of coupling ratio variation on the magnitude response assuming (a) random variation and (b) systematic variation.

the filter function with wavelength depending on whether the effective length increases or decreases, respectively.

Finally, random effective index variations are simulated by adding a random phase shift to each ring. The phase errors are assumed to be uniformly distributed. Significant changes in the filter function are noted with maximum shifts of $\Delta\phi_n = \pm 30°$. Figure 5-19 shows the results for the worst cases. The filter shape can be significantly impacted. If this variation arises from a change in effective index of refraction that is uniform around the ring, then the maximum phase change $\Delta\phi = \Delta\beta L = 30°$ corresponds to $\Delta n_e \approx 5 \times 10^{-6}$ for $L \approx 25$ mm and $\lambda = 1.55$ μm. A more realistic model is to treat the effective index variations statistically [10]. Assuming that the autocorrelation of $\Delta n_e(x)$ depends on the distance between two points on a ring, the autocorrelation can be approximated by a rectangular function:

$$R_{\Delta n}(\Delta x) = E[\Delta n_e(x)\Delta n_e(x + \Delta x)] = \sigma_{\Delta n}^2 \, \text{rect}\left(\frac{\Delta x}{L_{\text{corr}}}\right) \tag{40}$$

where L_{corr} is the effective correlation length and $\sigma_{\Delta n}^2 = E[\Delta n_e(x)^2]$ is the local index variation. The rectangular function is defined as follows:

$$\text{rect}(x) = \begin{cases} 1 & |x| \leq 0.5 \\ 0 & |x| > 0.5 \end{cases} \tag{41}$$

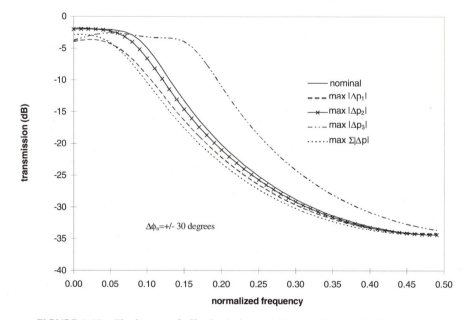

FIGURE 5-19   The impact of effective index variations on the magnitude response.

The relationship between the phase change and index variation is

$$\sigma_{\Delta\phi}^2 = \left(\frac{2\pi}{\lambda}\right)^2 \sigma_{\Delta n}^2 \int_0^L dx \int_0^L \text{rect}\left(\frac{x-x'}{L_{\text{corr}}}\right) dx' \tag{42}$$

which simplifies to the following expression:

$$\sigma_{\Delta\phi} = \frac{2\pi L}{\lambda} \sigma_{\Delta n} \sqrt{\frac{L_{\text{corr}}}{L}} \tag{43}$$

Alternatively, variation in the effective index can be viewed as shifting the center wavelength of each ring relative to the others. The standard deviation of the localized index fluctuations, $\sigma_{\Delta n}$, is related to fluctuations in the center wavelength, $\sigma_{\Delta\lambda}$, as follows:

$$\frac{\sigma_{\Delta\lambda}}{\Delta\lambda_{\text{FSR}}} = \frac{\sqrt{L L_{\text{corr}}}}{\lambda} \sigma_{\Delta n} \tag{44}$$

The phase variations depend on both local index fluctuations and on the correlation length, which is a function of the fabrication process. A coherence length, L_{coh}, of 27 m has been published for a CVD fabrication process [10]. It is related to the correlation length and local index variations by $L_{\text{coh}} = (\lambda^2/2\pi^2)1/(L_{\text{corr}}\sigma_{\Delta n}^2)$.

We can model the impact of core and index fluctuations using one of the waveguide models from Chapter 2. By applying Marcatili's model, variations in the propagation constant ($\Delta\beta = 2\pi\Delta n_e/\lambda$) for a rectangular waveguide were modeled assuming manufacturing deviations of $\pm 0.01\%$ for the core to cladding index difference and ± 0.1 μm for core size. A maximum change of $\Delta n_e \approx 2 \times 10^{-4}$ is predicted for a 6×6 μm^2 waveguide with a core-to-cladding index difference of $\Delta = 0.75\%$ at $\lambda = 1.55$ μm. This index change implies $L_{\text{corr}} \sim 0.1$ μm for $L_{\text{coh}} = 27$ m. This magnitude for L_{corr} provides an order of magnitude improvement over the case with uniform index variation. In practice, waveguide heaters may be included on each filter stage to compensate for phase errors after fabrication. Local index fluctuations increase with increasing Δ. For example, consider a 4.5×4.5 μm^2 core waveguide with $\Delta = 1.5\%$. With the same tolerances on core size and Δ as above, the calculated variation in Δn_e doubles. Therefore, effective index variations are the dominant concern for fabricating AR lattice filters. We see that uniform index changes scale linearly with L while random variations scale as $\sqrt{L}$. In either case, they become more problematic as the length of the feedforward or feedback path increases, i.e. for filters with smaller FSRs.

5.2.4 Bandpass Filter Design and Experimental Results

Bandpass filters have been fabricated using the AR ring lattice architecture in planar waveguides [11]. The design and measurement results are discussed for a second- and third-order filter. For the waveguide layout, the couplers must fit in the rings without adding significant excess loss, and the rings must have the same circumference. One solution is to use the ring configuration shown in Figure 5-20. Since different coupling ratios occur in the same ring, the end of the straight coupling region must intersect the curved region at two different vertical positions. To force the curved section to intersect tangent to the straight coupling region, an ellipse is defined for each half ring with major and minor axis a_N and b_N where N is the coupler number. If, instead, the coupling length were to form a chord that intersected the ring's circumference, then the tangent of the curved region and straight region would be at an angle large enough to contribute unacceptable losses for the longest coupling length. The two elliptical segments meet at the ring top and bottom where a straight section of length ΔL can be interposed to compensate the ring circumference so that all rings in a given filter have the same nominal circumference. Heaters are deposited on each ring to adjust the relative phases of the rings after fabrication. The heaters were 5 mm long with 2.5 mm contact pads as shown in the waveguide layout for the $N = 3$ AR filter in Figure 5-21a. Test couplers were placed adjacent to the filter to independently measure the fabricated coupling ratios.

Filter measurements for the AR magnitude responses after compensating for the fabrication induced phase errors are shown in Figures 5–22 and 5–23, respectively. A detailed discussion of the phase compensation algorithm is given in Chapter 7. The FSR was 64 pm (8 GHz), and the rings had an estimated loss of 0.25 dB/cm. The passband loss of 2.5 dB is significantly smaller than the 9 dB loss for the flat passband cascade response shown in Figure 5-4. The rejection for the lattice responses are within 1 dB of the predicted results based on the fabricated coupling ratio values. Both polarization responses are shown in the same wavelength region; however, for the same order, they are separated by ~0.43 nm due to the waveguide birefringence [12]. The phase errors for each polarization were slightly different, so the phase error could not be simultaneously compensated for both polarizations.

The associated ARMA magnitude responses are shown in Figures 5-24 and 5-25. Both ARMA responses are shown for the $N = 2$ filter. In Figure 5-24, it is evident

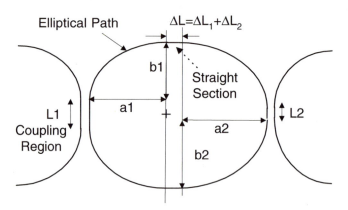

FIGURE 5-20   Coupler and ring configuration for AR lattice ring filters.

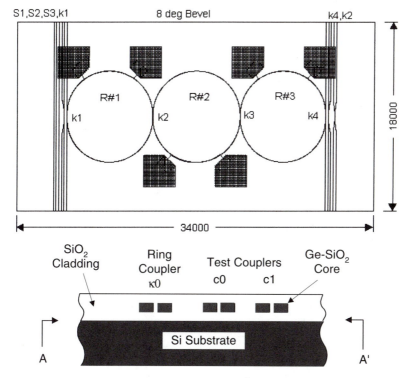

FIGURE 5-21   A third-order AR lattice planar waveguide filter: (a) top view showing the heater pads and (b) side view showing the substrate, core, and cladding areas.

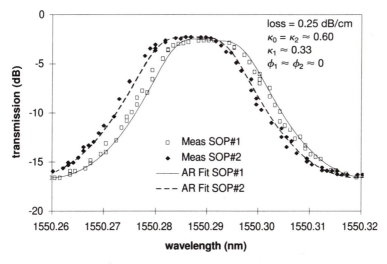

FIGURE 5-22   Measured AR magnitude response for a second-order filter showing both states of polarization (SOP), the test coupler values, and the phases for optimized heater settings.

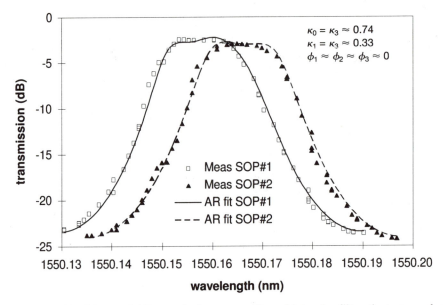

FIGURE 5-23   Measured AR magnitude response for a third-order filter, the test coupler values, and relative phases.

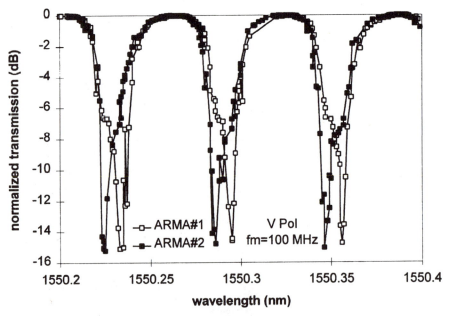

FIGURE 5-24   Measured ARMA magnitude response for a second-order filter.

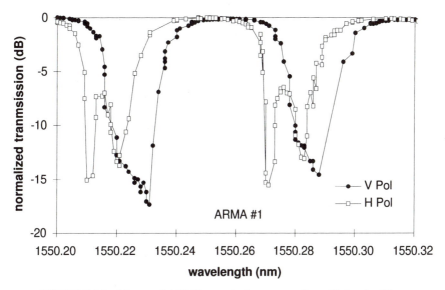

FIGURE 5-25 Measured ARMA magnitude response for a third-order filter.

that one zero is closer to the unit circle than the other. The wavelength of maximum rejection shifts between the ARMA responses as expected from the reciprocal relationship of their zeros. To design an ARMA response with a maximum stopband rejection, the zeros must be on the unit circle. For a lossy filter, the zeros are symmetric around γ, so only a maximum-phase design allows all the zeros to be located on the unit circle. The associated minimum-phase response will have all of its zeros within the unit circle. The rejection for a lossy filter, therefore, can only be optimized for one of the ARMA responses by design. Both polarization responses are shown for a single ARMA response of the $N = 3$ filter in Figure 5-25. The zero locations are slightly different for each polarization.

The group delay ARMA responses are shown in Figures 5-26 and 5-27. For the $N = 2$ filter, the minimum transmission in the rejection band is 12 to 15 dB below the maximum passband transmission, so the group delay measurement is more noisy in this region. The group delays for both polarizations of the $N = 2$ ARMA responses are negative in the rejection band. Negative group delay results from zeros that have a magnitude <1. Poles contribute positive group delay, but the poles are farther away from the unit circle so their contribution to the overall group delay is much smaller than the zeros'. For the $N = 3$ filter, the group delay for the vertical polarization is negative in the rejection band. For the horizontal polarization, which is not optimized, there is positive group delay as expected.

5.2.5 Gain Equalizer Design

Next, we investigate the design of a gain equalization filter using an AR response to best match the desired shape. The design goal is to find the lowest order filter

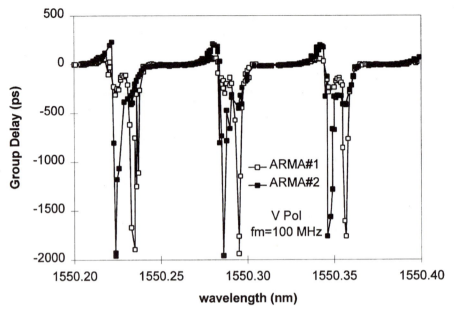

FIGURE 5-26   Measured ARMA group delay responses for the vertical polarization of a second-order filter.

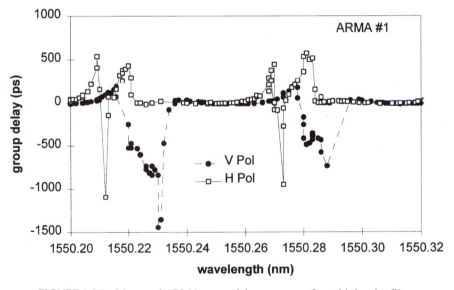

FIGURE 5-27   Measured ARMA group delay responses for a third-order filter.

that fits the desired equalization function to within 0.1 dB over a 20 nm range. The FSR is defined slightly wider than the wavelength range of interest. To use the least squares AR design method discussed in Chapter 3, we first take the inverse Fourier transform of the desired square magnitude response to obtain the autocorrelation function, which is complex in general with Hermitian symmetry. The autocorrelation function is input to a standard Levinson algorithm to obtain Γ_N and the coefficients of $A_N(z)$. Then, the fit $H_{\text{fit}}(z) = \Gamma_N/A_N(z)$ is normalized and compared to the desired response. Using this approach, a fifth-order filter was found to approximate the desired response to within the required tolerance. The spectral response is shown in Figure 5-28, and Table 5-5 lists the design values. The excess loss is quite low, <0.05 dB. Some of the resulting coupling ratios are very close to unity (for example, $\kappa_0 = 0.99998$), which are not practical to fabricate. Some alternatives to obtain more practical coupling ratios include: (1) cascade lower order AR sections and (2) design the filter to equalize several concatenated amplifiers instead of a single amplifier.

An important consideration for bandpass and gain equalization filters is that the filter's FSR must encompass a large number of channels. For a FSR on the order of 20 nm, a ring radius of 13 μm for $n_g = 1.45$ is needed, which requires a large core-to-cladding index difference. Rings with FSRs > 20 nm have been fabricated in Si/SiO$_2$ [13] and AlGaAs/GaAs [14] waveguides. Alternatively, thin film filters and Bragg gratings provide a way to realize an equivalent optical filter, as discussed in Section 5.4, while avoiding FSR limitations due to bend loss.

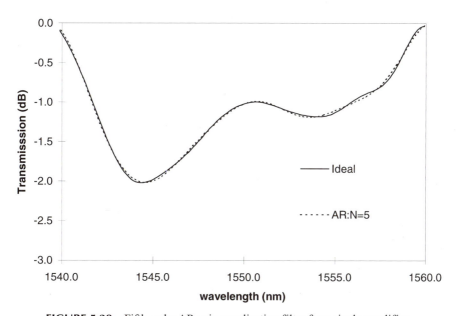

FIGURE 5-28   Fifth-order AR gain equalization filter for a single amplifier.

TABLE 5-5   Design Values for a Fifth-Order AR Gain Equalization Filter for a Single Amplifier

	$n = 0$	1	2	3	4	5
$\sqrt{\sigma}$	0.8816					
$\|a_n\|$	1.0000	0.0628	0.0593	0.0146	0.0061	0.0016
$\angle a_n$	0.0000	−2.4715	2.9015	1.9763	1.8083	−3.0443
$\|b_n\|$	0.4028	0.2005	0.1613	0.0429	0.0159	0.0041
$\angle b_n$	0.0000	−2.9007	2.7767	1.9706	1.9160	−3.0443
κ_n	1.0000	0.9997	0.9981	0.9735	0.9549	0.8377
ϕ_n		−1.8284	3.1063	2.3458	2.5751	0.1816

5.2.6   Dispersion Compensator Design

The ideal filter for compensating constant dispersion (i.e. $D \approx$ constant) has unity transmission across the passband and a linearly varying group delay. Since the AR response can only employ positive group delay from the poles to create the linear group delay variation, these two criteria cannot simultaneously be met. Both the magnitude and group delay vary over the passband producing a filter which is unsatisfactory for dispersion compensation. So, we consider the ARMA response. For now, we note that the ARMA response of chirped Bragg gratings in optical fibers are used for dispersion compensation applications. We shall find in Section 5.4 that the reflection response of a Bragg filter is equivalent to the ARMA response of an AR ring waveguide filter.

Since we are interested in the magnitude and group delay responses, we define an error for the ARMA response as follows:

$$E = \sum_n \xi [|H_{\text{fit}}(\nu_n)| - |H_d(\nu_n)|]^2 +$$

$$(1 - \xi)[\tau_{\text{fit}}(\nu_n) - \tau(0) - \tau_d(\nu_n) + \tau_d(0)]^2 \qquad (45)$$

where ξ is the weighting of the magnitude response fit relative to the group delay fit. The subscript "fit" indicates the response from the set of poles being optimized and "d" represents the desired response. The normalized frequency at discrete points is denoted by ν_n. Each group delay response is referenced to the delay at one frequency, for example $\nu_n = 0$, to avoid issues with different filter orders having different constant group delays. As an example, we take a V-shaped group delay with a slope of $d\tau/d\nu = \pm 12$ as the desired response and set the error weighting to $\xi = 0.5$. In addition, we define the passbands for $\nu_n = 0.1$–0.4 and 0.6–0.9, so that only frequencies in these ranges are included in the error. Figure 5-29 shows the resulting magnitude and group delay responses while the dispersion is plotted in Figure 5-30. The magnitude response is unity over the passbands, while the group delay response follows the desired linear behavior. The derivative of the group delay exhibits nonuniform ripples for the dispersion response. By optimizing the roots with re-

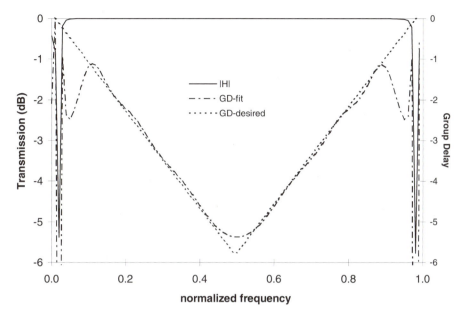

FIGURE 5-29 ARMA magnitude and normalized group delay response (relative to the maximum delay) for an eighth-order dispersion compensator

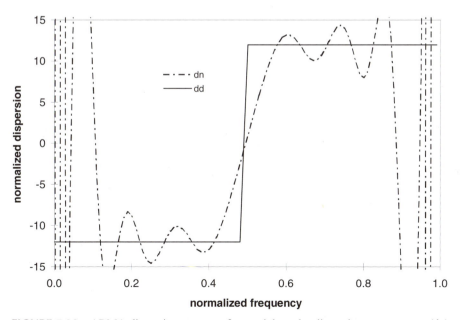

FIGURE 5-30 ARMA dispersion response for an eight-order dispersion compensator (dn) compared to the desired response (dd).

spect to the dispersion and magnitude response instead of the group delay and magnitude response, these oscillations can be reduced.

5.3 VERNIER OPERATION

Since the maximum FSR is limited by the bend radius, methods to extend the FSR are critical to increasing the range of applications for ring resonator based filters. In particular, Vernier operation has been employed to extend the FSR by staggering the passbands between concatenated stages [15]. Rings with unequal radii (i.e. $FSR_1 \neq FSR_2$) are combined so that the passbands overlap only once in several periods as shown in Figure 5-31. The overall FSR is equal to $N*FSR_1 = M*FSR_2$ where N and M are integers. The transfer function for the two ring architecture shown in Figure 5-32 is

$$H(z) = \frac{\sqrt{\kappa_1 \kappa_2 \kappa_3}}{1 - c_1 c_2 z^{-N} - c_2 c_3 z^{-M} + c_1 c_3 z^{-(N+M)}} \tag{46}$$

where $c_i = \sqrt{1 - \kappa_i}$ for $i = 1, 2, 3$. Figure 5-33 shows a comparison of the calculated magnitude responses for a ratio N:M of 1:2, 3:4, 7:8, and 11:12 based on the coupling ratios from [16]. Note the increasing secondary transmission peaks for

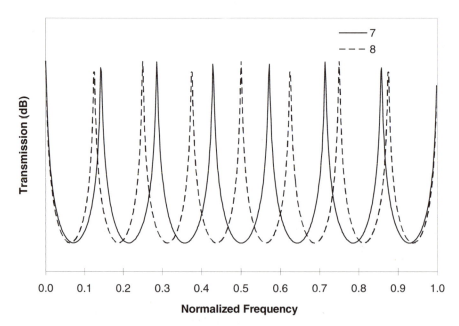

FIGURE 5-31 Calculated response using coupling ratios from [18] for individual rings with $N = 7$ and $M = 8$. Reprinted by permission. Copyright © 1994 IEEE.

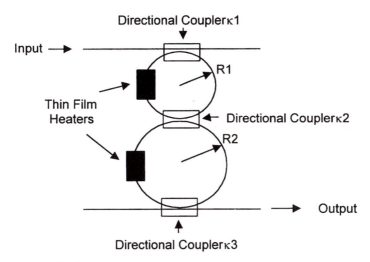

FIGURE 5-32 Double ring resonator architecture [19]. Reprinted by permission. Copyright © 1991 IEEE.

the ratios of larger integer values. The passband width of the major transmission peak is dominated by the largest ring; consequently, the response can be made sharper only at the expense of reducing the passband width. There is a design tradeoff between minimizing passband loss and maximizing sideband rejection for filters employing Vernier operation [17]. Although Vernier operation is clearly advantageous for increasing the FSR, its major disadvantages are: (1) small passband width, (2) decreasing rejection across the stopband with increasing integer ratios $N:M$, and (3) tradeoffs between maximum passband transmission and stopband rejection. An FDM transmission experiment using frequency shift keying (FSK) modulation, 10 GHz channel spacing, 8-channels and 622 Mb/s bitrate [18] was performed successfully using the double ring resonator shown in Figure 5-32 [19] for channel selection. An FSR of 100 GHz and FWHM = 500 MHz were achieved using $4.5 \times 4.5 \ \mu m^2$, $\Delta = 1.5\%$ waveguides and ring radii $R_1 = 3.33$ and $R_2 = 3.82$ mm.

Other architectures employing Vernier operation have been proposed. A triple coupler ring resonator (TCRR) design is shown in Figure 5-34. The device has two characteristic FSRs which depend on the feedback path lengths $L_1 = l_1 + l_3 + l_4$ and $L_2 = l_2 + l_3 + l_4$. The overall FSR is given by $N \times FSR_1 = M \times FSR_2$ where $FSR_i = c/n_g L_i$ for $i = 1, 2$. For N = 2 and M = 3, the FSR is double that of the smallest feedback path. The calculated spectral response for the add or drop channel is shown in Figure 5-35 and for the through-channels in Figure 5-36. Ring propagation losses of 0.17 dB and 0.26 dB were assumed for L_1 and L_2, respectively, in addition to excess losses of 0.01 dB per coupler. The passband transmittance of the dropped channels is maximized by setting $K_1 = K_2$ and $K_3 = 2 K_1 \sqrt{1 - K_1^2}$. The passband transmittance is improved by 5 dB over the single ring design while maintaining an extinc-

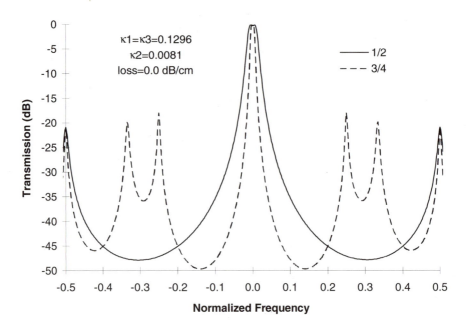

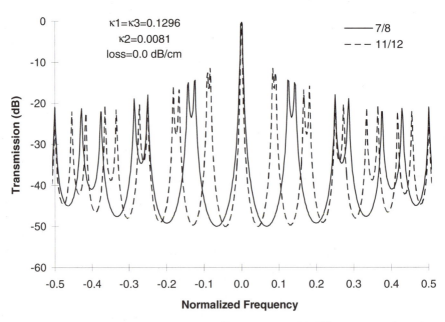

FIGURE 5-33   Comparison of double ring filter response with different FSR ratios *N/M*.

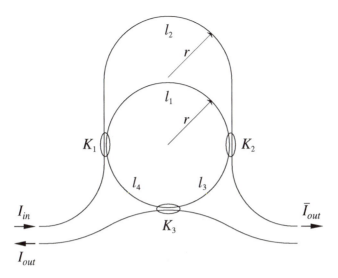

FIGURE 5-34 Schematic of a triple coupler ring resonator filter employing Vernier operation to increase the FSR with $l_2 = 2l_1$, $l_3 = l_4 = l_1/2$ and optimal values $K_1 = K_2$, $K_3 = 2K_1$
$\sqrt{1 - K_1^2}$ [15]. Reprinted by permission. Copyright © 1995 IEEE.

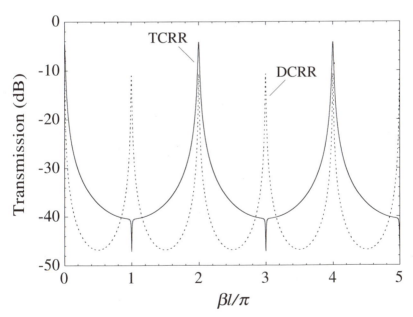

FIGURE 5-35 Calculated transmission for the add/drop response of the triple coupler ring
resonator (TCRR) with N = 2 and M = 3 compared to a single ring with two couplers, designated as a double coupler ring resonator (DCRR) [15]. The finesse for both filters was set to
100 for comparison. Reprinted by permission. Copyright © 1995 IEEE.

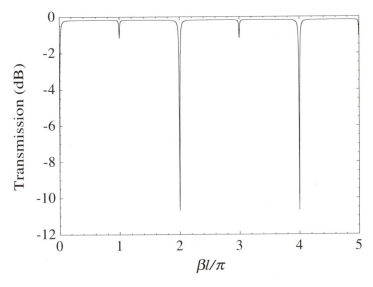

FIGURE 5-36   Calculated through-channel response for the TCRR with N = 2, M = 3, and finesse = 100 [15]. Reprinted by permission. Copyright © 1995 IEEE.

tion ratio of 36 dB. The extinction ratio falls below 5 dB for $N > 2$, and the passband width is relatively small.

5.4   REFLECTIVE LATTICE FILTERS

An analog to the single-stage ring filter discussed in Section 5.1 is a Fabry–Perot interferometer consisting of two partial reflectors with parallel surfaces and an intervening cavity. A glass slide or substrate with polished surfaces is an example of a Fabry–Perot cavity where the partial reflectors are realized by reflections at the air-to-glass interfaces. Much stronger partial reflectors are made by depositing thin film layers with alternating refractive indices on either side of the cavity. The reflection and transmission responses of thin film stacks of different dielectric materials are investigated using a Z-transform approach and the more conventional method employing characteristic matrices. Coupled cavities, which are equivalent to the lattice ring architecture, are discussed next. While thin film filters are typically made of materials with significantly different refractive indices, Bragg gratings use very low index contrasts to form a periodic index perturbation. The number of periods is much larger for Bragg gratings than the number of layers in thin film filters. The low index contrast and large number of periods make coupled-mode theory an attractive method for analyzing the spectral characteristics of Bragg gratings. Coupled-mode theory is presented in Section 5.4.2 for coupling between forward and backward propagating fundamental modes of a waveguide.

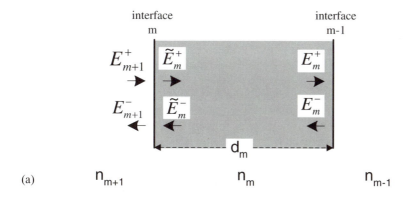

(a)

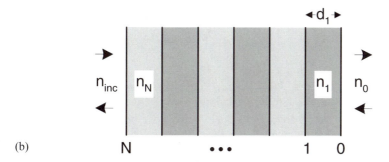

(b)

FIGURE 5-37   Thin film cavity schematic: (a) a single layer and (b) multiple layers.

5.4.1   Thin-Film Filters

A Z-transform description for the transmission and reflection of a stack of lossless, dielectric materials [5] as shown in Figure 5-37b is developed. The first step is to write the boundary conditions on the electric and magnetic fields of a transverse electromagnetic wave at an interface in matrix form. Figure 5-37a shows the interfaces and labeling for the electric field. The arrows indicate the field propagation direction, not the polarization. The layers are numbered in reverse order, like the coupled ring structure. The total transverse electric field must be identical on either side of the interface, so

$$\tilde{\mathcal{E}}_m^+ + \tilde{\mathcal{E}}_m^- = \mathcal{E}_{m+1}^+ + \mathcal{E}_{m+1}^- \tag{47}$$

where m and $m + 1$ designate the layer numbers on either side of the interface and $\mathcal{E}$ is the phasor for a monochromatic wave traveling in the positive or negative z direction as indicated by the superscript. The complete electric field description for a plane wave polarized in the $\hat{y}$ direction is $\mathcal{E}_m^{\pm} e^{j(\omega t \mp kz)} \hat{y}$. The tangential component of the magnetic field at the interface must be continuous since there are no surface currents, so the magnetic boundary condition is also a continuity relation.

$$\tilde{\mathcal{H}}_m^+ + \tilde{\mathcal{H}}_m^- = \mathcal{H}_{m+1}^+ + \mathcal{H}_{m+1}^- \tag{48}$$

Using Maxwell's equation for the curl of the electric field, $\nabla \times \vec{\mathcal{E}}_m^\pm = -j\omega\mu_0\vec{\mathcal{H}}_m^\pm$, and assuming that the derivatives of the fields with respect to x and y are zero, the magnetic field is related to the electric field as follows: $\mathcal{H}_m^\pm = \pm\varepsilon_0 n_m c\mathcal{E}_m^\pm\hat{x}$. In terms of the electric field, the magnetic boundary condition is

$$n_{m+1}(\mathcal{E}_{m+1}^+ - \mathcal{E}_{m+1}^-) = n_m(\tilde{\mathcal{E}}_m^+ - \tilde{\mathcal{E}}_m^-) \tag{49}$$

From equations (47) and (49), the following transfer matrix for an interface is obtained:

$$\begin{bmatrix} \mathcal{E}_{m+1}^+ \\ \mathcal{E}_{m+1}^- \end{bmatrix} = \Phi_m^i \begin{bmatrix} \tilde{\mathcal{E}}_m^+ \\ \tilde{\mathcal{E}}_m^- \end{bmatrix} \text{ where } \Phi_m^i = \frac{1}{\tau_m}\begin{bmatrix} 1 & \rho_m \\ \rho_m & 1 \end{bmatrix} \tag{50}$$

The transmission and reflection amplitude coefficients are $\tau_m = \tau_m^+ \equiv 2n_{m+1}/(n_m + n_{m+1})$ and $\rho_m = \rho_m^+ \equiv (n_{m+1} - n_m)/(n_m + n_{m+1})$, where the + superscript denotes that the incident wave is from the left. The relationships for a wave incident from the right are $\tau_m^- = 2n_m/(n_m + n_{m+1})$ and $\rho_m^- = -\rho_m^+$. Note that $\tau_m^+\tau_m^- - \rho_m^+\rho_m^- = 1$. The intensity reflection and transmission are given by $|R|^2 = |\rho|^2$ and $|T|^2 = 1 - |\rho|^2 = (n_0/n_{\text{inc}})|\tau|^2$, where the average intensity in each medium is given by $I_m^\pm = (n_m/2Z_0)|\mathcal{E}_m^\pm|^2$, and $Z_0 = \sqrt{\mu_0/\varepsilon_0}$ is the impedance of free space. It is useful in the Z-transform description to make the transmission coefficients symmetric [5], so the electric field is normalized as follows: $E_m^\pm \equiv \sqrt{n_m}\mathcal{E}_m^\pm$. Then, we define $t_m \equiv \sqrt{\tau_m^+\tau_m^-} = \sqrt{1 - \rho_m^2}$. The definition for ρ_m does not change as a result of this normalization.

The transfer matrix for the mth layer as shown in Figure 5-37a is derived next. The propagation in the $+z$ direction across one layer is given by $E_m^+ = e^{-jk_0 n_m d_m}\tilde{E}_m^+$ and in the opposite direction by $\tilde{E}_m^- = e^{-jk_0 n_m d_m}E_m^-$, where each layer has a width d_m and a propagation constant $k_0 n_m$. Next, the structure is simplified by assuming that each layer has the same unit delay, $T = n_m 2d_m/c$ where n_m is the group index of the mth layer. Small deviations in the path length from the nominal are included in the phase term ϕ_m for each layer. In matrix form, these relationships are

$$\begin{bmatrix} \tilde{E}_m^+ \\ \tilde{E}_m^- \end{bmatrix} = \Phi_m^p \begin{bmatrix} E_m^+ \\ E_m^- \end{bmatrix} \text{ where } \Phi_m^p \equiv \sqrt{e^{+j\phi_m}z}\begin{bmatrix} 1 & 0 \\ 0 & e^{-j\phi_m}z^{-1} \end{bmatrix} \tag{51}$$

where $z^{-1} = e^{-j\Omega T}$. By multiplying the interface matrix and the propagation matrix, $\Phi_m = \Phi_m^i\Phi_m^p$, the transfer matrix for a single layer is written as follows:

$$\begin{bmatrix} E_{m+1}^+ \\ E_{m+1}^- \end{bmatrix} = \Phi_m \begin{bmatrix} E_m^+ \\ E_m^- \end{bmatrix} \text{ where } \Phi_m \equiv \frac{\sqrt{z}e^{+j\phi_m}}{t_m}\begin{bmatrix} 1 & \rho_m e^{-j\phi_m}z^{-1} \\ \rho_m & e^{-j\phi_m}z^{-1} \end{bmatrix} \tag{52}$$

Note that by using t_m instead of τ_m, the determinant of Φ_m is unity. Transmission through multiple films is calculated by cascading matrices as follows:

$$\begin{bmatrix} E_{N+1}^+ \\ E_{N+1}^- \end{bmatrix} = \Phi_N(z) \cdots \Phi_1(z)\Phi_0^i(z) \begin{bmatrix} \tilde{E}_0^+ \\ \tilde{E}_0^- \end{bmatrix} \tag{53}$$

The overall transfer matrix is expressed in terms of two polynomials and their reverses as follows:

$$\Phi_{N0}(z) \equiv \Phi_N(z) \cdots \Phi_1(z)\Phi_0^i(z) = \frac{\sqrt{e^{j\phi_{\text{tot}}z^N}}}{\sigma_N} \begin{bmatrix} A_N(z) & B_N^R(z) \\ B_N(z) & A_N^R(z) \end{bmatrix} \tag{54}$$

where $\sigma_N \equiv \Pi_{n=0}^N t_n$. The reverse polynomial of $A_N(z)$ is defined by $A_N^R(z) = e^{-j\phi_{\text{tot}}z^{-N}}A_N^*(z^{*-1})$ where $\phi_{\text{tot}} = \Sigma_{n=1}^N \phi_n$, and similarly for $B_N(z)$. The amplitude transmission and reflection responses of the filter in Z-transform notation are

$$T_N^+(z) = \left.\frac{\tilde{E}_0^+}{E_{N+1}^+}\right|_{\tilde{E}_0^-=0} = \frac{\sqrt{e^{-j\phi_{\text{tot}}z^{-N}}}\sigma_N}{A_N(z)} \tag{55}$$

$$T_N^-(z) = \left.\frac{E_{N+1}^-}{\tilde{E}_0^-}\right|_{E_{N+1}^+=0} = \frac{\sqrt{e^{-j\phi_{\text{tot}}z^{-N}}}\sigma_N}{A_N(z)} \tag{56}$$

$$R_N^+(z) = \left.\frac{E_{N+1}^-}{E_{N+1}^+}\right|_{\tilde{E}_0^-=0} = \frac{B_N(z)}{A_N(z)} \tag{57}$$

$$R_N^-(z) = \left.\frac{\tilde{E}_0^+}{\tilde{E}_0^-}\right|_{E_{N+1}^+=0} = -\frac{B_N^R(z)}{A_N(z)} \tag{58}$$

The transmission responses are identical, independent of loss and whether the incident wave is from the left or right. The reflection responses have identical magnitude responses only for the lossless case. Let the phase of $A_N(\omega)$ and $B_N(\omega)$ be $\phi_A(\omega)$ and $\phi_B(\omega)$, then the reflection phases are $\phi_R^+(\omega) = \phi_B(\omega) - \phi_A(\omega)$ and

$$\phi_R^-(\omega) = -\phi_B(\omega) - \phi_A(\omega) - \omega N - \phi_{\text{tot}} - \pi \tag{59}$$

The transmission phase is related to the reflection phases as follows:

$$\phi_T(\omega) = -(\omega N + \phi_{\text{tot}})/2 - \phi_A(\omega) = [\phi_R^+(\omega) + \phi_R^-(\omega) + \pi]/2 \tag{60}$$

In general, the group delays of the forward and reverse reflection responses are different. For a symmetric filter, however, one cannot distinguish the forward and reverse reflection responses because $B_N^R(z) = -B_N(z)$, which implies that $\phi_B(\omega) = -\frac{1}{2}[\omega N + \phi_{\text{tot}} + \pi]$. Therefore, the group delay of $R_N^+(\omega)$ and $R_N^-(\omega)$ for a symmetric filter are equivalent and equal to the group delay for $T_N^+(\omega)$ and $T_N^-(\omega)$.

The determinant of $\Phi_{N0}(z)$ is equal to the product of determinants of $\Phi_n(z)$, which gives the following relationship.

$$A_N(z)A_N^R(z) - B_N(z)B_N^R(z) = \sigma_N^2 e^{-j\phi_{tot}}z^{-N} \tag{61}$$

In the frequency domain, Eq. (61) is a statement of power conservation.

$$1 = \frac{\sigma_N^2}{|A_N(\omega)|^2} + \frac{|B_N(\omega)|^2}{|A_N(\omega)|^2} = |T|^2 + |R|^2 \tag{62}$$

For a single cavity, the transfer matrix is given as follows:

$$\begin{bmatrix} E_2^+ \\ E_2^- \end{bmatrix} = \Phi_1(z)\Phi_0^i \begin{bmatrix} \tilde{E}_0^+ \\ \tilde{E}_0^- \end{bmatrix} = \frac{\sqrt{e^{+j\phi_1}z}}{t_0 t_1} \begin{bmatrix} 1 + \rho_0\rho_1 e^{-j\phi_1}z^{-1} & \rho_0 + \rho_1 e^{-j\phi_1}z^{-1} \\ \rho_1 + \rho_0 e^{-j\phi_1}z^{-1} & \rho_0\rho_1 + e^{-j\phi_1}z^{-1} \end{bmatrix} \begin{bmatrix} \tilde{E}_0^+ \\ \tilde{E}_0^- \end{bmatrix} \tag{63}$$

The transfer functions are

$$T_1^+(z) = T_1^-(z) = \frac{t_0 t_1 \sqrt{e^{-j\phi_1}z^{-1}}}{1 + \rho_0\rho_1 e^{-j\phi_1}z^{-1}} \tag{64}$$

$$R_1^+(z) = \frac{\rho_1 + \rho_0 e^{-j\phi_1}z^{-1}}{1 + \rho_0\rho_1 e^{-j\phi_1}z^{-1}} \tag{65}$$

$$R_1^-(z) = -\frac{\rho_0 + \rho_1 e^{-j\phi_1}z^{-1}}{1 + \rho_0\rho_1 e^{-j\phi_1}z^{-1}} \tag{66}$$

Equations (64)-(66), after substituting $z^{-1} = e^{-j\omega} = e^{-j\Omega T}$, give the frequency response of a Fabry–Perot cavity. Note that the zeros of the reflection responses are on the unit circle when the partial reflectors are identical, independent of their magnitude. Expressions for the passband width, maximum-to-minimum transmission ratio, and finesse can be derived from the square magnitude of Eq. (64). The transfer functions for the single-stage ring resonator filter discussed in Section 5.2.1 are repeated below for comparison.

$$H_{AR}(z) = \frac{-s_0 s_1 \sqrt{e^{-j\phi_1}z^{-1}}}{1 - c_0 c_1 e^{-j\phi_1}z^{-1}} \text{ and } H_{ARMA\#1}(z) = \frac{c_1 - c_0 e^{-j\phi_1}z^{-1}}{1 - c_0 c_1 e^{-j\phi_1}z^{-1}} \tag{67}$$

The absolute value of the reflection coefficient is analogous to the coupler through-port transmission. Note that $-1 \le \rho_m \le 1$, but $0 \le c_m \le 1$. If ρ_0 and ρ_1 have opposite signs, then Eqs. (65) and (67) have the same form. The recursion relations for the thin film and ring resonator filters for the lossless case are given by

Thin Film Recursions   Ring Recursions

$A_0^{TF}(z) = 1$ and $B_0^{TF}(z) = \rho_0$ $A_0(z) = 1$ and $B_0(z) = c_0$

$$A_m^{TF}(z) = A_{m-1}^{TF}(z) + \rho_m z^{-1} e^{-j\phi_m} B_{m-1}^{TF}(z) \qquad A_m(z) = A_{m-1}(z) - c_m z^{-1} e^{-j\phi_m} B_{m-1}(z)$$

$$B_m^{TF}(z) = \rho_m A_{m-1}^{TF}(z) + z^{-1} e^{-j\phi_m} B_{m-1}^{TF}(z) \qquad B_m(z) = c_m A_{m-1}(z) - z^{-1} e^{-j\phi_m} B_{m-1}(z)$$

for $m = 1, 2, \ldots, N$.

We now examine the reflection and transmission for a stack of alternating layers of high and low refractive index materials, where each layer is a quarter wavelength thick at the design center wavelength λ_c. In this case, $n_m d_m = \lambda_c/4$, $\omega = \Omega T = \pi \lambda_c/\lambda$, and $\phi_m = 0$. One analysis approach is to multiply the transfer matrices for a high and low index layer and then apply Chebyshev's identity (given in Chapter 4) to obtain the response for N periods. A more common approach is to express the transfer matrix in an alternate form, known as the characteristic matrix for a thin film. For the characteristic matrix approach, the boundary conditions relating the total electric and magnetic fields at each interface are written in matrix form as follows [20]

$$\begin{bmatrix} \mathcal{E}_{m+1} \\ Z_0 \mathcal{H}_{m+1} \end{bmatrix} = \begin{bmatrix} \cos\varphi_m & \dfrac{j \sin\varphi_m}{n_m} \\ jn_m \sin\varphi_m & \cos\varphi_m \end{bmatrix} \begin{bmatrix} \mathcal{E}_m \\ Z_0 \mathcal{H}_m \end{bmatrix} = M_m \begin{bmatrix} \mathcal{E}_m \\ Z_0 \mathcal{H}_m \end{bmatrix} \tag{68}$$

where $\mathcal{E}_m = \mathcal{E}_m^+ + \mathcal{E}_m^-$, $\mathcal{H}_m = \mathcal{H}_m^+ + \mathcal{H}_m^-$, and $\varphi_m = (2\pi/\lambda)n_m d_m$. The characteristic matrix for the mth layer is designated by M_m. For quarter-wavelength thick layers, $\varphi_m = (\pi/2)(\lambda_c/\lambda)$ and the characteristic matrix simplifies to

$$\begin{bmatrix} 0 & \dfrac{j}{n_m} \\ jn_m & 0 \end{bmatrix}$$

at $\lambda = \lambda_c$. For a half-wavelength thick layer at $\lambda = \lambda_c$, the characteristic matrix is

$$\begin{bmatrix} -1 & 0 \\ 0 & -1 \end{bmatrix}$$

Such layers are referred to as absentee layers, since they have no impact on the response at the design center wavelength. For non-normal incidence, the same characteristic matrix is used except that the refractive indices are replaced by $n_m \cos\alpha_m$ if the electric field is polarized perpendicular to the plane of incidence or by $n_m/\cos\alpha_m$ for parallel polarization, where α_m is the angle of incidence with respect to the normal in the mth layer. In addition, $\varphi_m = (2\pi/\lambda)n_m d_m \cos\alpha_m$.

To determine the transmission and reflection coefficients using the characteristic matrix approach, we need to relate the forward and backward propagating electric fields to the total fields. The transformation for the fields in the mth layer is given by [20]

$$\begin{bmatrix} \mathcal{E}_m \\ Z_0 \mathcal{H}_m \end{bmatrix} = \begin{bmatrix} 1 & 1 \\ n_m & -n_m \end{bmatrix} \begin{bmatrix} \mathcal{E}_m^+ \\ \mathcal{E}_m^- \end{bmatrix} \tag{69}$$

The overall transfer matrix, Q, is obtained from the characteristic matrix as follows:

$$\begin{bmatrix} \mathcal{E}_{N+1}^{+} \\ \mathcal{E}_{N+1}^{-} \end{bmatrix} = \begin{bmatrix} 1 & 1 \\ n_{N+1} & -n_{N+1} \end{bmatrix}^{-1} \begin{bmatrix} M_{11} & M_{12} \\ M_{21} & M_{22} \end{bmatrix} \begin{bmatrix} 1 & 1 \\ n_0 & -n_0 \end{bmatrix} \begin{bmatrix} \tilde{\mathcal{E}}_0^{+} \\ \tilde{\mathcal{E}}_0^{-} \end{bmatrix}$$

$$= \begin{bmatrix} Q_{11} & Q_{12} \\ Q_{21} & Q_{22} \end{bmatrix} \begin{bmatrix} \tilde{\mathcal{E}}_0^{+} \\ \tilde{\mathcal{E}}_0^{-} \end{bmatrix} \tag{70}$$

The transmission and reflection are then defined from the coefficients of Q, just as Φ_{N0} is used in Eqs. (55)–(58), as follows:

$$T^{+}(\lambda) = \frac{\mathcal{E}_0^{+}}{\mathcal{E}_{N+1}^{+}} = \frac{1}{Q_{11}(\lambda)} \quad \text{and} \quad R^{+}(\lambda) = \frac{\mathcal{E}_{N+1}^{-}}{\mathcal{E}_{N+1}^{+}} = \frac{Q_{21}(\lambda)}{Q_{11}(\lambda)} \tag{71}$$

A very useful property of the characteristic matrix approach is that any symmetrical combination of layers can be expressed by an equivalent matrix of the same form [20,22]. A filter consisting of alternating layers of high and low index materials is written in terms of symmetrical elements consisting of $L/2$ H $L/2$ or $H/2$ L $H/2$, where L and H represent a quarter-wave thickness at λ_c of low and high index material, respectively. The coefficients, M_{ij}, of the characteristic matrix for the symmetrical layer structure are calculated. Then, an equivalent index is defined by $N = +\sqrt{M_{21}/M_{12}}$ and an equivalent thickness by $\Gamma = \cos^{-1} M_{11}$. For P periods of the symmetrical layer, the characteristic matrix is

$$\begin{bmatrix} \cos \Gamma & j \dfrac{\sin \Gamma}{N} \\ jN \sin \Gamma & \cos \Gamma \end{bmatrix}^{P} = \begin{bmatrix} \cos P\Gamma & j \dfrac{\sin P\Gamma}{N} \\ jN \sin P\Gamma & \cos P \Gamma \end{bmatrix} \tag{72}$$

The stopband corresponds to regions of high reflectance where $|M_{11} + M_{22}|/2 > 1$. In the stopband, both N and Γ are imaginary. The cos and sin functions are then written as cosh and sinh of real variables.

The peak reflectance for a stack consisting of N periods is [21]

$$|R(\lambda_{\text{max}})|^2 = \tanh^2 \left[N \ln \frac{n_H}{n_L} + \frac{1}{2} \ln \frac{n_0}{n_{inc}} \right] \tag{73}$$

where n_H = high refractive index material and n_L = low refractive index. The reflectance depends on the ratio of the high and low indices, and it goes to unity in the limit as the number of periods goes to infinity. The reflection bandwidth is given by [22]

$$\frac{\Delta\lambda}{\lambda} = \frac{4}{\pi} \sin^{-1} \left(\frac{n_H - n_L}{n_H + n_L} \right) \tag{74}$$

The reflection bandwidth depends on the index difference. Using an analogy to

TABLE 5-6   Refractive Indices for Some Thin Film Materials from [22]

Material	Refractive Index $1 \leqslant \lambda \leqslant 2 \ \mu m$
SiO_2	1.44
SiO	1.85
TiO_2	2.25
Ta_2O_5	2.09
Si	3.49

electronic band structures, the frequency range over which light is reflected is also referred to as a photonic bandgap. Examples of materials used for thin film filters and their refractive indices are listed in Table 5-6 [22]. For alternating layers of titanium dioxide and silica, a 400 nm reflection bandwidth is predicted by Eq. (74) for $\lambda_c = 1550$ nm.

A narrowband transmission resonance at λ_c is created by inserting an odd multiple of quarter-wavelength layers between two partial reflectors to form a cavity. If each reflector consists of M periods, then the overall layer structure is written as

$$\left[\frac{L}{2} H \frac{L}{2} \right]^M L \left[\frac{L}{2} H \frac{L}{2} \right]^M$$

The single L in the center combines with the adjacent $L/2$ layers to create an absentee half-wave layer and likewise for subsequent pairs of layers. Thus, the structure looks like an absentee layer at λ_c, and the transmission is 100%. By cascading several of these structures, multi-cavity filters are realized, which can be used to achieve wider passbands and sharper rolloffs than a single stage. A calculation of various bandpass filters using 1, 2 and 3 cavities with 10 periods for each reflector is shown in Figure 5-38 assuming $n_L = 1.45$ and $n_H = 1.92$. An advantage of thin film filters is that the choice of substrate thermal expansion can be engineered to reduce the temperature dependence of the filter [23]. A temperature dependence of 0.001 nm/°C has been realized [24]. For normal incidence, the polarization dependence is also negligible.

Coupled-cavity filters with much larger cavity thicknesses can be realized by cascading coated substrates with partial reflectors as shown in Figure 5-39. Dowling and MacFarlane [25] first applied Z-transform analysis techniques to such filters. Each reflector consists of many layers, but is treated as a single, wavelength independent element over the FSR of interest. The unit delay is proportional to the distance between reflectors, L_m, as opposed to a quarter-wave thickness as in the analysis of a thin film stack. Each reflector has an associated interface matrix given by Eq. (54) and restated as follows in the frequency domain:

$$\Phi_m^i(\omega) = \frac{\sqrt{e^{j(\omega N + \phi_{\text{tot}})}}}{\sigma_m} \begin{bmatrix} A_m(\omega) & B_m^R(\omega) \\ B_m(\omega) & A_m^R(\omega) \end{bmatrix} \tag{75}$$

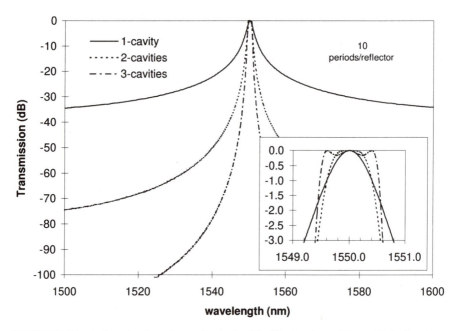

FIGURE 5-38   A 1-cavity, 2-cavity, and 3-cavity thin film bandpass filter with the bandpass region enlarged in the inset.

The subscript m denotes the mth partial reflector and N is the number of layers making up the stack. The coefficients of Φ_m^i are evaluated at the filter's center frequency and treated as a constant over the FSR. By rewriting Φ_m^i in terms of the partial transmission and reflectance, we obtain

$$\Phi_m^i = \frac{1}{T_m^+}\begin{bmatrix} 1 & -R_m^- \\ R_m^+ & \Theta_m \end{bmatrix} \tag{76}$$

where $\Theta_m = e^{-j2\Phi_m^A}$, $\Phi_m^A = \arg\{A_m(\omega_c)\}$, and the transmittance and reflectances are given by Eq. (55)–(58). The phase associated with Θ_m, T_m^+ and $R_m^\pm$ distinguishes Eq.

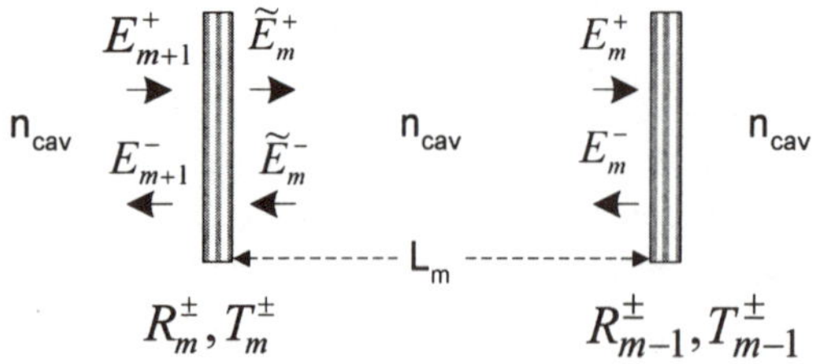

FIGURE 5-39   Reflective optical AR lattice filter with partial reflectors.

(76) from Eq. (52). The new expression for the reflectance of a Fabry–Perot cavity, corresponding to Eq. (65), is

$$R^+(z) = \frac{R_1^+ + e^{-j\phi_1}R_0^+\Theta_1 z^{-1}}{1 - e^{-j\phi_1}R_0^+R_1^- z^{-1}} \tag{77}$$

The phases of R_0^+ and R_1^- impact the location of the pole and zero within the FSR, and their group delay contributes to the effective cavity length. An effective reflectance plane is defined for each partial reflector. The offset of the reflectance plane from the end of the stack is included as part of the total cavity length. The transfer matrix for a single stage is then calculated using $\Phi_m = \Phi_m^i \Phi_m^p$ where the delay matrix Φ_m^p is given by Eq. (51) and $z^{-1} = e^{-j\Omega T}$, where T is the nominal roundtrip delay. Multi-stage AR filters with arbitrary pole locations can be realized by designing the partial reflectors and cavity phases using step-down recursion relations in an algorithm similar to the one discussed for ring filters. Flat-top passband coupled-cavity filters have been designed [25], and filters to give a prescribed pulse train for encoding pulses have been demonstrated [26]. In the latter case, a 200 fs pulse was filtered to obtain an 800 GHz pulse burst with a prescribed amplitude, or code. Several three-cavity designs were demonstrated using 125 μm thick glass substrates coated to achieve different reflectance values.

5.4.2 Bragg Gratings

For a periodic index modulation with a small index contrast, $n_L \approx n_H$, the reflection and transmission are accurately modeled using coupled-mode theory. Coupled-mode theory is widely used, and many physical insights are gained from it regarding the filter characteristics of Bragg gratings. Using the thin film as an analogy, each half period is viewed as a weak partial reflector. When the wavelength is such that the partial reflectances add constructively, a narrowband reflection results as illustrated in Figure 5-40. The center wavelength of the reflection is called the Bragg wavelength, which is related to the grating period by $\Lambda = \lambda_B/2n_e$ where Λ is the grating period and the Bragg wavelength is denoted by λ_B. The index variation may be induced by UV exposure of a photosensitive material or etched. Etched gratings are used extensively in semiconductors for fabricating distributed feedback (DFB) and distributed Bragg reflector (DBR) lasers. The UV-induced Bragg gratings written in optical fibers and planar waveguides are important for realizing low loss, low polarization dependent, narrowband filters. We focus on UV-induced gratings for this discussion, since their detailed filter characteristics are very important for WDM system applications. The induced index changes are less than $\Delta n = 0.01$. The filters are typically several millimeters to several centimeters long, consisting of thousands of periods. Since the interference mechanism is the same as for thin film filters, we can describe Bragg gratings using Z-transforms. From the Z-transform description given in Section 5.4.1, we know that the transmission response is AR and that the reflection response is ARMA. In this section, we extend our understanding of discrete AR/ARMA filters to the continuous case by developing the coupled-mode solutions.

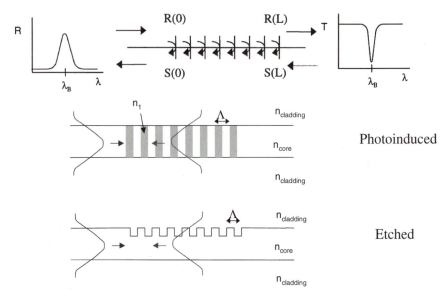

FIGURE 5-40   Bragg gratings: coherent interference of partial reflectances creates a band-pass reflection response and a stopband transmission response. The periodic effective index bandstop is created by inducing an index change using external fields or etching.

The fabrication of UV-induced gratings is briefly discussed to provide a motivation for how the coupled-mode equations are stated. A phase mask designed to minimize diffraction into the zeroth order and having a period of Λ_{PM} is placed near the waveguide [27] as shown in Figure 5-41. The ± 1 diffraction orders occur at an angle α to the normal given by $\sin\alpha = \lambda_{UV}/\Lambda_{PM}$. The two beams interfere in the near field of the mask to produce a pattern with half the period of the phase mask, $\Lambda = \lambda_{UV}/2 \sin \alpha = \Lambda_{PM}/2$. Ultraviolet exposure raises the refractive index. Exposure with a Gaussian-shaped UV beam creates an index change as shown in Figure 5-43a that contains a nonzero dc index change across the profile equal to half the ac index change. By scanning the beam along the waveguide at a constant speed, a uniform index profile as shown in Figure 5-42a is created. In this case, neither the dc index nor the ac index profile depend on z, where z is the direction parallel to the waveguide. The periodic index change for UV induced gratings is conveniently written as [28]

$$\delta n(x, y, z) = \delta\bar{n}(x, y, z)\left\{1 + \upsilon\cos\left[\frac{2\pi}{\Lambda}z + \phi(z)\right]\right\} \qquad (78)$$

where $\delta\bar{n}$ is the dc index change, the period of the grating is given by Λ, υ is the modulation index, and chirping of the grating period is included via the term $\phi(z)$. In many cases, only the core is photosensitive, so the index change is assumed to be

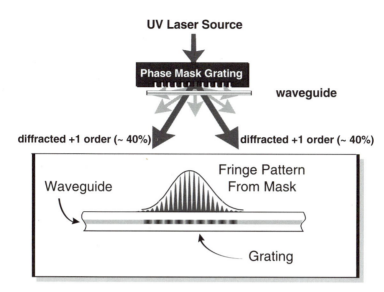

FIGURE 5-41   A typical UV-induced grating fabrication setup using the diffraction from a phase mask to create the interference pattern.

uniform over the core and zero otherwise. A more general expression for the index perturbation allows the ac and dc index profiles to be specified independently.

$$\delta n(x, y, z) = \delta n_{dc}(x, y, z) + \delta n_{ac}(x, y, z) \cos\left[\frac{2\pi}{\Lambda}z + \phi(z)\right] \tag{79}$$

The dc index profile can be UV induced by sweeping the beam across the waveguide without the phase mask while varying the beam speed or UV power. Alternatively, the waveguide width can be varied by design to achieve the desired dc effective index profile. When the grating period is nonuniform, Λ represents a constant period, for example the average, while the local period is given by $\Lambda(z) = \Lambda + \Delta\Lambda(z)$. For a linear chirp with the grating centered at $z = 0$, $\Delta\Lambda(z) = Cz/2n_e$ and $d\phi/dz = -2\pi Cz/n_e\Lambda^2$ where $C = d\lambda_B/dz$.

In a similar manner to the derivation in Chapter 2 for the directional coupler, we solve for the fields with the index perturbation in terms of the field solutions without the perturbation. The derivation is limited to coupling between the forward and reverse traveling fundamental mode of a waveguide. The transverse electric field is described by

$$\vec{E}(x, y, z, t) = [A(z)e^{-j\beta z} + B(z)e^{+j\beta z}]\vec{e}_t(x, y)e^{+j\omega t} \tag{80}$$

where $A(z)$ and $B(z)$ are the amplitudes of the forward and reverse traveling waves, respectively, z is the propagation distance along the waveguide, and $\vec{e}_t(x, y)$ is the transverse mode profile. For coupling to occur, the difference in propagation constants of the forward and reverse modes must be matched by the grating spatial frequency as follows: $\beta - (-\beta) = 4\pi n_e/\lambda = m(2\pi/\Lambda)$, where m is the grating order. We shall only consider a first-order grating, so $m = 1$ and the Bragg wavelength condition is $\Lambda = \lambda_B/2n_e$, as previously stated. The coupled-mode equations are derived by substituting Eq. (80) into the Helmholtz equation, $\nabla^2\vec{E}_t + k^2 n_0^2 \vec{E}_t = 0$, as derived in Chapter 2. The term n_0^2 is replaced by $(n_0 + \delta n)^2 \approx n_0^2 + 2n_0\delta n$, where $n_0(x, y)$ is the unperturbed refractive index and $\delta n(x, y, z)$ is given by Eq. (78) or (79). For an etched grating, the index change $\delta n(x, y, z)$ is expanded in a Fourier series in z. For a nonchirped symmetric grating, $\delta n(z) = \Sigma_p \delta n_p \cos(2\pi p z/\Lambda)$ so that $\delta n_{dc} = \delta n_0$ and $\delta n_{ac} = \delta n_1$ in Eq. (79). The mode amplitudes are assumed to vary slowly, so the terms with second derivatives of $A(z)$ and $B(z)$ are neglected. The remaining terms are grouped into two equations with exponents $\exp(+j\delta)$ and $\exp(-j\delta)$, where

$$\delta \equiv \beta - \frac{\pi}{\Lambda} = 2\pi n_e\left(\frac{1}{\lambda} - \frac{1}{\lambda_B}\right) \tag{81}$$

Finally, the equations are multiplied by $\vec{e}_t^*(x, y)$ and integrated over x and y. Two coupling coefficients are defined after the integration, which are proportional to the ac and dc index perturbations. The dc coupling coefficient is $\hat{\sigma} \equiv \delta + \sigma - (1/2)(d\phi/dz)$ where $\sigma = (2\pi/\lambda)\delta\overline{n}_e$ and

$$\delta\overline{n}_e \equiv \delta\overline{n} \int\int_A |\vec{e}_t(x, y)|^2 dxdy \Big/ \int\int_{A_\infty} |\vec{e}_t(x, y)|^2 dxdy \tag{82}$$

The effective index change $\delta\overline{n}_e$ is proportional to the overlap of the field with the index perturbation. The average index change, $\delta\overline{n}$, is assumed to be uniform across an area A. The ac coupling coefficient is

$$\kappa = \kappa^* = \frac{\pi}{\lambda} v\delta\overline{n}_e \tag{83}$$

The following coupled-mode equations result which describe the change in amplitude of the forward and reverse waves [28]:

$$\frac{dR}{dz} = -j\hat{\sigma}R(z) - j\kappa S(z) \tag{84}$$

$$\frac{dS}{dz} = +j\hat{\sigma}S(z) + j\kappa^*R(z)$$

where $R(z) \equiv A(z)e^{-j(\delta z - \phi/2)}$ and $S(z) \equiv B(z)e^{+j(\delta z - \phi/2)}$. The dc coupling coefficient depends on the deviation of the wavelength from the Bragg wavelength, on the average index change, and on the grating chirp. Consequently, an average index change can be used to create an effective chirp and vice versa since they both enter the coupled-mode equations in the same way.

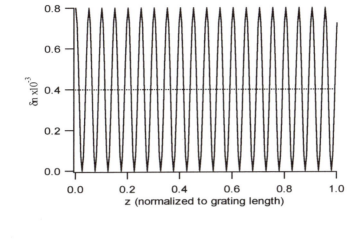

(a)

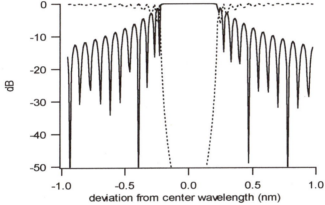

(b)

FIGURE 5-42 (a) A uniform refractive index profile and (b) the transmission and reflection spectrum for a 10 mm long Bragg grating with $\delta\bar{n}_e = 0.4 \times 10^{-3}$.

For a uniform grating, ϕ and $\delta\bar{n}_e$ are constants, and the coupled-mode equations have an analytic solution. The amplitude reflectance for a uniform grating of length L is given by

$$\rho = \frac{S(0)}{R(0)}\bigg|_{S(L)=0} = \frac{-j\kappa\,\sinh(\gamma L)}{\gamma\cosh(\gamma L) + j\hat{\sigma}\sinh(\gamma L)} \tag{85}$$

where $\gamma = \sqrt{\kappa^2 - \hat{\sigma}^2}$. The maximum reflection occurs at $\hat{\sigma} = 0$, corresponding to a wavelength of $\lambda_{max} = (1 + \delta\bar{n}/n_e)\lambda_B$. The peak reflection is given by $R = |\rho|^2 = \tanh^2(\kappa L)$, which has the same form as Eq. (73) for the periodic thin film case. Inside the bandgap, the fields are exponentially decaying, while outside they vary sinusoidally. The reflection and transmission spectrum for a 10 mm long uniform grating with $\delta\bar{n}_e = 0.4 \times 10^{-3}$ and $v = 1$ are shown in Figure 5-42 for $\lambda_B = 1550$ nm.

To quantify the reflection bandwidth, we define it by the first zeros of the reflection spectrum [28].

$$\frac{\Delta \lambda}{\lambda_B} = \frac{v \delta \overline{n}_e}{n_g} \sqrt{1 + \left(\frac{\lambda_B}{L v \delta \overline{n}_e} \right)^2} \qquad (86)$$

Bragg gratings are classified in terms of their reflectance as weak or strong gratings. Strong gratings have a reflectance very close to 100% at the Bragg wavelength. The weak grating regime is described by [28] $v \delta \overline{n}_e \ll \lambda_B / L$. Its reflection bandwidth is given by $\Delta \lambda / \lambda = 2/N$, where $\Delta \lambda$ is the bandwidth and N is the number of periods. The bandwidth is inversely proportional to the grating length for weak gratings, and the group delay is proportional to the length ($\tau_g = L/v_g$ for a uniform grating). For weak gratings, the reflection spectrum is proportional to the scaled Fourier transform of the coupling strength $|\rho(\delta)| = |\int \kappa(z) e^{j\phi(z)} e^{-j2\delta z} dz|$ [29]. A consequence of this approximation is that the reflection response behaves like a MA filter instead of an ARMA response. This behavior has been used to demonstrate a square passband filter response with an approximately constant group delay, i.e. a filter which closely approximates the dispersionless properties of symmetric MA filters [30]. The drawback is that the peak reflection is ~50%; consequently, the remainder of the power is transmitted at the Bragg wavelength. In the strong grating regime, $v \delta \overline{n}_e \gg \lambda_B / L$ and the reflection bandwidth is proportional to the index variation given by the ac index change, $\Delta \lambda / \lambda = v \delta \overline{n}_e / n_g$. Increasing the length of a strong grating makes the band edge sharper but does not change the bandwidth. For a strong uniform grating, the group delay is $\tau_g = 1/k v_g$.

To analyze a nonuniform grating, one can solve the coupled-mode equations or segment the grating into approximately uniform sections and multiply the solutions of each section. In the latter case, the solution for a uniform grating is conveniently written in a transfer matrix form for easy concatenation. The transfer matrix which relates the input and output electric fields [31,32] for each section of length Δz is given as follows:

$$\begin{bmatrix} R_{n+1} \\ S_{n+1} \end{bmatrix} = \begin{bmatrix} F_{11}^n & F_{12}^n \\ F_{21}^n & F_{22}^n \end{bmatrix} \begin{bmatrix} R_n \\ S_n \end{bmatrix} \qquad (87)$$

where

$$F_{11}^n = \cosh(\gamma_n \Delta z) + j \frac{\hat{\sigma}_n}{\gamma_n} \sinh(\gamma_n \Delta z) \qquad (88)$$

$$F_{12}^n = +j \frac{\kappa_n}{\gamma_n} \sinh(\gamma_n \Delta z) \qquad (89)$$

$$F_{21}^n = -j \frac{\kappa_n}{\gamma_n} \sinh(\gamma_n \Delta z) \qquad (90)$$

$$F_{22}^n = \cosh(\gamma_n \Delta z) - j\frac{\hat{\sigma}_n}{\gamma_n}\sinh(\gamma_n \Delta z) \qquad (91)$$

The product of transfer matrices, $F = F^N F^{N-1} \cdots F^1$, yields the overall response where

$$\begin{bmatrix} R(0) \\ S(0) \end{bmatrix} = \begin{bmatrix} F_{11} & F_{12} \\ F_{21} & F_{22} \end{bmatrix} \begin{bmatrix} R(L) \\ S(L) \end{bmatrix}$$

The grating's amplitude reflectance is given by $\rho = F_{21}/F_{11}$.

Flat passband filters with sharp transition bands and large stopband rejections have been demonstrated in fibers [33]. A key to realizing narrowband reflection fil-

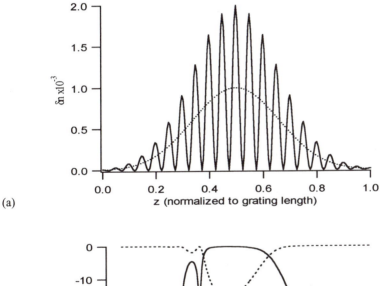

(a)

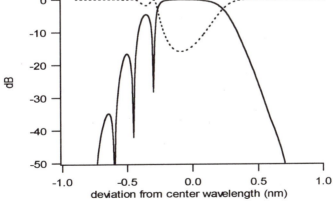

(b)

FIGURE 5-43   (a) A Gaussian refractive index profile and (b) the transmission and reflection spectrum for a 10 mm long Bragg grating with a maximum $\delta\bar{n}_e = 1.0 \times 10^{-3}$.

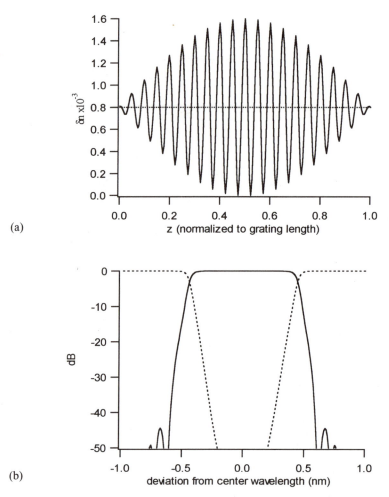

FIGURE 5-44 (a) A raised cosine refractive index profile and (b) the transmission and reflection spectrum for a 10 mm long Bragg grating with a maximum $\delta\overline{n}_e = 0.8 \times 10^{-3}$.

ters for WDM applications is control over the coupling strength profile. Tapering, or apodization, of the coupling strength decreases the sidelobes that are present for uniform coupling. A comparison of various profiles is shown in Figures 5-42 through 5-44. The uniform profile has the largest sidelobes. Its average index is constant along the length of the grating, and both the ac and dc index functions are symmetric, thus resulting in a symmetric spectral response. The Gaussian profile has a dc index change across the grating length equal to half the maximum index change as shown in Figure 5-43. The ripples on the short wavelength side are attributable to a Fabry–Perot cavity created by the dc index change. By plotting the center Bragg wavelength and the reflection bandwidth along the length of the grating, as shown in Figure 5-45, it is evident that the short wavelengths experience a re-

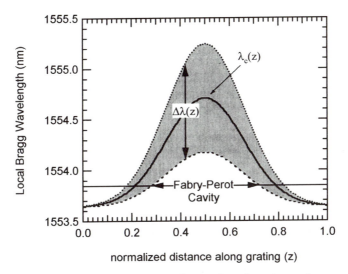

FIGURE 5-45 Photonic bandgap diagram for a Gaussian grating.

flectance at each end of the grating with a cavity in the middle [34]. Even though
the index variation of the Gaussian grating is symmetrical, its magnitude response
is not, which is attributed to the non-zero dc index change across the grating. Finally, by compensating the grating strength so that there is both an apodization of the
ac index and no change in the dc index, such as for the raised cosine profile shown

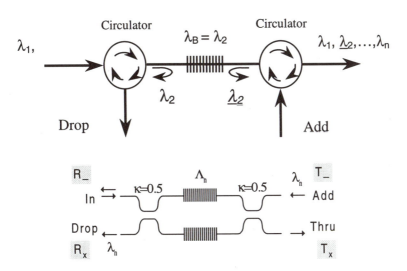

FIGURE 5-46 An add-drop filter using Bragg gratings: (a) a single grating with a circulator on each end and (b) an MZI with identical Bragg gratings in each arm.

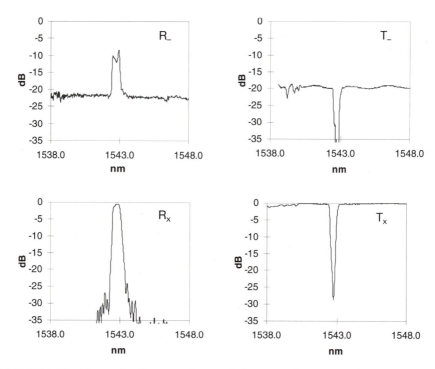

FIGURE 5-47 Measured reflection and transmission spectral responses for a planar wave-guide MZI with Bragg gratings in each arm showing the bar (–) and cross (X) states.

in Figure 5-44, the sidelobes are greatly reduced and a symmetric spectral response results.

Since the reflected signal is traveling in the same waveguide as the input, a method to separate it from the input is needed. One solution is to use a circulator on both the drop and add sides as shown in Figure 5-46a. An alternative, which is particularly suitable for implementation in planar waveguides, is to write a pair of identical gratings in an MZI [35,36]. The reflected signal at the Bragg wavelength is then output at the cross-port as shown in Figure 5-46b. Identical gratings are written across the MZI arms in a single UV exposure [37]. Measured spectral response

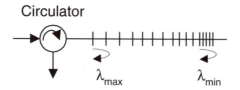

Circulator

FIGURE 5-48 Chirped Bragg grating for dispersion compensation.

curves for each port are shown in Figure 5-47 for apodized gratings [38]. The pass-band loss for the drop (or add) port is <0.5 dB. Variations in the 3 dB couplers from their nominal value degrades the achievable isolation in the transmission and reflection bar-ports.

Chirped Bragg gratings are used for dispersion compensation [39]. The advantages of UV-induced chirped gratings are their low loss and negligible nonlinearity compared to dispersion compensating fibers. To compensate the +17 ps/nm-km dispersion of conventional fiber, the longer wavelengths must experience a shorter group delay. Compensation is accomplished with a negative chirp as shown in Figure 5-48. For linear chirped gratings with a chirp rate of C nm/mm and neglecting the impact of index profile apodization, the dispersion is approximated by $D \approx 2n_e/cC \approx 10/C\,(ps/nm)$ where $n_e = 1.5$. It is the reflected, or ARMA, response that is most useful for dispersion compensation since an almost uniform reflectance can be achieved across the passband. A very small chirp rate is needed to produce large dispersions, and the required grating length scales with the bandwidth to be compensated. For example, a 17.5 cm long grating with a 0.06 nm/cm chirp is required to compensate 100 km of conventional fiber (1700 ps/nm) over a 1 nm bandwidth [40]. Apodization of the grating strength is required to reduce ripples in the group delay response, but it also decreases the reflection bandwidth [41]. For comparison to the uniform profile, a positive-tanh apodization profile requires 19.2 cm/nm to compensate a dispersion of 1700 ps/nm [41]. Typically, one grating covers one

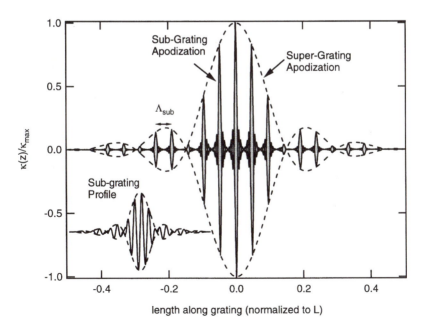

FIGURE 5-49 Relative coupling strength for a sampled grating. Both the sub-grating and super-grating profiles are apodized with a sinc function times $\cos(\pi z/L)$.

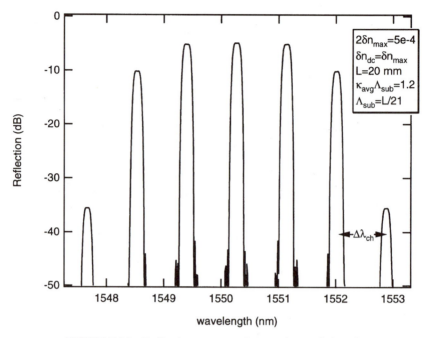

FIGURE 5-50   Reflection spectrum for a weak sampled grating.

channel; however, sampled gratings [42] as well as long gratings have been reported for compensating multiple channels. A 1.3 meter filter consisting of thirteen 10 cm gratings was fabricated to compensate for 100 km of conventional fiber over a 10 nm bandwidth [43].

By modulating the coupling strength, a sampled or superstructure grating can be made [44–46]. A sampled grating refers to one whose coupling strength is modulated in a binary fashion to produce multiple passbands; whereas, a superstructure profile may include tapering within each period of the sample (sub-grating) and tapering of the overall profile (super-grating). As an example, consider the index profile shown in Figure 5-49 for a weak grating. The reflection spectrum obtained from coupled mode theory is shown in Figure 5-50. The channel spacing is given by $\Delta\lambda_{ch}$ = $\lambda_B^2/2n_g\Lambda_{sub}$, where Λ_{sub} is the period of the sub-gratings. The width of the envelope factor for the reflectance response is inversely proportional to the width of the sub-grating apodization profile. A narrower sub-grating profile yields more reflected channels since its Fourier transform is wider. The super-grating apodization profile determines the spectral response for each passband. In this example, both the sub-grating and super-grating profiles are sinc functions multiplied by a window cosine to truncate them to a finite length. In an experimental demonstration, a uniform super-grating profile and a sinc-shaped sub-grating profile were used to fabricate a 10 cm grating [47]. A reflection response with 16 channels and passband widths of 16 pm was demonstrated.

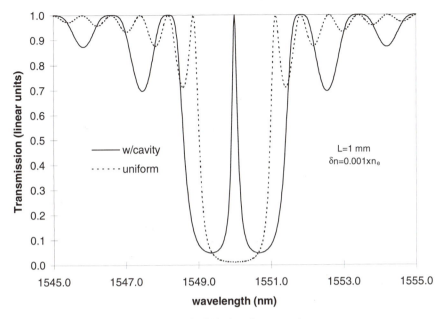

FIGURE 5-51 A $\lambda/4$ shifted cavity spectral response.

As previously discussed for thin films, quarter-wave shifted cavities can produce very narrow bandpass filters in transmission [48]. A bandpass transmission response is advantageous since it avoids the need for circulators or equivalent methods to separate the reflected signal from the input signal. A 1 mm long uniform grating with and without a $\lambda/4$-shifted cavity is shown in Figure 5-51 assuming an index difference of $\delta\bar{n}_e/n_e = 0.001$. The response was calculated using the matrix defined in Eqs. (87)–(91) and a delay matrix,

$$ F_{\text{delay}} = \begin{bmatrix} e^{+j\phi} & 0 \\ 0 & e^{-j\phi} \end{bmatrix} $$

where $\phi = (2\pi n_e/\lambda)(\lambda_c/4n_e)$. The number of cavities can be increased and the length of the reflectors chosen to provide a flatter passband response [49–51]. The stopband width is limited by the index change, so large δn's are needed for large stopband widths. High index difference structures integrated in waveguides offer the opportunity for wide photonic bandgaps and very compact devices, since only a few layers are required for each reflector. Foresi et al. [52] reported a one-dimensional photonic bandgap filter using a silicon waveguide as shown in Figure 5-52a, which supports only one polarization. The index of silicon is 3.4, so a quarter wavelength at $\lambda_c = 1550$ nm is $\lambda_c/4n_e \approx 0.1$ μm. X-ray and e-beam lithography were used to define the features. The filter consisted of four circular holes with 0.10 μm radii on a period of 0.4 μm located on either side of a $\lambda/4$ shifted cavity. The measured trans-

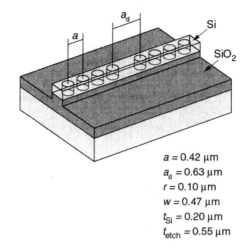

$a = 0.42\ \mu m$
$a_d = 0.63\ \mu m$
$r = 0.10\ \mu m$
$w = 0.47\ \mu m$
$t_{Si} = 0.20\ \mu m$
$t_{etch} = 0.55\ \mu m$

(a)

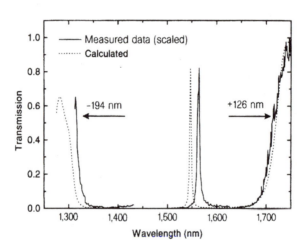

(b)

FIGURE 5-52 Integrated waveguide 1-D photonic bandgap phase-shifted filter: (a) waveguide structure and (b) measured results compared to theory [52]. Reprinted by permission.

mission response has a narrow passband with a FWHM = 5.9 nm and a 400 nm rejection band as shown in Figure 5-52b.

The reflection bandwidth for a uniform grating is limited by the maximum index change to around 20 nm for UV-induced gratings. The reflection bandwidth can be increased by chirping the grating period. A chirped grating Fabry–Perot cavity was demonstrated with a 140 nm reflection bandwidth and an FSR = 11.3 nm [53]. Chirped gratings are also useful for filtering wavelength bands instead of individual channels. To realize an ultra-wide erbium-doped fiber amplifier [54], the incoming signal was split into two 40 nm bands using a chirped fiber Bragg grating. After amplifying each band separately, the bands were recombined with a second chirped grating. Each grating had a bandwidth in excess of 40 nm and a nearly rectangular shape. The transition from the −15 dB point of the reflection spectra to the −15 dB point of the transmission spectra on one side of the grating was achieved in 1 nm [55].

REFERENCES

1. C. Madsen and J. Zhao, "Planar Waveguide AR Lattice Filters: Advantages and Challenges," Integrated Photonics Research Topical Meeting. Boston, MA, April 29-May 1, 1996, pp. 167–170.

2. B. Moslehi, J. Goodman, M. Tur, and H. Shaw, "Fiber-Optic Lattice Signal Processing," *Proc. IEEE,* vol. 72, no. 7, pp. 909–930, 1984.

3. K. Jackson, S. Newton, B. Moslehi, M. Tur, C. Cutler, J. Goodman, and H. Shaw, "Optical Fiber Delay-Line Signal Processing," *IEEE Trans. Microwave Theory Techn.,* vol. 33, no. 3, pp. 193–209, 1985.

4. C. Madsen and J. Zhao, "A General Planar Waveguide Autoregressive Optical Filter," *J. Lightw. Technol.,* vol. 14, no. 3, pp. 437–447, 1996.

5. S. Orfanidis, Optimum Signal Processing: An Introduction, 2nd. Ed., New York: McGraw-Hill, 1988.

6. E. Dowling and D. MacFarlane, "Lightwave Lattice Filters for Optically Multiplexed Communication Systems," *J. Lightw. Technol.,* vol. 12, no. 3, pp. 471–486, 1994.

7. A. Oppenheim and R. Schafer, Digital Signal Processing, 2nd. ed., Englewood, NJ: Prentice-Hall, 1975.

8. T. Miyashita, S. Sumida, and S. Sakaguchi, "Integrated Optical Devices Based on Silica Waveguide Technologies," SPIE Vol. 993 Integrated Optical Circuit Engineering VI, pp. 288–294, 1988.

9. R. Orta, P. Savi, R. Tascone, and D. Trinchero, "Synthesis of Multiple-Ring-Resonator Waveguides," *IEEE Photonics Technol. Lett.,* vol. 7, no. 12, pp. 1447–1449, 1995.

10. R. Adar, C.H. Henry, M.A. Milbrodt, and R.C. Kistler, "Phase Coherence of Optical Waveguides," *J. Lightw. Technol,* vol. 12, pp. 603–606, 1994.

11. C. Madsen and J. Zhao, "Post-fabrication Optimization of an Autoregressive Planar Waveguide Lattice Filter," vol. 36, no. 3, *J. Appl. Opt.,* pp. 642–647, 1997.

12. M. Kawachi, "Silica Wavegides on Silicon and Their Application to Integrated-Optic Components," *Opt. Quantum Electron.,* vol. 22, pp. 391–416, 1990.

13. J. Foresi, B. Little, G. Steinmeyer, E. Thoen, S. Chu, H. Haus, E. Ippen, L. Kimerling, and W. Greene, "Si/SiO$_2$ micro-ring resonator optical add/drop filters," CLEO Conf. Baltimore, MD, May 18–23, 1997, pp. CPD22–2.

14. D. Rafizadeh, J. Zhang, S. Hagness, A. Taflove, K. Stair, and S. Ho, "Nanofabricated waveguide-coupled 1.5-μm microcavity ring and disk resonators with high Q and 21.6-nm free spectral range," CLEO Conf. Baltimore, MD, May 18–23, 1997, pp. CPD23–2.

15. G. Barbarossa, A. Matteo, and M. Armenise, "Theoretical Analysis of Triple-Coupler Ring-Based Optical Guided-Wave Resonator," *J. Lightw. Technol.,* vol. 13, no. 2, pp. 148–157, 1995.

16. S. Suzuki, K. Oda, and Y. Hibino, "Integrated-Optic Double-Ring Resonators with a Wide Free Spectral Range of 100 GHz," *J. Lightw. Technol.,* vol. 8, no. 13, pp. 1766–1771, 1995.

17. Y. Ja, "Vernier Operation of Fiber Ring and Loop Resonators," *Fiber and Integrated Optics,* vol. 14, pp. 225–244, 1995.

18. K. Oda, S. Suzuki, H. Takahashi, and H. Toba, "An Optical FDM Distribution Experiment Using a High Finesse Waveguide-Type Double Ring Resonator," *IEEE Photonics Technol. Lett.,* vol. 6, no. 8, pp. 1031–1034, 1994.

19. K. Oda, N. Takato, and H. Toba, "A Wide-FSR Waveguide Double-Ring Resonator for Optical FDM Transmission Systems," *J. Lightw. Technol.,* vol. 9, no. 6, pp. 728–736, 1991.

20. A. Thelen, Design of Optical Interference Coatings. New York: McGraw-Hill, 1989.

21. P. Yeh, Optical Waves in Layered Media. New York: John Wiley, 1988.

22. H. Macleod, Thin-film Optical Filters. New York: McGraw-Hill, 1989.

23. H. Takashashi, "Temperature stability of thin-film narrow-bandpass filters produced by ion-assisted deposition," *Appl. Opt.,* vol. 34, no. 4, pp. 667–675, 1995.

24. M. Scobey and D. Spock, "Passive DWDM components using MicroPlasma optical interference filters," in Technical Digest, Optical Fiber Communications1996, pp. 242–243.

25. E. Dowling and D. MacFarlane, "Lightwave Lattice Filters for Optically Multiplexed Communication Systems," *J. Lightw. Technol.,* vol. 12, no. 3, pp. 471–486, 1994.

26. V. Narayan, E. Dowling, and D. MacFarlane, "Design of Multi-Mirror Structures for High Frequency Bursts and Codes of Ultrashort Pulses," *IEEE J. Quantum Electronics,* vol. 30, no. 7, pp. 1671–1680, 1994.

27. K. Hill, B. Malo, F. Bilodeau, D. Johnson, and J. Albert, "Bragg gratings fabricated in monomode photosensitive optical fiber by UV exposure through a phase mask," *Applied Physics Letters,* vol. 62, no. 10, pp. 1035–1037, 1993.

28. T. Erdogan, "Fiber Grating Spectra," *J. Lightw. Technol.,* vol. 15, no. 8, pp. 1277–1294, 1997.

29. H. Kogelnik, "Filter Response of Nonuniform Almost-Periodic Structures," *Bell System Techn. J.,* vol. 55, no. 1, pp. 109–126, 1976.

30. M. Ibsen, M. Durkin, M. Cole, and R. Laming, "Optimized Square Passband Fiber Bragg Grating Filter With In-band Flat Group Delay Response," *Electron. Lett.,* vol. 34, no. 8, pp. 800–801, 1998.

31. G. Bjork and O. Nilsson, "A New Exact and Efficient Numerical Matrix Theory of Complicated Laser Structures: Properties of Asymmetric Phase-Shifted DFB Lasers," *J. Lightw. Technol.,* vol. 5, no. 1, pp. 140–146, 1987.

32. M. Yamada and K. Sakuda, "Analysis of Almost-Periodic Distributed Feedback Slab Waveguides Via a Fundamental Matrix Approach," *Appl. Opt.,* vol. 26, no. 16, pp. 3474–3478, 1987.

33. T. Strasser, P. Chandonnet, J. DeMarco, C. Soccolich, J. Pedrazzani, D. DiGiovanni, M. Andrejco, and D. Shenk, "UV-induced Fiber Grating OADM devices for efficient bandwidth utilization," Optical Fiber Conference, San Jose, CA, Feb. 25-Mar. 1, no. PD8, pp. 1–4, 1996.

34. V. Mizrahi and J. Sipe, "Optical Properties of Photosensitive Fiber Phase Gratings," *J. Lightw. Technol.,* vol. 11, no. 10, pp. 1513–1517, 1993.

35. D. Johnson, K. Hill, F. Bilodeau, and S. Faucher, "New design concept for a narrowband wavelength-selective optical tap and combiner," *Electron. Lett.,* vol. 23, no. 13, pp. 668–669, 1987.

36. R. Kashyap, G. Maxwell, and B. Ainslie, "Laser-Trimmed Four-Port Bandpass Filter Fabricated in Single-Mode Photosensitive Ge-Doped Planar Waveguide," *IEEE Photonics Technol. Lett.,* vol. 5, no. 2, pp. 191–194, 1993.

37. T. Erdogan, T. Strasser, M. Milbrodt, E. Laskowski, C. Henry, and G. Kohnke, "Integrated-Optical Mach–Zehnder Add-Drop Filter Fabricated by a Single UV-Induced Grating Exposure," Optical Fiber Conference. San Jose, CA, Feb., 1996.

38. C. Madsen, J. DeMarco, C. Henry, E. Laskowski, R. Scotti, and T. Strasser, "Apodized UV-induced Gratings in Planar Waveguides for Compact Add-Drop Filters," in Optical Society of America Technical Digest Vol. 17, Bragg Gratings, Photosensitivity, and Poling in Glass Fibers and Waveguides: Applications and Fundamentals. Williamsburg, VA, Oct. 26–28, 1997, pp. 262–264.

39. F. Ouellette, "Dispersion cancellation using linearly chirped Bragg grating filters in optical waveguides," *Opt. Lett.,* vol. 12, pp. 847–849, 1987.

40. K. Ennser, M. Zervas, and R. Laming, "Optimization of Apodized Linearly Chirped Fiber Gratings for Optical Communications," *IEEE J. Quantum Electron.,* vol. 34, no. 5, pp. 770–778, 1998.

41. D. Atkinson, W. Loh, J. O'Reilly, and R. Laming, "Numerical study of 10-cm chirped-fiber grating pairs for dispersion compensation at 10 Gb/s over 600 km of nondispersion shifted fiber," *IEEE Photon. Technol. Lett.,* vol. 8, pp. 1085–1087, 1996.

42. F. Ouellette, P. Krug, T. Stephens, G. Dhosi, and B. Eggleton, "Broadband and WDM dispersion compensation using chirped sampled fibre Bragg gratings," *Electron. Lett.,* vol. 3, no. 11, pp. 899–901, 1995.

43. R. Kashyap, A. Ellis, D. Malyon, H. Froehlich, A. Swanton, and D. Armes, "Eight wavelength *x* 10 Gb/s simultaneous dispersion compensation over 100 km single-mode fiber using a single 10 nanometer bandwidth, 1.3 meter long, super-step-chirped fiber bragg grating with a continuous delay of 13.5 nanoseconds," Proc. European Conf. on Optical Communications (ECOC), p. ThB.3.2, 1996.

44. P. Russell, "Optical superlattices for modulation and deflection of light," *J. Appl. Phys.,* vol. 59, pp. 3344–3355, 1986.

45. P. Russell, "Bragg resonance of light in optical superlattices," *Phys. Rev. Lett.,* vol. 56, pp. 596–599, 1986.

46. B. Eggleton, P. Krug, L. Poladian, and F. Ouellette, "Long periodic superstructure Bragg gratings in optical fibers," *Electron. Lett.,* vol. 30, pp. 1620–1622, 1994.

47. M. Ibsen, M. Durkin, M. Cole, and R. Laming, "Sinc-sampled fiber Bragg grating for identical multiwavelength operation," in Vol. 2 1998 OSA Technical Digest Series (Optical Society of America, Washington, DC, 1998), Optical Fiber Communication Conference, San Jose, CA, Feb. 22–27, 1998, pp. 5–6.

48. R. Alferness, M. Joyner, M. Divino, M. Martyak, and L. Bull, "Narrowband grating resonator filters in InGaAsP/InP waveguides," *Appl. Phys. Lett.,* vol. 46, no. 3, pp. 125–127, 1986.

49. R. Zengerle and O. Leminger, "Phase-shifted Bragg-grating filters with improved transmission characteristics," *J. Lightw. Technol.,* vol. 13, no. 12, pp. 2354–2358, 1995.

50. L. Wei and J. Lit, "Phase-shifted Bragg grating filters with symmetrical structures," *J. Lightw. Technol.,* vol. 15, no. 8, pp. 1405–1409, 1997.

51. F. Bakhti and P. Sansonetti, "Design and realization of multiple quarter-wave phase-shifts UV-written bandpass filters in optical fibers," *J. Lightw. Technol.,* vol. 15, no. 8, pp. 1433–1437, 1997.

52. J. Foresi, P. Villeneuve, F. Ferrara, E. Thoen, G. Steinmeyer, S. Fan, J. Joannopoulos, L. Kimerling, H. Smith, and E. Ippen, "Photonic-bandgap microcavities in optical waveguides," *Nature,* vol. 390, 13 November, pp. 143–145, 1997.

53. G. Town, K. Sugden, J. Williams, I. Bennion, and S. Poole, "Wide-Band Fabry–Perot-Like Filters in Optical Fiber," *IEEE Photonics Technol. Lett.,* vol. 7, no. 1, pp. 78–80, 1995.

54. A. Srivastava, Y. Sun, J. Sulhoff, C. Wolf, M. Zirngibl, R. Monnard, A. Chraplyvy, A. Abramov, R. Espindola, T. Strasser, J. Pedrazzani, A. Vengsarkar, J. Zyskind, J. Zhou, D. Ferrand, P. Wysocki, J. Judkins, and Y. Li, "1 Tb/s transmission of 100 WDM 10 Gb/s channels over 400 km of TrueWave Fiber," in OSA Technical Digest Series, Vol. 2, Optical Fiber Communications Conference. San Jose, CA, 1998, p. PD10.

55. T. Strasser, "Fiber Grating Devices for WDM Communications Systems," in Optical Society of America Technical Digest Vol. 17. Bragg Gratings, Photosensitivity, and Poling in Glass Fibers and Waveguides: Applications and Fundamentals. Williamsburg, VA, Oct. 26–28, 1997, pp. 254–255.

PROBLEMS

1. For a Fabry–Perot interferometer, derive expressions for (a) the full-width half maximum (FWHM) of the transmission response, (b) cavity finesse $F = FSR/FWHM$, and (c) maximum to minimum transmission ratio using Eqs. (64)–(66) with the substitution $z^{-1} = e^{-j\Omega T}$.

2. Design a reflective lattice filter using the lowest order ARMA response that fits the desired gain equalization response to within 0.2 dB. Use the fifth-order AR design as the desired response. Provide the cavity thicknesses, partial reflectances, and relative phases.

3. For Vernier operation where there is no feedback path length of one unit delay, show that the poles must satisfy $\Sigma_{n=1}^{N} p_n = 0$ and $\Sigma_{n=1}^{N} p_n^{-1}$. Hint: use the sum of all optical paths.

4. Calculate the poles and zeros of the ARMA response for a three-layer thin film stack with a layer structure of $L_\infty |H|L|H|L_\infty$ where L_∞ denotes that the outside layers are infinite in extent.

   (a) Use titanium dioxide and silica. Let each layer be a quarter-wave thick at $\lambda = 1550$ nm.

   (b) Using the same materials, introduce a non-zero average phase across the layers of $\phi_n = \{0, 0.5, 0\} \pi$.

(c) Repeat part (a) using tantalum pentoxide and titanium dioxide. Compare the pole and zero locations between the large index contrast in part (a) and the low index contrast in part (c).

5. Design a partial reflector with an amplitude reflectance of $\rho = 50\%$ over 1520–1560 nm. Use only 2 materials in the design. Specify the refractive index and thickness of each layer.

CHAPTER SIX

Multi-Stage ARMA Filters

This chapter covers general ARMA optical filter architectures that allow the pole and zero locations to be specified independently, in contrast to the ARMA responses discussed in Chapter 5. The poles and zeros can be realized as separate sections by cascading multi-stage AR and MA filters; however, such a cascade arrangement may introduce undesirable excess loss. To minimize passband loss, an ARMA lattice architecture is advantageous. A single-stage lattice ARMA filter is described in Section 6.1 to provide physical insight into the filtering action of a ring within an MZI. Then, we discuss a general multi-stage ARMA lattice architecture in Section 6.2. A special type of ARMA filter is an all-pass filter, which ideally has a constant magnitude response. All-pass filters are well-known in analog and digital filter theory and practice. Optical all-pass filters are explained in Section 6.3, and applications for dispersion compensation are presented. All-pass filters can also be used in an interferometer to efficiently realize Butterworth, Chebyshev, elliptic and other optimal bandpass filters. The design of bandpass responses using all-pass filters is the subject of Section 6.4. General multi-stage ARMA optical filters are very new; so, we rely mainly on theoretical descriptions and simulations in contrast to the numerous fabrication results presented in Chapters 4 and 5 for MA and AR optical filters.

6.1 A MAXIMALLY FLAT ARMA FILTER

By combining a ring resonator inside an MZI as shown in Figure 6-1 [1,2], an ARMA response with a maximally flat passband can be obtained. The filter response differs from that of a typical asymmetric MZI because the ring introduces a frequency dependent, nonlinear phase in one arm instead of a linear phase. Let the

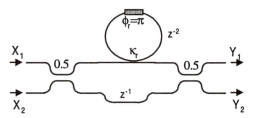

FIGURE 6-1   A maximally flat ARMA filter.

delay of the ring be given by $z_r^{-1} = e^{-j\Omega T_r}$ and the delay in the through-path arm of the MZI be denoted by $z_t^{-1} = e^{-j\Omega T_t}$ where T_r and T_t are the delays of the ring circumference and added through-path, respectively. To simplify the notation, we assume that $n_e \approx n_g$. For now, we shall use Ω instead of the normalized angular frequency ω, and explicitly indicate which delay is being used. Neglecting loss, the following transfer function for the ring is obtained by summing the contributions from all optical paths:

$$H_r(z_r) = \frac{e^{-j\phi}[\rho e^{+j\phi} - z_r^{-1}]}{1 - \rho e^{-j\phi}z_r^{-1}} \tag{1}$$

where $\rho = \sqrt{1 - \kappa_r}$ and $\phi = \phi_r$ determine the pole magnitude and phase and κ_r is the power coupling ratio of the ring. Equation (1) can also be obtained by setting $\kappa_1 = 1 - \rho^2$ and $\kappa_0 = 0$ in the expression for the single-stage, two-coupler ring given in Chapter 3. The ring has its antiresonance condition at $\Omega T_r + \phi = \pi(2m + 1)$ where m is an integer. For a ring with loss, the antiresonance condition equates to the frequency with maximum transmission, while the minimum transmission occurs at resonance where $\Omega T_r + \phi = 2\pi m$. The ring's phase response is defined by

$$\Phi_r(\Omega) = \tan^{-1}\left\{ \frac{\mathrm{Im}[H_r(z)]}{\mathrm{Re}[H_r(z)]} \right\}_{z = e^{j\Omega T_r}}$$

As a function of the pole location, the nonlinear phase response is given by

$$\Phi_r(\Omega) = \tan^{-1}\left\{ \frac{(1 - \rho^2)\sin(\Omega T_r + \phi)}{2\rho - (1 + \rho^2)\cos(\Omega T_r + \phi)} \right\} \tag{2}$$

The phase response for two different pole magnitudes is compared in Figure 6-2 with $\phi = 0$ and $\omega = \Omega T_r$ modulo 2π. The $\rho = 0$ case corresponds to a linear phase with a non-normalized slope of T_r. For the $\rho = 0.9$ case, the phase changes abruptly at resonance. Off-resonance, the phase changes much more slowly. The $\rho = 0.9$ case intersects the linear slope at $\omega + \phi = 0$ and $\omega + \phi = \pi$, so the phase difference for the MZI response is independent of the pole magnitude at these frequencies, as evident from Eq. (2).

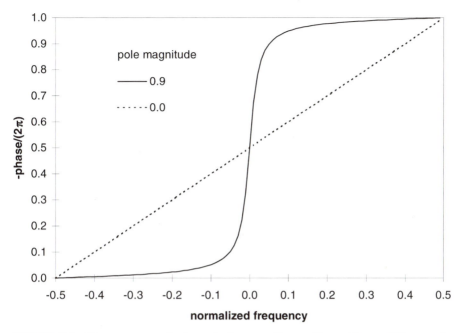

FIGURE 6-2   Phase response Φ_r for a single-stage all-pass filter with two different pole magnitudes assuming that $\phi = 0$.

Assuming that the MZI has perfect coupling ratios of $\kappa = 0.5$, the square magnitude responses for port X_1 to Y_1 and port X_1 to Y_2 are given by

$$|H_{11}(\Omega)|^2 = \sin^2\left(\frac{\Delta\Phi(\Omega)}{2}\right) \tag{3}$$

$$|H_{12}(\Omega)|^2 = \cos^2\left(\frac{\Delta\Phi(\Omega)}{2}\right) \tag{4}$$

where $\Delta\Phi(\Omega) = \Phi_r(\Omega) - \Omega T_t$. When $\Delta\Phi(\Omega)$ is an even multiple of π, $|H_{11}| = 0$. For an odd multiple, $|H_{11}| = 1$. The typical asymmetric MZI response, with a phase difference given by $\Delta\Phi(\Omega) = \Omega(T_r - T_t)$, is obtained when $\rho = 0$ ($\kappa_r = 1$). The phases for each arm of the MZI are illustrated in Figure 6-3 for $\phi = -\pi$, $T_r = 2T_t$, and a normalized angular frequency of $\omega = \Omega T_t$. The passband and stopband center frequencies occur where the phase difference is an integer multiple of π. To flatten the passband response, a constant phase difference around the passband center frequency is required. The solution is to set the slope of $\Phi_r(\Omega)$ equal to the group delay of the through-path arm, T_t, as follows [2]:

$$\tau_r(\Omega_c) = -\frac{d\Phi_r(\Omega)}{d\Omega}\bigg|_{\Omega = \Omega_c} = \frac{d[\Omega T_t]}{d\Omega} = T_t \tag{5}$$

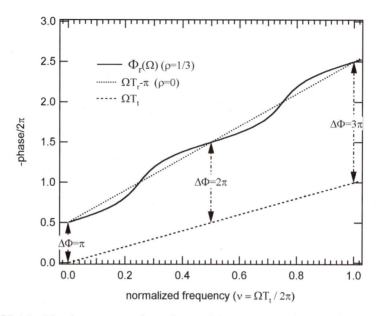

FIGURE 6-3   The phase response for each arm of the MZI. Two values of ρ for the ring resonator are shown.

where Ω_c is the passband center frequency. An explicit formula for the ring's group delay as a function of the pole location is

$$\tau_r(\Omega) = \frac{(1 - \rho^2)}{1 + \rho^2 - 2\rho \cos(\Omega T_r + \phi)} T_r \tag{6}$$

On resonance, the group delay is

$$\tau_r(\Omega)|_{\Omega T_r + \phi = 2\pi m} = \left(\frac{1 + \rho}{1 - \rho}\right) T_r \tag{7}$$

which has a value between T_r and ∞. The phase and group delay change rapidly around the resonant frequency. At antiresonance, the ring's group delay is

$$\tau_r(\Omega)|_{\Omega T_r + \phi = (2m+1)\pi} = \left(\frac{1 - \rho}{1 + \rho}\right) T_r \tag{8}$$

which has a value between 0 and T_r. The phase and group delay change slowly off-resonance, making it a good choice for passband flattening. Note that T_t must be less than T_r so that ρ can be chosen to satisfy Eq. (5). The ring's antiresonance condition is shifted to $\Omega T_r = 2\pi m$ by setting $\phi = \pm \pi$. Off-resonance, equating the group delay in each arm provides the following relationship:

$$T_t = \left(\frac{1 - \rho}{1 + \rho}\right) T_r \tag{9}$$

For $T_t = T_r/2$, the solution for the pole magnitude is $\rho = 1/3$, so the critical coupling ratio is $\kappa_c = 8/9$. The ring's phase response with the critical coupling ratio is illustrated in Figure 6-3. The magnitude response is shown in Figure 6-4. Both the bar- and cross-ports have flat passband responses. The responses for a conventional, asymmetrical MZI are also shown for comparison.

The Z-transform is now examined for this special architecture. Let $z = z_t$, $z^2 = z_r$, and $\phi = \pi$. The transfer matrix is then written as follows:

$$\frac{1}{A(z)}\begin{bmatrix} c_{2t} & -js_{2t} \\ -js_{2t} & c_{2t} \end{bmatrix}\begin{bmatrix} A^R(z) & 0 \\ 0 & z^{-1}A(z) \end{bmatrix}\begin{bmatrix} c_{1t} & -js_{1t} \\ -js_{1t} & c_{1t} \end{bmatrix} \tag{10}$$

where $A(z) = 1 + \rho z^{-2}$, $c_{nt} = \sqrt{1 - \kappa_{nt}}$, and $s_{nt} = \sqrt{\kappa_{nt}}$ for $n = 1,2$. A simplified expression is obtained by assuming that the couplers are identical with $\kappa_{1t} = \kappa_{2t} = 0.5$:

$$H_{11}(z) = \frac{1}{2}\left\{\frac{\rho - z^{-1} + z^{-2} - \rho z^{-3}}{1 + \rho z^{-2}}\right\} \tag{11}$$

$$H_{12}(z) = \frac{-j}{2}\left\{\frac{\rho + z^{-1} + z^{-2} + \rho z^{-3}}{1 + \rho z^{-2}}\right\} \tag{12}$$

The magnitude response for $H_{12}(z)$ is identical to the Butterworth 3rd-order filter with a cutoff at $\omega_c = \pi/2$, which has the following Z-transform:

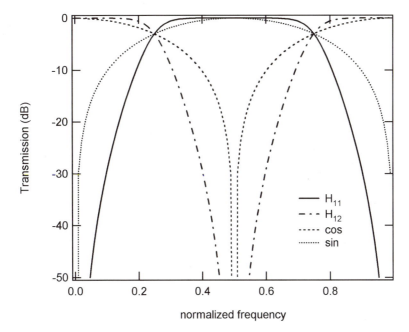

FIGURE 6-4   Amplitude response for a maximally flat ARMA filter compared to an asymmetric MZI response denoted by cos and sin.

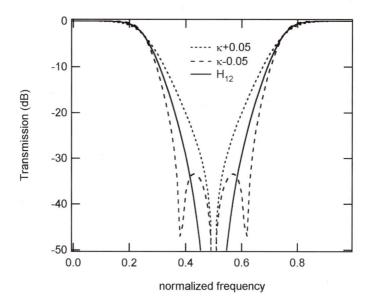

FIGURE 6-5   Sensitivity of maximally flat ARMA response to coupling ratio variations.

$$H_B^3(z) = \frac{0.1667 + 0.5z^{-1} + 0.5z^{-2} + 0.1667z^{-3}}{1 + 0.3333z^{-2}} \tag{13}$$

The critical coupling ratio is $\rho = \sqrt{1 - \kappa}$ or $\kappa = 8/9$, as before. Thus, the architecture in Figure 6-1 realizes a 3rd-order, maximally flat response.

A maximally flat ARMA filter was fabricated using silica planar waveguides with $\Delta = 0.7\%$ and a minimum bend radius of 5 mm [1]. The FSR was 10 GHz, and the passband width at the 0.5 dB points was 4 GHz, 1.8 times wider than a conventional asymmetric MZI. The cross-talk loss was 13.2 dB. The dependence of the spectral response on fabrication variations is illustrated by substituting $\kappa = \kappa_c \pm 0.05$ for the ring coupling ratio. The resulting amplitude response is shown in Figure 6-5. The stopband rejection is no longer monotonic, i.e. the zeros are no longer clustered at $z = -1$, which is characteristic of a Butterworth design. In Section 6.4.1, it is shown how higher order Butterworth filters with varying cutoff frequencies as well as Chebyshev and elliptic filters can be realized.

6.2   A GENERAL ARMA LATTICE ARCHITECTURE

Jinguji [3] demonstrated that a general multi-stage ARMA lattice architecture, shown in Figure 6-6, can be constructed from building blocks similar to the single stage described in Section 6.1. We derive the Z-transform description and the synthesis algorithm next. Then, design examples are given for bandpass filters and gain equalizers.

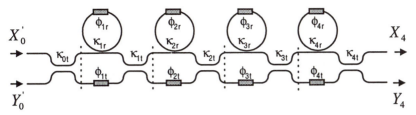

FIGURE 6-6   Multi-stage general ARMA lattice filter schematic illustrated for a fourth-order filter.

6.2.1   The Z-Transform Description

To analyze the lattice ARMA architecture, we begin by describing the transfer function for a single stage. The single stage consists of an MZI with nominally equal arm lengths as shown in Figure 6-7. One arm contains a ring resonator with a unit delay. There are four design variables per stage: the coupling ratio κ_{nt} at one end of the MZI, the coupling ratio κ_{nr} in the ring resonator, the phase controller ϕ_{nt} in one arm of the MZI, and the phase controller ϕ_{nr} in the ring resonator. The transfer matrix for a single stage Φ_n is defined by

$$\begin{bmatrix} X_n \\ Y_n \end{bmatrix} = \Phi_n \begin{bmatrix} X_{n-1} \\ Y_{n-1} \end{bmatrix} \tag{14}$$

The matrix Φ_n is given by the product of the individual matrices for one coupler, propagation through the ring in one arm, and propagation through a phase delay in the other arm. So,

$$\Phi_n = \begin{bmatrix} c_{nt} & -js_{nt} \\ -js_{nt} & c_{nt} \end{bmatrix} \begin{bmatrix} \dfrac{-e^{-j\phi_{nr}}A_n^R(z)}{A_n(z)} & 0 \\ 0 & e^{-j\phi_{nt}} \end{bmatrix} \tag{15}$$

where $A_n(z) = 1 - \rho_n z^{-1}$ and $A_n^R(z) = z^{-1}A_n^*(z^{*-1}) = -\rho_n + z^{-1}$. For this derivation, we include the magnitude and phase in defining the pole for the nth ring, $\rho_n = c_{nr}e^{-j\phi_{nr}}$.

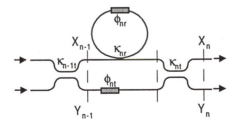

FIGURE 6-7   Single-stage ARMA filter: a symmetric MZI with a ring resonator in one arm.

The trigonometric definitions of the coupling ratios are repeated for use later in the derivation, $c_{nt} = \cos(\theta_{nt}) = \sqrt{1 - \kappa_{nt}}$ and $s_{nt} = \sin(\theta_{nt}) = \sqrt{\kappa_{nt}}$. Equation (15) is rewritten as

$$\Phi_n = \frac{1}{A_n(z)} \begin{bmatrix} -c_{nt}A_n^R(z)e^{-j\phi_{nr}} & -js_{nt}A_n(z)e^{-j\phi_{nt}} \\ js_{nt}A_n^R(z)e^{-j\phi_{nr}} & c_{nt}A_n^R(z)e^{-j\phi_{nt}} \end{bmatrix} \tag{16}$$

For an N-stage filter, the transfer matrix is equal to the product of individual matrices.

$$\Phi_{N0} = \prod_{n=0}^{N} \Phi_n = \frac{1}{G_N(z)} \begin{bmatrix} H_N(z) & jF_N^R(z) \\ jF_N(z) & H_N^R(z) \end{bmatrix} \tag{17}$$

where the equality on the right defines two numerator polynomials, $H_N(z)$ and $F_N(z)$, and the denominator polynomial $G_N(z)$. The first matrix is defined by

$$\begin{bmatrix} X_0 \\ Y_0 \end{bmatrix} = \Phi_0 \begin{bmatrix} X'_0 \\ Y'_0 \end{bmatrix} \text{ where } \Phi_0 = \begin{bmatrix} c_{0t} & -js_{0t} \\ -js_{0t} & c_{0t} \end{bmatrix} \tag{18}$$

The prime denotes the input as shown in Figure 6-6. For an input on X'_0, the bar- and cross-port transfer functions are given by $H_N(z)/G_N(z)$ and $jF_N(z)/G_N(z)$, respectively.

For filter design, we are interested in step-down recursion relations for reducing the filter order. Equation (17) is generalized for any order by replacing N by n where $n = 0, 1, \ldots, N$. The coefficients of the polynomials are

$$H_n(z) = h_0^n + h_1^n z^{-1} + \cdots + h_n^n z^{-n} \tag{19}$$

$$F_n(z) = f_0^n + f_1^n z^{-1} + \cdots + f_n^n z^{-n} \tag{20}$$

$$G_n(z) = g_0^n + g_1^n z^{-1} + \cdots + g_n^n z^{-n} \tag{21}$$

The reverse polynomials are defined as follows:

$$H_n^R(z) = (-1)^n z^{-n} \exp\left(-j \sum_{m=1}^{n} (\phi_{mt} + \phi_{mr})\right) H_n^*(z^{*-1}) \tag{22}$$

$$F_n^R(z) = (-1)^n z^{-n} \exp\left(-j \sum_{m=1}^{n} (\phi_{mt} + \phi_{mr})\right) F_n^*(z^{*-1}) \tag{23}$$

Equations (22) and (23) require the knowledge of all $\phi_{mt} + \phi_{mr}$ in order to define the reverse polynomials from the forward polynomials of $H_n(z)$ and $F_n(z)$. However, the roots of the reverse polynomials are located at mirror image points about the unit circle from the roots of the forward polynomial, so the reverse polynomials are easily defined to within a multiplicative constant. For the zeroth-stage, $\phi_{0r} = \phi_{0t} = 0$ and $\kappa_{0r} = 0$. Step-up recursion relations are defined for each polynomial as follows:

$$\frac{1}{G_n(z)}\begin{bmatrix} H_n(z) & jF_n^R(z) \\ jF_n(z) & H_n^R(z) \end{bmatrix} = \frac{1}{G_{n-1}(z)}\Phi_n(z)\begin{bmatrix} H_{n-1}(z) & jF_{n-1}^R(z) \\ jF_{n-1}(z) & F_{n-1}^R(z) \end{bmatrix} \tag{24}$$

The step-down relation for $G_{n-1}(z)$ is found by inspection of Eq. (24) to be

$$G_{n-1}(z) = \frac{G_n(z)}{A_n(z)} \tag{25}$$

The remaining recursions are found by inverting the matrix $A_n(z)\Phi_n(z)$.

$$H_{n-1}(z) = \frac{e^{+j\phi_{nr}(z)}}{A_n^R(z)}\tilde{H}_n(z) \tag{26}$$

$$F_{n-1}(z) = \frac{e^{+j\phi_{nt}(z)}}{A_n(z)}\tilde{F}_n(z) \tag{27}$$

where

$$\tilde{H}_n(z) = -c_{nt}H_n(z) + s_{nt}F_n(z) \tag{28}$$

$$\tilde{F}_n(z) = s_{nt}H_n(z) + c_{nt}F_n(z) \tag{29}$$

Since Eqs. (26) and (27) reduce the order of the polynomial, $\tilde{H}_n(z)$ and $\tilde{F}_n(z)$ must contain $A_n^R(z)$ and $A_n(z)$ as roots, respectively. By evaluating $\tilde{F}_n(z)$ at the root of $A_n(z)$, the following relationship is obtained for the nth through-port coupling ratio:

$$\theta_{nt} = \tan^{-1}\left(-\frac{F_n(\rho_n)}{H_n(\rho_n)}\right) \tag{30}$$

To define a valid coupling ratio, θ_{nt} must be real, so the ratio $F_n(\rho_n)/H_n(\rho_n)$ must be a real number [3]. This condition will be used in the synthesis algorithm discussed in Section 6.2.2. Only one more value, ϕ_{nt}, is required to completely define the nth stage so that the step-down recursion relations can be used. The through-path phase, ϕ_{nt}, is found by applying the real value requirement from Eq. (30) to the $n-1$ polynomials. The argument of the ratio of Eq. (27) to Eq. (26) evaluated at ρ_{n-1} gives the following equation:

$$\arg\left\{-\frac{F_{n-1}(\rho_{n-1})}{H_{n-1}(\rho_{n-1})}\right\} = \phi_{nt} - \phi_{nr} + \arg\left\{-\frac{A_n^R(\rho_{n-1})\tilde{F}_n(\rho_{n-1})}{A_n(\rho_{n-1})\tilde{H}_n(\rho_{n-1})}\right\} = 0 \tag{31}$$

The resulting solution for ϕ_{nt} is

$$\phi_{nt} = \phi_{nr} - \arg\left\{-\frac{A_n^R(\rho_{n-1})\tilde{F}_n(\rho_{n-1})}{A_n(\rho_{n-1})\tilde{H}_n(\rho_{n-1})}\right\} \tag{32}$$

For the final recursion, a value for ρ_0 must be defined to evaluate Eq. (32). We let $\rho_0 = 1$ ($\kappa_{0r} = 0$) which gives $A_0^R(z)/A_0(z) = 1$.

6.2.2 Synthesis Algorithm

For the design of ARMA filters, we assume that the numerator $H_N(z)$ and the denominator $G_N(z)$ polynomials are given. The roots of $G_N(z)$ define the ring parameters completely, i.e. both ϕ_{nr} and κ_{nr} for $n = 1, \ldots, N$. For optical filter synthesis, we need to determine a valid $F_N(z)$ polynomial. The determinant of $G_N(z)\Phi_{N0}(z)$ is

$$\det\{G_N(z)\Phi_{N0}(z)\} = H_N(z)H_N^R(z) + F_N(z)F_N^R(z) \tag{33}$$

Equation (33) is equal to the product of determinants of the individual matrices, $A_n(z)\Phi_n(z)$.

$$\det\{A_n(z)\Phi_n(z)\} = -A_n(z)A_n^R(z)e^{-j(\phi_{nt}+\phi_{nr})} \tag{34}$$

The product of Eq. (34) for $n = 1, \ldots, N$ is

$$\prod_{n=1}^{N} \det\{A_n(z)\Phi_n(z)\} = (-1)^N \exp\left(-j\sum_{n=1}^{N}(\phi_{nt} + \phi_{nr})\right)G_N(z)G_N^R(z) \tag{35}$$

where

$$G_N^R(z) = \prod_{n=1}^{N} A_n^R(z) = z^{-N}\prod_{n=1}^{N} A_n^*(z^{*-1}) \tag{36}$$

By substituting Eq. (35) into Eq. (33), we find that

$$F_N(z)F_N^R(z) = (-1)^N \exp\left(-j\sum_{n=1}^{N}(\phi_{nt} + \phi_{nr})\right)G_N(z)G_N^R(z) - H_N(z)H_N^R(z) \tag{37}$$

Equation (37) requires *a priori* knowledge of the ϕ_{nt}'s and ϕ_{nr}'s. By substituting the definitions for the reverse polynomials into Eq. (37), the following equation is found, for which all of the values on the right hand side are known:

$$F_N(z)F_N^*(z^{*-1}) = G_N(z)G_N^*(z^{*-1}) - H_N(z)H_N^*(z^{*-1}) \tag{38}$$

Note that $F_M^*(z^{*-1})$, $G_M^*(z^{*-1})$, and $H_M^*(z^{*-1})$ are non-causal, having powers of z instead of z^{-1}. Since we are only interested in the roots of the right hand side, this non-causality does not matter. The roots of the right hand side occur in pairs about the unit circle, which follows from the definition of the forward and reverse polynomials. Any combination of one root from each pair may be chosen for the roots of $F_N(z)$, so there are 2^N possibilities. For example, one choice is to pick all of the minimum-phase roots. A multiplicative factor, $\alpha = |\alpha|e^{-j\phi_\alpha}$, must be determined to define $F_N(z)$ completely. By evaluating Eq. (38) at one of the roots of $G_N(z)$, for example, ρ_N, the following expression results:

$$|\alpha|^2 \, F_N'(\rho_N)F_N'^R(\rho_N) = -H_N(\rho_N)H_N^R(\rho_N) \tag{39}$$

where $F_N(\rho_N) = \alpha F_N'(\rho_N)$. The phase is determined by the requirement that the ratio $F_N(\rho_N)/H_N(\rho_N)$ be real; therefore,

$$\phi_\alpha = \arg\{F_N'(\rho_N)\} - \arg\{H_N(\rho_N)\} \tag{40}$$

With the equations defined so far, we are able to design optical filters with general ARMA responses. Since the filters are passive, the transmission must be ≤ 1. To assure this, the first step in the design process is to scale the given polynomials, $H_N(z)$ and $G_N(z)$, with a constant multiplier, ζ, to satisfy the following equation:

$$\max\left\{ \frac{\zeta H_N(z)}{G_N(z)} \right\}_{|z|=1} \leq 1 \tag{41}$$

The design steps are summarized as follows assuming that $H_N(z)$ and $G_N(z)$ are given and that $G_N(z)$ has all of its poles within the unit circle:

1. Find ζ to normalize the peak transmission to unity
2. Calculate a valid $F_N(z)$ using Eqs. (38), (39) and (40).
3. Find the roots of $G_N(z)$, which define $\rho_n = c_{nr}e^{-j\phi_{nr}}$ for $n = 1, \dots, N$
4. Use Eqs. (30) and (32) to determine θ_{nt} and ϕ_{nt} for $n = N$
5. Calculate the next lower order polynomials using Eqs. (25), (26) and (27)
6. Repeat steps 4 and 5 until $n = 0$.

A direct implementation of the step-down recursion relations defined in Eqs. (25), (26) and (27) would require root finding for three equations on each iteration. A better approach suggested by Jinguji [3] is to first write expressions for each term of the next lower order polynomials. The terms of each polynomial are defined in Eqs. (19)–(21). Now, consider the restatement of Eq. (26) as follows:

$$(-\rho_n^* + z^{-1})\begin{bmatrix} h_0^{n-1} \\ h_1^{n-1} \\ \vdots \\ h_{n-1}^{n-1} \\ 0 \end{bmatrix} = -\rho_n^*\begin{bmatrix} h_0^{n-1} \\ h_1^{n-1} \\ \vdots \\ h_{n-1}^{n-1} \\ 0 \end{bmatrix} + \begin{bmatrix} 0 \\ h_0^{n-1} \\ h_1^{n-1} \\ \vdots \\ h_{n-1}^{n-1} \end{bmatrix} = e^{+j\phi_{nr}}\left\{ -c_{nt}\begin{bmatrix} h_0^n \\ h_1^n \\ \vdots \\ h_{n-1}^n \\ h_n^n \end{bmatrix} + s_{nt}\begin{bmatrix} f_0^n \\ f_1^n \\ \vdots \\ f_n^{n-1} \\ f_n^n \end{bmatrix}\right\} \quad (42)$$

The first coefficient of $H_{n-1}(z)$ is written explicitly as follows:

$$h_0^{n-1} = \frac{e^{+j\phi_{nr}}}{\rho_n^*}[c_{nt}h_0^n - s_{nt}f_0^n] = -\frac{e^{+j\phi_{nr}}}{\rho_n^*}\tilde{H}_0^n \quad (43)$$

and the remaining terms are

$$h_k^{n-1} = \frac{1}{\rho_n^*}[h_{k-1}^{n-1} - e^{+j\phi_{nr}}\tilde{h}_k^n] \quad (44)$$

for $k = 1, \ldots, n - 1$. An implementation problem is encountered if $\rho_n = 0$. By rewriting Eq. (44) in terms of h_{k-1}^{n-1} instead, this problem can be avoided as shown in Eq. (45).

$$h_{k-1}^{n-1} = \rho_n^* h_k^{n-1} - e^{+j\phi_{nr}}\tilde{h}_k^n \quad (45)$$

where $k = n, \ldots, 1$ and $h_k^{n-1} = 0$. By using the same approach, the following expressions for f_k^{n-1} and g_k^{n-1} are obtained.

$$f_{k-1}^{n-1} = \rho_n f_{k-1}^{n-1} + e^{+j\phi_{nt}}\tilde{f}_k^n \quad (46)$$

$$g_k^{n-1} = g_k^n + \rho_n g_{k-1}^{n-1} \quad (47)$$

for $k = 0, \ldots, n - 1$ and $f_{-1}^{n-1} = g_{-1}^{n-1} = 0$.

A special case arises during the evaluation of Eq. (32) when $\rho_{n-1} = \rho_n$, because $A_n(\rho_{n-1}) = 0$. In this case, the term-by-term definition of Eq. (46) is used for the numerator of the n–1st version of Eq. (30), and $H_N(z)$ is evaluated as before. The following simplification is obtained for $F_{n-1}(z)$ [3]:

$$F_{n-1}(\rho_n) = \sum_{k=0}^{n-1} f_k^{n-1} z^{-k}\bigg|_{z=\rho_n} = e^{+j\phi_{nt}}\sum_{k=0}^{n-1}(n-k)\tilde{f}_k^n \rho_n^{-k} \quad (48)$$

using the expression $f_k^{n-1} = e^{+j\phi_{nt}}\sum_{m=0}^k \rho_n^{k-m}\tilde{f}_m^n$. The special version of Eq. (32) is:

$$\phi_{nt} = \phi_{nr} - \arg\left\{\frac{A_n^R(\rho_{n-1})\sum_{k=0}^{n-1}(n-k)\tilde{f}_k^n \rho_{n-1}^{-k}}{\tilde{H}_n(\rho_{n-1})}\right\} \quad (49)$$

To avoid problems evaluating Eqs. (32) or (49) when $\rho_{n-1} = 0$, the numerator and denominator are multiplied by z^{n-1} before substituting ρ_{n-1}. The following simplification of Eq. (49) results from this process [3]:

$$\phi_{nt} = \phi_{nr} - \arg\left\{ -\frac{\tilde{f}_{n-1}^n}{\tilde{h}_n^n} \right\} \tag{50}$$

6.2.3 Design Examples

6.2.3.1 Bandpass Filter   As an example, an ARMA bandpass filter is designed to separate one out of 10 channels per FSR. Since we want to obtain a filter with the fewest number of stages, the elliptic design is chosen. It is optimal in the Chebyshev sense, i.e., equiripple, in both the passband and stopband. The following filter characteristics were chosen as requirements: a passband ripple of 0.1 dB, a stopband maximum of −30 dB, and a cutoff of 0.1 × FSR. A fourth-order elliptic filter meets the design criteria. For comparison, a tenth-order Butterworth filter is required to meet the same conditions. The magnitude responses for the elliptic and Butterworth filters are shown in Figure 6-8. The group delay and dispersion are plotted in Figures 6-9 and 6-10, respectively. In this example, the elliptical filter has the smaller passband group delay and dispersion. The waveguide design values obtained from the step-down recursion relations are listed in

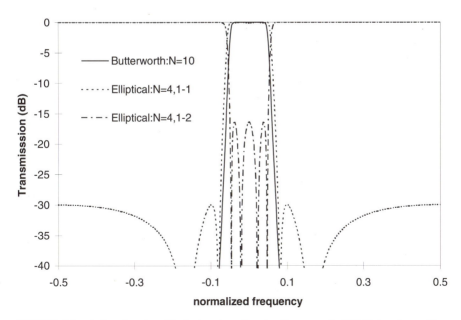

FIGURE 6-8   A fourth-order elliptical and tenth-order Butterworth ARMA bandpass filter magnitude response.

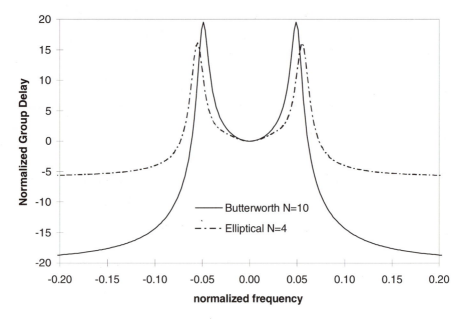

FIGURE 6-9   A comparison of ARMA bandpass filter group delay responses.

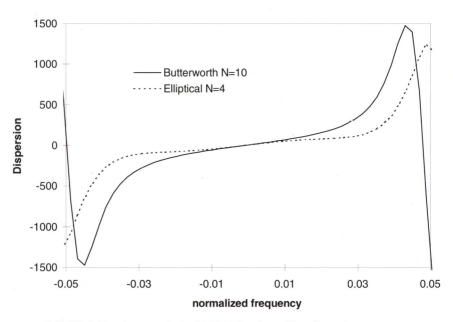

FIGURE 6-10   A comparison of ARMA bandpass filter dispersion responses.

TABLE 6-1   Design Values for a Fourth-Order ARMA Elliptical Bandpass Filter

	$n = 0$	1	2	3	4
$|h_n|$	0.0329	0.0916	0.1253	0.0916	0.0329
$\angle h_n$	0.0000	3.1416	0.0000	3.1416	0.0000
$|g_n|$	1.0000	3.3560	4.3466	2.5608	0.5781
$\angle g_n$	0.0000	3.1416	0.0000	3.1416	0.0000
$|f_n|$	0.7600	2.9599	4.4011	2.9599	0.7600
$\angle f_n$	1.5720	−1.5697	1.5718	−1.5699	1.5716
κ_{tn}	0.5010	1.0000	0.0000	1.0000	0.5000
ϕ_{tn}		1.2012	2.4457	−2.1629	−0.8314
κ_{rn}		0.1035	0.1035	0.3551	0.3551
ϕ_{rn}		−0.3480	0.3480	−0.1944	0.1944

Table 6-1. The coupling ratios for the through-path are [0.5, 1.0, 0.0, 1.0, 0.5]. The presence of 0% or 100% coupling simplifies the layout of the filter to a single MZI with several rings on each arm as shown schematically in Figure 6-11. The through-path coupling ratios retain this pattern even when the cutoff, passband ripple, or stopband rejection are varied. For the fourth-order elliptical filter, the denominator roots are: $0.9468\angle\pm0.3480$ and $0.8030\angle\pm0.1944$. Complex conjugate root pairs are placed on opposite sides of the MZI. The waveguide layout is greatly simplified in comparison to the general lattice architecture. The underlying properties which result in this simplification are examined in Section 6.4.1.

6.2.3.2   Gain Equalization Filter   Next, we design an ARMA filter having an equal number of poles and zeros to fit a desired gain equalization function (the same one as used in Chapters 4 and 5) to within 0.1 dB over a 20 nm range. For a starting point, the output of an AR model is used for the pole locations, and the zeros are set at $z = 0$. An optimization routine is then used to find the best fit in a least squares sense for a given order ARMA response. The shortest ARMA filter to meet the above criterion was third-order. The magnitude response is shown in Figure 6-12, and the waveguide design values are given in Table 6-2. Note that the coupling ratios are quite large, with a maximum of 0.9927. More realistic values can be obtained by allowing the peak transmission to be <1. Recall that the minimum order

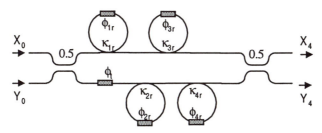

FIGURE 6-11   Simplified layout for a fourth-order ARMA elliptic bandpass filter.

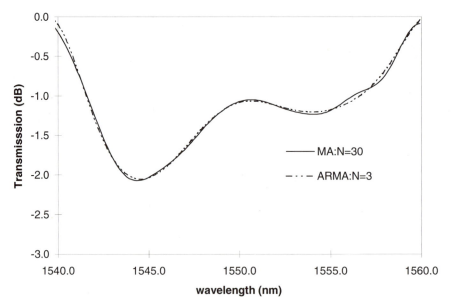

FIGURE 6-12   An ARMA third-order gain equalization filter.

was 5 for an AR filter and 6 for an MA filter. The ability to combine both poles and zeros in one filter allows a lower order filter to be used to meet the same requirements.

6.3   ALL-PASS FILTERS

A special type of ARMA filter is an all-pass filter. The frequency response of an all-pass filter is given by $A(\omega) = e^{j\Phi(\omega)}$. It has a unity magnitude response, and its phase

TABLE 6-2   Design Values for a Third-Order ARMA Gain Equalization Filter

	$n = 0$	1	2	3
$|h_n|$	0.8775	0.3211	0.0104	0.0042
$\angle h_n$	0.0000	3.0792	0.1732	−0.3200
$|g_n|$	1.0000	0.4157	0.0405	0.0183
$\angle g_n$	0.0000	−3.1031	2.3238	0.3115
$|f_n|$	0.4077	0.3508	0.1048	0.0380
$\angle f_n$	−0.0231	−3.0688	2.3307	0.4268
κ_{tn}	0.9927	0.0475	0.0342	0.7985
ϕ_{tn}		−0.5817	2.0954	−6.1892
κ_{rn}		0.8087	0.9531	0.9628
ϕ_{rn}		0.2363	−0.5626	−3.1267

can be tailored to approximate any desired response. Its Z-transform is expressed as the ratio of two polynomials as follows:

$$A_N(z) = e^{j\zeta}\frac{d_N^* + d_{N-1}^* z^{-1} + \cdots + z^{-N}}{1 + d_1 z^{-1} + \cdots + d_N z^{-N}} = e^{j\zeta}\frac{\displaystyle\prod_{k=1}^{N}(-z_k^* + z^{-1})}{\displaystyle\prod_{k=1}^{N}(1 - z_k z^{-1})} = \frac{D_N^R(z)}{D_N(z)} \qquad (51)$$

The zero locations are mirror images about the unit circle from the pole locations, thus providing a unity magnitude response for all frequencies. The numerator coefficients can be determined directly from the denominator by reversing the order of the coefficients and taking their complex conjugate. A constant phase factor, ζ, is included for generality; it does not change the magnitude or the group delay responses.

6.3.1   Optical All-Pass Filters

Optical all-pass filters can be realized by cascading multi-stage AR and MA sections [4]. Although a constant magnitude response will be achieved for a lossless filter, it will not have unity magnitude. A different approach is needed for a low loss implementation. A lossless ring resonator and a reflective cavity, as shown in Figure 6-13, are single-stage optical all-pass filters since their transfer function is given by

$$H(z) = -e^{-j\phi}\frac{(-\rho e^{+j\phi} + z^{-1})}{1 - \rho e^{-j\phi} z^{-1}} = -e^{-j\phi}\frac{D^R(z)}{D(z)} \qquad (52)$$

where $z^{-1} = e^{-j\Omega T}$, $T = c/n_g L$ the feedback path length is L, and the cavity group in-

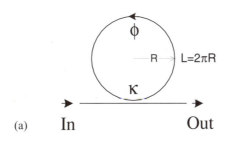

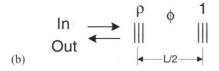

FIGURE 6-13   Single-stage optical all-pass filters: (a) a ring resonator and (b) a reflective cavity.

dex is n_g. For the ring resonator, $\kappa = 1 - \rho^2$. The reflective cavity shown in Figure 6-13b is a Gires-Tournois interferometer (GTI) [5]. The last reflector has an amplitude reflectance of $|\rho| = 1$. The FSR of the all-pass filter is $FSR = c/n_g L$ where $n_g = n_e - \lambda(dn_e/d\lambda)$.

Multi-stage all-pass filters can be realized by cascading single stages or using lattice architectures [6] as shown in Figure 6-14. For the cascade architectures, the synthesis algorithm is simply to take each pole of the desired response, z_n, and set

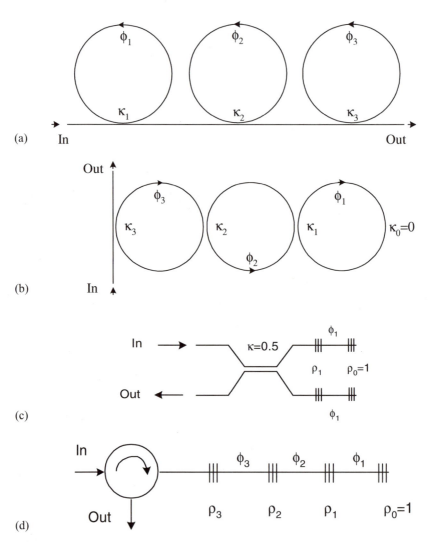

FIGURE 6-14   All-pass optical filter architectures: (a) ring cascade, (b) ring lattice, (c) single-stage cavity in a Mach–Zehnder for cascading, and (d) cavity lattice with circulator to separate the input and output.

the phase and reflection coefficient (or coupling ratio) accordingly. The reflection coefficient and coupling ratio are given by $\rho_n = |z_n|$ and $\kappa_n = 1 - |z_n|^2$, respectively. The cavity phase is $\phi_n = -\arg(z_n)$. As discussed in Chapter 5, the phase of the reflector is included in ϕ_n, and its group delay is included in the cavity's roundtrip or feedback path delay. For reflective cavities, extending the single stage response to multiple stages requires a method to separate the reflected output from the input. A circulator can be used between each stage as shown in Figure 6-14d, or identical filters can be written in the arms of a MZI to separate the output as shown in Figure 6-14c. The reflectances must be wavelength independent over the FSR so that the polynomial $D(z)$ has constant coefficients. Thin films with alternating layers of high and low index or Bragg gratings can be used to realize the reflectors.

Lattice architectures can also be used to realize multi-stage all-pass filters as shown in Figure 6-14b and d. The ring architecture is identical to the ring lattice architecture discussed in Chapter 5 except that $\kappa_0 = 0$. Using this substitution, the recursion relations and transfer functions obtained for synthesizing AR lattice filters are applicable to all-pass filter synthesis.

Lossless all-pass filters offer the flexibility to approximate a desired phase response arbitrarily closely by increasing the number of stages without changing the filter's magnitude response [6]. In contrast, the phase response cannot be controlled independently of the magnitude response for MA or AR filters. Three examples are given to illustrate the use of optical all-pass filters for engineering a desired phase response. The first example addresses fiber dispersion compensation. In the second example, the non-linear phase response of an elliptic filter is compensated across its passband. In the third example, the use of all-pass filters to realize a wavelength-dependent delay line is explored.

6.3.2   Fiber Dispersion Compensation

Dispersion compensating applications were first demonstrated for single-stage all-pass filters. A GTI was used to compensate for dispersion and laser chirp [7], and a fiber ring resonator was used to compensate for fiber dispersion [8]. A difficulty with the single-stage approach is that the group delay response and the bandwidth over which a desired response can be approximated are limited. By using multi-stage all-pass filters, a desired response can be more closely approximated, and it can be achieved over a broader portion of the FSR compared to single-stage filters. Compensation of linear group delay ($D \cong$ constant) from a transmission fiber is addressed in this example. A fourth- and sixth-order all-pass filter were designed to provide a constant dispersion of $12D_n$, where $D_n = d\tau_n/d\nu_n$ and the normalized frequency is defined relative to a starting frequency of f_0, $\nu_n = (f - f_0)/\text{FSR}$. The all-pass filter was designed to be symmetrical so that on one half of the FSR the dispersion is positive, and on the other half it is negative. The roots of $D_N(z)$ were found by minimizing the difference between the desired and all-pass filter dispersion responses. Unlike the filters discussed in Chapters 4 and 5, it was not necessary to trade off error between the magnitude and dispersion (or group delay) response in the optimization process. The results are shown in Figure 6-15. As the filter order is

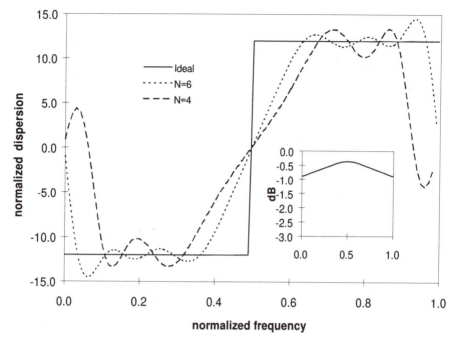

FIGURE 6-15   Dispersion for a 4th and a 6th-order all-pass filter. The amplitude response for the 6th-order filter with a loss of 0.1 dB/stage is shown in the inset.

increased, the error between the desired dispersion and the filter dispersion decreases. Table 6-3 shows the design values for the sixth-order filter. For the lattice design, the last coupling ratios are very close to unity. The cascade design, however, has practical coupling ratios.

From Chapter 3, the dispersion in ps/nm is given by $D = d\tau/d\lambda = -(T^2c/\lambda^2)D_n$. The quadratic dependence on the unit delay arises because the dispersion is propor-

TABLE 6-3   Design Values for a Sixth-Order All-Pass Filter for Linear Dispersion Compensation

$D(z)$	κ_l	$|\rho|$	κ_c	ϕ_c
1.000	0.0000	1.0000	0.8760	−1.8245
−1.1673	0.41731	0.7633	0.8364	−0.9364
0.6516	0.82732	0.4155	0.8130	−0.0221
−0.2625	0.96766	0.1798	0.8760	1.8245
0.0846	0.99638	0.0602	0.8364	0.9364
−0.0232	0.99965	0.0188	0.8130	0.0221
0.0038	0.99999	0.0038		

tional to the delay and inversely proportional to the FSR. As an example, let the FSR = 25 GHz (T = 40 ps). Then, for D_n = 12, the dispersion is 2,400 ps/nm which is sufficient to compensate 140 km of standard fiber with a dispersion of +17 ps/nm-km. For a ring implementation, the corresponding bend radius is 1.3 mm, assuming n_g = 1.45. For a sixth-order all-pass filter, the device size is on the order of 3 × 16 mm². Thus, all-pass filters have the potential to make very compact filters that will compensate many channels with their periodic frequency response.

Since the desired group delay response is required only over the channel passband width, FSRs on the order of the channel spacing are sufficient. In contrast to the tens of nanometers required for passband filter applications, dispersion compensation is a viable application for ring-based filters with currently available low loss, planar waveguide fabrication methods. Rings are also advantageous because they avoid the issue of separating the incoming and reflected signals associated with reflective cavities.

So far, lossless filters have been assumed. For practical filters, a finite feedback path loss must be considered. A 0.1 dB loss per feedback path was assumed in a simulation of the sixth-order all-pass dispersion compensating filter. The resulting magnitude response had a change in loss of 0.5 dB, as shown in the inset of Figure 6-15. Although the introduction of loss results in a frequency dependent magnitude response, such small propagation losses can be tolerated since the induced magnitude changes are gradual with respect to frequency. It is also possible to use gain to negate the loss; however, the gain must be controlled so that the filter does not become unstable and lase. Loss moves the pole and zero locations to γp_k and γz_k, respectively. When the loss becomes unacceptably high, a constant magnitude response may still be realized using the general ARMA filter architecture discussed in Section 6.2, where the pole and zero locations can be set independently. In this case, the poles and zeros are placed by design at p_k/γ and $1/p_k\gamma$, respectively. When the filter is fabricated, the loss moves them to their desired locations.

A two-stage dispersion compensating filter was fabricated using ring resonators in silica planar waveguides with Δ = 1.2%. The minimum bend radius was limited to 2.2 mm. The group delay and magnitude responses are shown in Figure 6-16. The filter had a FSR = 12.5 GHz and a 4.5 GHz passband width over which a linear group delay was achieved. Over the passband width, the group delay ripple was ±5 ps. The loss in the rings introduced a wavelength-dependent variation in the loss of the filter of 2 dB peak-to-peak across the FSR and slightly less across the passband. A dispersion of −4251 ps/nm was obtained, which is sufficient to compensate for 250 km of standard fiber.

6.3.3 Filter Dispersion Compensation

The AR or ARMA bandpass filters have significant group delay variation near steep band edges, since the filters are not linear-phase. At bit rates where the dispersion is significant compared to the bit period, the filter dispersion reduces the useable passband width and increases the minimum spacing of the adjacent channels [9,10]. Multi-stage optical all-pass filters have the capability to compensate

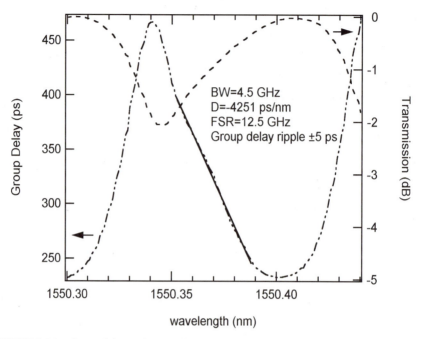

FIGURE 6-16 Group delay and spectral loss measurements on a 2-stage all-pass ring planar filter.

the non-linear phase without impacting the magnitude response. To demonstrate this capability, an all-pass filter is designed to compensate the group delay variation for the fifth-order elliptic filter shown in Figure 6-17. The group delay relative to its value at the middle of the passband is plotted. The elliptic filter's magnitude response in dB is shown in the inset. The following parameters were used for the elliptic filter design: a cutoff of 0.1 × FSR, stopband rejection of 50 dB and a passband ripple of 0.1 dB. The vertical markers indicate the magnitude response FWHM. Near the passband edges, the group delay varies substantially. To compensate the filter dispersion, a fourth-order all-pass filter was designed. The roots of $D_N(z)$ were found by minimizing the difference between the desired and all-pass filter group delay responses. The group delay is equalized over a much larger portion of the passband than before, as shown in Figure 6-17. The filter design values are given in Table 6-4. Each root of $D(z)$ gives the coupler and phase values for one stage of the cascade design vis-a-vis Eq. (52). For the lattice architecture, the step-down recursion relations from Chapter 5 are used to determine the filter design values. Since $N(z)$ and $D(z)$ have real coefficients, the phases for the lattice design are all zero.

The effects of the elliptic filter dispersion and all-pass filter compensation on a pulse are illustrated by a time-domain analysis [6]. A Gaussian pulse is examined after transmitting through the elliptic filter and subsequently after transmitting

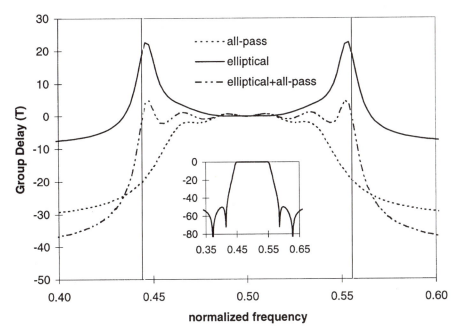

FIGURE 6-17   Group delay of a 5th-order elliptic bandpass filter and a 4th-order all-pass phase compensating filter.

through the all-pass filter. The Gaussian pulse is chosen such that its power spectrum suffers negligible loss upon filtering. A FWHM of $0.08 \times$ FSR was used for the pulse spectrum compared to a FWHM = $0.11 \times$ FSR for the elliptic filter. Since the elliptic filter has an approximately parabolic group delay over most of its passband, the filtered pulse undergoes mainly cubic dispersion effects (asymmetry and an oscillatory trailing part) and additional broadening of the main pulse on the rising edge side as shown in Figure 6-18. The all-pass filter substantially improves the pulse profile, by suppressing some of the trailing edge (resulting in less energy in

TABLE 6-4   Design Values for a Fourth-Order All-Pass Filter for Compensation of an Elliptic Filter's Nonlinear Phase Response

$N(z)$	$D(z)$	κ_r	ρ	κ_r	ϕ_r
0.6488	1.0000	0.0000	1.0000	0.1852	2.9180
2.8520	3.5403	0.0130	0.9935	0.2038	3.0694
4.7444	4.7444	0.0276	0.9861	0.1852	−2.9180
3.5403	2.8520	0.0811	0.9586	0.2038	−3.0694
1.0000	0.6488	0.5791	0.6488		

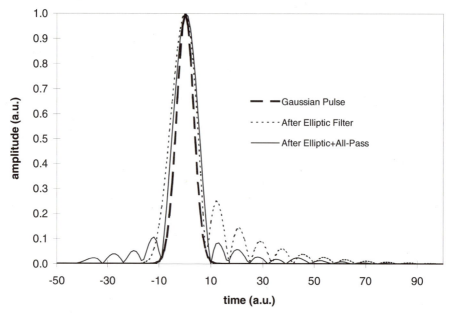

FIGURE 6-18   Gaussian pulse propagation through the elliptic and all-pass filters.

the pulse wings) and narrowing the main pulse, making it more like the original input Gaussian pulse.

6.3.4   Wavelength Dependent Delays

Next, the application of all-pass filters to realize wavelength-dependent delay lines is discussed. The group delay for a single-stage all-pass filter is wavelength dependent. As the pole approaches the unit circle, the group delay increases and the bandwidth decreases. This behavior can be used to vary the delay of a signal through the filter by moving the filter resonances to the signal wavelength or away from it. A constant group delay response can be achieved over a portion of the FSR by using two or more stages and staggering the pole (and zero) locations. Consider the group delay of four all-pass filters, each containing two staggered stages as shown in Figure 6-19 for the dotted curve. The pole magnitudes are 0.7, and the relative phase difference between the poles in each filter is 0.43 radians. To increase the maximum delay at a given wavelength, two or more of the filters are tuned to the same center frequency, as shown by the solid curve in Figure 6-19 where all of the filters are set to the same center frequency. The tuning is accomplished by changing the phase of each filter. Note that the maximum group delay is over 30 times the unit delay, whereas the maximum number of unit delays in the example is 8. Thus, a magnification of 3.75 over the non-resonant group delay is achieved for the structure. A larger group delay amplification factor can be obtained by increasing the pole mag-

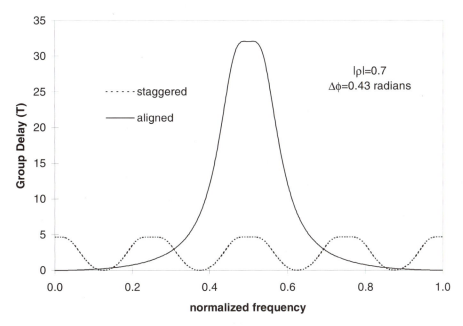

FIGURE 6-19   Group delay for four concatenated all-pass filters with 2 stages per filter for two cases: (1) all filters are equally spaced across the FSR (solid), and (2) all filters are aligned at a single frequency (dotted).

nitudes; however, a decrease in the bandwidth over which the delay is approximately constant will also occur. For example, a pole with a magnitude of 0.98 has a multiplication factor of $100 \times T$; however, the bandwidth over which the delay is > 90% of the maximum delay is only 0.2% of the FSR. Similar cavity enhanced group delay has been investigated using a Bragg grating in transmission at wavelengths which are just off-resonance [11]. Applications for variable delay lines include multiplexing low speed channels up to a high speed stream in time division multiplexing systems (TDM) [12]. The wavelength dependence can also be used to provide a differential delay between channels on a WDM system for applications where fiber dispersion or chirped gratings might otherwise be used, for example with chirped pulsed WDM sources [13]. All-pass filters offer a potentially more compact and tunable solution compared to other variable delay alternatives, but have the disadvantage of trading larger delays for smaller bandwidths.

Dispersion compensation and delay applications for all-pass filters use the filter response at or near resonance. We now address the impact of loss on a single-stage filter at resonance. A lossy all-pass filter's minimum transmission occurs at resonance and is given by

$$T_{\min} = \left| \frac{\rho - \gamma}{1 - \rho\gamma} \right|^2 \qquad (53)$$

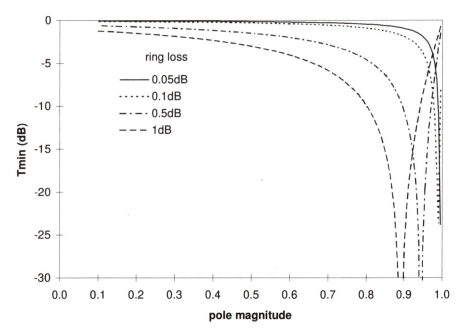

FIGURE 6-20   Transmission loss on-resonance as a function of pole magnitude and ring loss.

The minimum transmission is plotted as a function of pole magnitude in Figure 6-20. Clearly, it is not necessary to have lossless resonators, but losses should be minimized to allow a large range of pole magnitudes to be implemented without an associated large, wavelength dependent loss. Note that designs with smaller pole magnitudes can be implemented with significantly larger resonator losses and still maintain reasonably low transmission losses. Designs with smaller pole magnitudes are beneficial from another perspective; they are less sensitive to small changes in the pole magnitude that might be introduced during fabrication. *For time delay or dispersion compensation applications with lossy all-pass filters, the minimum FSR should be employed so that smaller pole magnitudes can be used and thus minimize the impact of loss and fabrication tolerances on the filter response!*

6.4   BANDPASS FILTERS

An important class of filters are the optimum bandpass filter designs, which include the Butterworth, Chebyshev, and elliptic filters that were discussed in Chapter 3. These filters are easily designed to meet desired bandpass and stopband requirements. The general lattice architecture described in Section 6.2 can realize such responses; however, a simpler architecture is now discussed [14].

The simplified architecture results because optimum bandpass filters can be realized as the sum or difference of two all-pass functions [15]. The frequency response of each all-pass function is written as follows:

$$A_1(\omega) = e^{j\Phi_1(\omega)} \text{ and } A_2(\omega) = e^{j\Phi_2(\omega)} \tag{54}$$

The sum and difference of A_1 and A_2 lead to the following expressions for the magnitude response of two new functions, G and H, respectively:

$$|G(\omega)|^2 = \left| \frac{1}{2}[A_1(\omega) + A_2(\omega)] \right|^2 = \cos^2\left(\frac{\Phi_1(\omega) - \Phi_2(\omega)}{2} \right) \tag{55}$$

$$|H(\omega)|^2 = \left| \frac{1}{2}[A_1(\omega) - A_2(\omega)] \right|^2 = \sin^2\left(\frac{\Phi_1(\omega) - \Phi_2(\omega)}{2} \right) \tag{56}$$

When both all-pass filters have the same phase, their sum is maximum. When their phases differ by π, the sum is zero but the difference is maximum. The resulting functions, G and H, are power complementary, i.e.

$$|G(\omega)|^2 + |H(\omega)|^2 = 1 \tag{57}$$

For filter synthesis, it is convenient to work with Z-transforms. We define $G(z)$ and $H(z)$ as

$$G(z) = \frac{1}{2}[A_1(z) + A_2(z)] = \frac{P(z)}{D(z)} \tag{58}$$

$$H(z) = \frac{1}{2}[A_1(z) - A_2(z)] = \frac{Q(z)}{D(z)} \tag{59}$$

Note that $G(z)$ and $H(z)$ share a common denominator polynomial $D(z)$, but have different numerator polynomials, $P(z)$ and $Q(z)$.

A simple decomposition algorithm is available for a class of functions which have polynomials with real coefficients and a numerator polynomial $P(z)$ with even symmetry. Even symmetry refers to $p_n = p_{N-n}$ and odd symmetry to $p_n = -p_{N-n}$. In addition, the function $H(z)$ must be power complementary to $G(z)$, and $Q(z)$ must be a linear-phase polynomial. The details of the synthesis method depend on whether $P(z)$ has even or odd order. If $P(z)$ has odd order, then the all-pass functions have order r and $N - r$, where r is determined by the decomposition algorithm, discussed in Section 6.4.1.1. If $P(z)$ has even order, then $A_1(z)$ and $A_2(z)$ are complex conjugates and have $N/2$ poles (and zeros) each.

The sum and difference are easily implemented in an optical filter using directional couplers. Let the directional coupler through- and cross-port transmission be $c = \sqrt{1 - \kappa}$ and $s = -j\sqrt{\kappa}$, respectively, where κ is the power coupling ratio. By

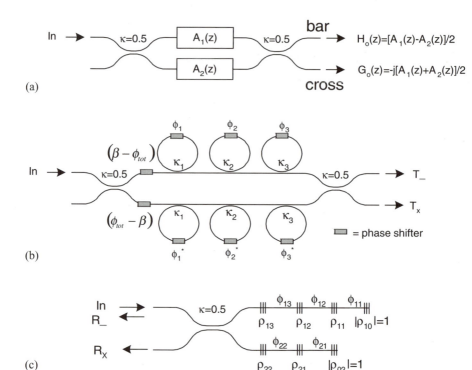

FIGURE 6-21 All-pass decomposition architectures: (a) schematic for a bandpass filter realized with two all-pass filters, (b) a cascade ring implementation showing the phase terms explicitly, and (c) a coupled cavity implementation using reflectors.

setting $\kappa = 50\%$, the sum and difference of the two all-pass functions are realized in the cross and bar transmission ports as shown in Figure 6-21a. The optical filter transfer functions are given by $G_o(z) = (-j/2)[A_1(z) + A_2(z)]$ and $H_o(z) = (1/2)[A_1(z) - A_2(z)]$. A bandpass architecture using multi-stage transmissive and reflective all-pass filters is shown in Figure 6-21b and c, respectively. The sum and difference are realized at the R_x and R_- outputs, respectively, for the reflective filters. By introducing the sum and difference of two all-pass functions, filters with two power complementary ARMA responses are achievable for thin film and grating-based devices.

The all-pass decomposition architecture allows the classical filters (Butterworth, Chebyshev, and elliptic) to be implemented exactly using thin films, Bragg gratings, or ring resonators as the basic filtering elements while eliminating the need to have a feed-forward path for each stage as in the general ARMA lattice filter. The transmissive architecture requires $N + 2$ couplers and $N + 2$ phase shifters to realize an Nth-order filter. In contrast, the general ARMA lattice architecture requires $2N + 1$ couplers and $2N$ phase shifters.

6.4.1 All-Pass Decomposition for 2 × 2 Filters

6.4.1.1 Optimum 2 × 2 Bandpass Filters An even and odd order elliptic filter are designed as examples to illustrate the all-pass decomposition algorithm. The details of the decomposition are the same as for digital filters [15]. It is assumed that there are no common factors between $P(z)$ and $Q(z)$. The power complementary condition of Eq. (57) is extended by analytic continuation to the z-plane using the following replacements:

$$P(\omega) \to P(z) \text{ and } P^*(\omega) \to P^R(z) = z^{-N}P^*(z^{*-1}) \tag{60}$$

and similarly for $Q(z)$ and $D(z)$. The power conservation expression in the z-domain is:

$$P(z)P^R(z) + Q(z)Q^R(z) = D(z)D^R(z) \tag{61}$$

Even-Order All-Pass Decomposition:

For an even order filter, both $P(z)$ and $Q(z)$ are symmetric. Assuming that they have real coefficients, $P^R(z) = P(z)$ and $Q^R(z) = Q(z)$ are substituted in Eq. (61) to yield

$$P^2(z) + Q^2(z) = D^R(z)D(z) \tag{62}$$

which can be factored as follows:

$$[P(z) + jQ(z)][P(z) - jQ(z)] = D^R(z)D(z) \tag{63}$$

where $D^R(z) = z^{-N}D^*(z^{*-1})$. The roots of $D(z)$ occur in complex conjugate pairs, and there are no real roots. If $D(z)$ had a real root at z_r, it would require that $P(z_r) = Q(z_r) = 0$ since both $P(z)$ and $Q(z)$ have real coefficients. Such a common factor between $P(z)$ and $Q(z)$ is ruled out in the preliminary statement of the problem. Half of the roots from the right hand side of Eq. (63) are assigned to the sum of $P(z)$ and $jQ(z)$, and their complex conjugates are assigned to the difference as follows:

$$P(z) + jQ(z) = e^{j\beta} \prod_{k=1}^{N/2}(1 - z_k z^{-1})(-z_k + z^{-1}) \tag{64}$$

$$P(z) - jQ(z) = e^{-j\beta} \prod_{k=1}^{N/2}(1 - z_k^* z^{-1})(-z_k^* + z^{-1}) \tag{65}$$

where β is a real constant. The minimum-phase roots of Eqs. (64) and (65) define $D(z)$ since the poles must lie within the unit circle for a stable filter. So,

$$D(z) = \prod_{k=1}^{N/2}(1 - z_k z^{-1})(1 - z_k^* z^{-1}) \tag{66}$$

By dividing Eqs. (64) and (65) by Eq. (66), expressions for the sum and difference of $G(z)$ and $jH(z)$ are obtained.

$$G(z) + jH(z) = e^{j\beta} \prod_{k=1}^{N/2} \frac{-z_k + z^{-1}}{1 - z_k^* z^{-1}} = A_1(z) \tag{67}$$

$$G(z) - jH(z) = e^{-j\beta} \prod_{k=1}^{N/2} \frac{-z_k^* + z^{-1}}{1 - z_k z^{-1}} = A_2(z) \tag{68}$$

The all-pass functions are defined to within a complex constant, $e^{j\beta}$. Expressions for $G(z)$ and $H(z)$ are:

$$G(z) = \frac{1}{2}[A_1(z) + A_2(z)] \tag{69}$$

$$H(z) = \frac{1}{2j}[A_1(z) - A_2(z)] \tag{70}$$

The constant β is defined as follows, once $Q(z)$ is determined:

$$e^{j\beta} \prod_{k=1}^{N/2} \frac{1 - z_k}{1 - z_k^*} = G(1) + jH(1) \tag{71}$$

To determine $Q(z)$, the roots of the right hand side of Eq. (72) must be found.

$$Q^2(z) = z^{-N} D(z^{-1}) D(z) - P^2(z) \tag{72}$$

Let the right hand side be denoted by $R(z) = \sum_{k=0}^{2N} r_k z^{-k}$, then the coefficients of $Q(z)$ are given by the following recursion relation:

$$q_n = q_{N-n} = \frac{r_n - \sum_{k=1}^{n-1} q_k q_{n-k}}{2q_0} \quad \text{for } n \geq 2 \tag{73}$$

where $q_0 = \sqrt{r_0}$, $q_1 = r_1/2q_0$.

As an example, an eighth-order elliptic filter is designed to have a 30 dB stopband rejection for both $G(\omega)$ and $H(\omega)$ and a cutoff of $0.1 \times$ FSR. The stopband rejection requirement for one response implies a passband ripple of 0.004 dB for the other response. The polynomials $P(z)$ and $D(z)$ were obtained from a digital filter design package. Then, $Q(z)$ was determined using Eqs. (72) and (73). The polynomials $P(z)$, $Q(z)$ and $D(z)$ are listed in Table 6-5. Note that $Q(z)$ has even symmetry. The denominator polynomials for the all-pass functions, which are both fourth-order, were then calculated from the minimum-phase roots of Eqs. (64) and (65). The denominator polynomials $D_1(z)$ and $D_2(z)$ are given in Table 6-5 where $A_1(z) =$

TABLE 6-5 Polynomial Coefficients for the Eighth-Order Elliptic Filter

$n(z^{-n})$	$P(z)$	$Q(z)$	$D(z)$	$D_1(z)$	$D_2(z)$
0	0.028905	0.656656	1.000000	1	1
1	−0.190293	−5.092601	−6.929052	−3.4645 − 0.1483i	−3.4645 + 0.1483i
2	0.577952	17.435711	21.270372	4.6227 + 0.3638i	4.6227 − 0.3638i
3	−1.059785	−34.415530	−37.744402	−2.8027 − 0.3076i	−2.8027 + 0.3076i
4	1.286524	42.831531	42.316488	0.6516 + 0.0864i	0.6516 − 0.0864i
5	−1.059785	−34.415530	−30.676715		
6	0.577952	17.435711	14.036958	$\beta = -1.39492$	
7	−0.190293	−5.092601	−3.705598		
8	0.028905	0.656656	0.432032		

$D_1^R(z)/D_1(z)$ and $A_2(z) = D_2^R(z)/D_2(z)$. The phase factor β is also shown. Figure 6-22 shows the magnitude response for the cross- and bar-port transmission. The coupling ratios are shown for both the cascade and lattice implementations in the inset of Figure 6-22. The coupling ratios are identical for A_1 and A_2, while the phases are complex conjugates of each other. A schematic of a cascade ring configuration is shown in Figure 6-21b for an even ordered filter. As seen for the single ring case in Eq. (52), there is a phase constant equal to the sum of the individual phases, i.e. ϕ_{tot}

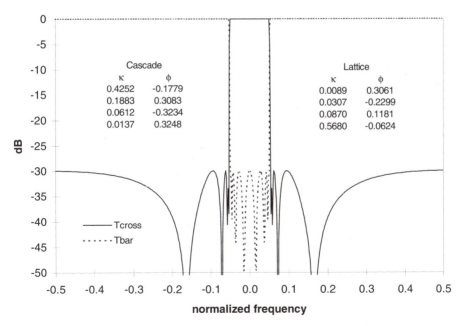

FIGURE 6-22 An eighth-order elliptic bandpass filter. Coupling ratios and phases for a cascade and lattice waveguide all-pass filter implementation are shown.

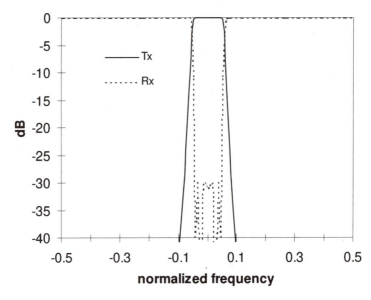

FIGURE 6-23   An optimized eighth-order AR filter with a 30 dB equiripple stopband.

$= \Sigma_{n=1}^{N} \phi_n$, associated with each all-pass transfer function that is not equal to β in general. Therefore, the phase terms β-ϕ_{tot} and ϕ_{tot}-β must be included in the upper and lower arms, respectively.

To demonstrate the difference between the elliptic filter response and an AR response, the filter response of the eighth-order elliptic bandpass filter is compared to an optimized AR response that also has a 30 dB stopband rejection for its power complementary response. The AR response represents the transmission response for a thin film filter. The optimized, eighth-order AR response is shown in Figure 6-23. Assuming a 40 nm FSR, the filter transition width from the passband to the stopband, defined at the 30 dB down points of the power complementary responses, is 1.4 nm for the AR response and 0.14 nm for the elliptic filter. By using all-pass filters in an MZI, an order of magnitude improvement in the filter rolloff is achieved! The all-pass filter decomposition architecture is particularly advantageous because the all-pass filter order that must be fabricated is half the total filter order, i.e. two fourth-order all-pass filters are needed instead of one eighth-order filter. For filters that demultiplex wavelength bands covering several WDM channels, a sharp rolloff is important so that a minimum guard band (or dead space) is required between bands.

Additional functionality is afforded by the all-pass decomposition architecture. By changing the phase in one arm of the MZI by π, the spectral responses switch between the outputs. The passband response becomes the stopband and vice-versa. Potential applications include wavelength dependent switching and modulation.

Odd-Order All-Pass Decomposition:

For an odd filter order, $Q(z)$ has odd symmetry, so $Q_R(z) = -Q(z)$. The following expression results from the power conservation property:

$$P(z)P(z) - Q(z)Q(z) = D^R(z)D(z) \tag{74}$$

Let

$$R(z) = Q^2(z) = P^2(z) - D^R(z)D(z) \tag{75}$$

Then, the following recursion relation is used to find the coefficients of $Q(z)$:

$$q_n = -q_{N-n} = \frac{r_n - \sum\limits_{k=1}^{n-1} q_k q_{n-k}}{2q_0} \quad \text{for } n \geq 2 \tag{76}$$

where $q_0 = \sqrt{r_0}$, $q_1 = r_1/2q_0$. Factoring the left hand side of Eq. (74) yields

$$[P(z) + Q(z)][P(z) - Q(z)] = D^R(z)D(z) \tag{77}$$

Since the coefficients of $P(z)$ and $Q(z)$ are real, the roots of $P(z)+Q(z)$ and $P(z)-Q(z)$ must be real or occur in complex conjugate pairs. Let the minimum-phase roots of $[P(z) + Q(z)]$ be denoted by $\{z_1, z_2, \ldots, z_r\}$ and the maximum phase roots by $\{1/z_{r+1}^*, \ldots, 1/z_N^*\}$. Then, the roots for the sum and difference of $P(z)$ and $Q(z)$ are given by

$$P(z) + Q(z) = \prod_{k=r+1}^{N} (-z_k^* + z^{-1}) \prod_{k=1}^{r}(1 - z_k z^{-1}) \equiv D_1^R(z)D_2(z) \tag{78}$$

$$P(z) - Q(z) = \prod_{k=r+1}^{N} (1 - z_k z^{-1}) \prod_{k=1}^{r}(-z_k^* + z^{-1}) \equiv D_1(z)D_2^R(z) \tag{79}$$

where $D(z) = D_1(z)D_2(z)$. Dividing Eqs. (78) and (79) by $D_1(z)D_2(z)$ yields the all-pass functions

$$A_1(z) = \frac{D_1^R(z)}{D_1(z)} = \prod_{k=r+1}^{N} \frac{-z_k^* + z^{-1}}{1 - z_k z^{-1}} = G(z) + H(z) \tag{80}$$

$$A_2(z) = \frac{D_2^R(z)}{D_2(z)} = \prod_{k=1}^{r} \frac{-z_k^* + z^{-1}}{1 - z_k z^{-1}} = G(z) - H(z) \tag{81}$$

The polynomials $G(z)$ and $H(z)$ are defined in terms of the all-pass functions as follows:

TABLE 6-6   Polynomial Coefficients for the Seventh-Order Elliptic Filter

z^{-n}	$P(z)$	$Q(z)$	$D(z)$	$D_1(z)$	$D_2(z)$
0	0.020704	0.653633	1.000000	1.000000	1.000000
1	−0.090034	−4.442218	−5.960679	−2.496684	−3.463995
2	0.148467	13.068580	15.456718	2.159719	4.648499
3	−0.078983	−21.569928	−22.568665	−0.632930	−2.848647
4	−0.078983	21.569928	20.018424		0.674337
5	0.148467	−13.068580	−10.778055		
6	−0.090034	4.442218	3.259370	$\beta = 0$	
7	0.020704	−0.653633	−0.426808		

$$G(z) = \frac{1}{2}[A_1(z) + A_2(z)] \tag{82}$$

$$H(z) = \frac{1}{2}[A_1(z) - A_2(z)] \tag{83}$$

For an odd filter order, $\beta = 0$.

To demonstrate an odd-order filter design, a seventh-order elliptic filter example is given. Table 6-6 lists the polynomials $P(z)$ and $D(z)$ obtained for the same pass-

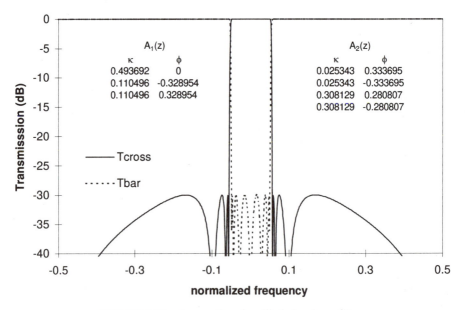

FIGURE 6-24   A seventh-order elliptic bandpass filter.

band ripple and rejection band loss requirement as the eighth-order example. The polynomial $Q(z)$ was calculated using Eqs. (75) and (76). The denominator polynomials for the all-pass functions were then calculated from the minimum-phase roots of Eqs. (78) and (79). For this example, $D_1(z)$ is third-order while $D_1(z)$ is fourth-order. Figure 6-24 shows the magnitude response and the coupling and phase design parameters for $A_1(z)$ and $A_2(z)$, assuming a ring cascade implementation. The cross-port magnitude response corresponding to $|G(\omega)|^2$ has a zero at $\omega = \pi$ in contrast to the even-order magnitude response.

So far, lossless filters were assumed. In the presence of loss, the all-pass property is broken. Each numerator zero moves to γ/ρ, whereas each pole moves to $\gamma\rho$; so, they are no longer reciprocals of each other. Using the eighth-order design, a loss of 0.2 dB/feedback path was simulated. The impact on the magnitude response is shown in Figure 6-25. The peak transmission of the cross-port $|G(\omega)|^2$ is reduced, but the 30 dB stopband rejection of the bar-port $|H(\omega)|^2$ is retained.

6.4.1.2 Multiple Passband Filters   The 2×2 design is now extended to allow multiple passbands per FSR so that several wavelengths, not necessarily adjacent, can be selected for adding or dropping in a similar fashion to the MA frequency selector [16] described in Chapter 4. The multi-wavelength selector uses two all-pass filters, $A_1(\omega)$ and $A_2(\omega)$, to construct two bandpass functions, $H_1(\omega)$ and $H_2(\omega)$. Each function is represented in the Z-domain as a ratio of polynomials as follows:

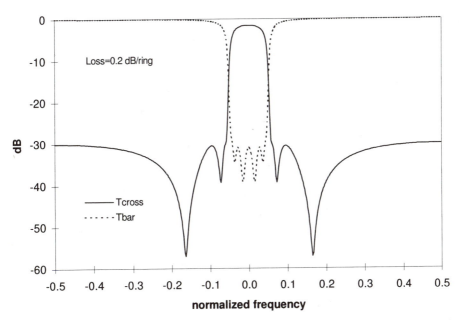

FIGURE 6-25   An eighth-order elliptic filter with 0.2 dB loss/feedback path.

$$H_1(z) = \frac{N_1(z)}{D(z)} \text{ and } H_2(z) = \frac{N_2(z)}{D(z)} \tag{84}$$

$$A_1(z) = \frac{D_1^R(z)}{D_1(z)} \text{ and } A_2(z) = \frac{D_2^R(z)}{D_2(z)} \tag{85}$$

where $D(z) = D_1(z)D_2(z)$. The numerator polynomials of the bandpass filters, $N_1(z)$ and $N_2(z)$, were previously assumed to have either even or odd symmetry. We now remove the symmetry requirement so that a multiple passband response, with each passband arbitrarily located within the FSR, can be obtained. The decomposition into all-pass functions is written as follows:

$$H_1(z) = \frac{1}{2}[A_1(z) + A_2(z)] \tag{86}$$

$$H_2(z) = \frac{1}{2}[A_1(z) - A_2(z)] \tag{87}$$

Given a desired $H_1(z)$ response, the $H_2(z)$ response can be found from power conservation. A simple way to determine $H_1(z)$ and $H_2(z)$ is by finding suitable zeros. If we multiply Eq. (57) (the statement of power conservation) by $|D(\omega)|^2$ and divide by $|N_1(\omega)|^2$, the following relationship is obtained [17]:

$$1 + \frac{|N_2(\omega)|^2}{|N_1(\omega)|^2} = \frac{|D(\omega)|^2}{|N_1(\omega)|^2} = \frac{1}{|H_1(\omega)|^2} \tag{88}$$

In the Z-domain, $K(z) = N_2(z)/N_1(z)$ is called the characteristic function. By evaluating the characteristic function on the unit circle, the magnitude response for $H_1(\omega)$ is defined. Similarly, by dividing Eq. (57) by $|N_2(\omega)|^2$ instead of $|N_1(\omega)|^2$, the magnitude response of $H_2(\omega)$ is obtained. The characteristic function for $H_2(\omega)$ is $1/K(z)$. The important result is that the characteristic function, which is defined solely by the numerator polynomials, defines the magnitude responses for both $H_1(\omega)$ and $H_2(\omega)$. Since 100% transmission is desired at certain frequencies, the zeros of the opposite response are located on the unit circle at these frequencies. When choosing the filter order for a particular desired response, it is important to consider that the change in the phase response of each all-pass filter over one FSR is limited by [18]

$$\Delta\Phi = \int_0^{2\pi} \tau(\omega)d\omega = 2\pi N \tag{89}$$

where N is the order of the all-pass filter. Thus, a reasonable starting point for the filter order is at least one stage per transition from stopband to passband or vice-versa.

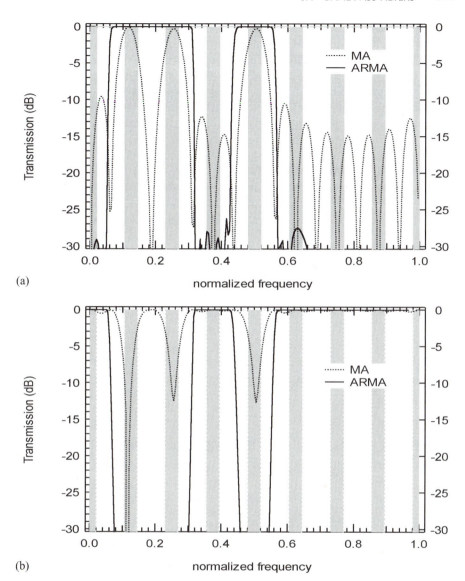

(a)

(b)

FIGURE 6-26 Comparison of MA and ARMA 8-channel selector: (a) dropping channels 2, 3 and 5, and (b) passing channels 1, 4, 6, 7 and 8.

As an example, a 16th-order filter was designed to drop channels 2, 3 and 5 out of 8 channels spaced uniformly across the FSR. An initial guess at the zero locations for $N_1(z)$ and $N_2(z)$ was made by distributing the zeros between the stopbands. The solution was refined by numerical optimization to yield the spectral responses shown in Figure 6-26 for $|H_1(\omega)|$ and $|H_2(\omega)|$. For comparison, the response for a

16th-order MA filter was also calculated and optimized. The desired response was a 30 dB stopband rejection and a passband width equal to 30% of the channel spacing. The channel widths are indicated by the shaded regions in Figure 6-26. For the drop-response, the ARMA selector exhibits a much flatter, wider passband and significantly better rejection. For the through-response, the MA filter offers very little rejection, which would limit its use as a multichannel add/drop filter. The ARMA selector's through-channel response, on the other hand, yields good rejection.

To obtain the design parameters for an optical ring or cavity implementation, the all-pass responses are needed. It is important to note that only the magnitudes of $H_1(\omega)$ and $H_2(\omega)$ are determined by Eq. (88), leaving a constant phase multiplier for each response undetermined. Since the all-pass filters are defined in terms of the sum and difference of $H_1(\omega)$ and $H_2(\omega)$, we can neglect any common phase term, but the relative phase must be determined. Let ϕ represent the phase difference, so the all-pass functions are now defined as follows:

$$A_1(z) = H_1(z) + e^{j\phi}H_2(z) \tag{90}$$

$$A_2(z) = H_1(z) - e^{j\phi}H_2(z) \tag{91}$$

From Section 6.4.1.1, we know that if $N_2(z)$ has odd order and symmetry, $\phi = 0$. If it has even order and symmetry, then $\phi = \pi/2$. For the general case, we must determine ϕ. Note that $|A_1(\omega)| = |A_2(\omega)| = 1$. Setting the magnitudes of Eqs. (90) and (91) equal to each other at an arbitrary frequency, say $z = 1$, results in the following expression:

$$|N_1(1) + e^{j\phi}N_2(1)| = |N_1(1) - e^{j\phi}N_2(1)| \tag{92}$$

Let the phase of $N_1(1)$ and $N_2(1)$ be represented by η_1 and η_2, then the condition for ϕ is given by

$$\phi = \eta_1 - \eta_2 + \frac{\pi}{2}(2m + 1) \tag{93}$$

where m is an integer. The roots of the denominator polynomials can now be determined. They are defined by the sum and difference of the numerator polynomials as follows:

$$N_1(z) + e^{j\phi}N_2(z) = D_1^R(z)D_2(z) \tag{94}$$

$$N_1(z) - e^{j\phi}N_2(z) = D_1(z)D_2^R(z) \tag{95}$$

For Eq. (94), the minimum-phase roots are used to define $D_2(z)$ and the maximum phase roots to define $D_1^R(z)$. A scale factor is then determined for $N_1(z)$ and $N_2(z)$ to make $|H_1(\omega)|^2 + |H_2(\omega)|^2 = 1$. The denominator of $A_1(z)$ and $A_2(z)$ are fully defined as well as the roots of the numerators. These roots completely define the magnitude for the all-pass response; however, their phase must also be determined. The phase is determined by evaluating Eqs. (90) and (91) at a particular frequency, for example $z = 1$.

$$A_1(1) = e^{j\alpha} \frac{D_1^R(1)}{D_1(1)} = \frac{N_1(1)}{D(1)} + e^{j\phi} \frac{N_2(1)}{D(1)} \tag{96}$$

$$A_2(1) = e^{-j\alpha} \frac{D_2^R(1)}{D_2(1)} = \frac{N_1(1)}{D(1)} - e^{j\phi} \frac{N_2(1)}{D(1)} \tag{97}$$

where α is the phase to be determined. The design process for a multi-wavelength selector is summarized as follows:

1. find the zeros of $N_1(z)$ and $N_2(z)$ which approximate the desired magnitude response
2. calculate the relative phase ϕ for $N_1(\omega)$ and $N_2(\omega)$ using Eq. (93)
3. determine the roots of $D_1(z)$ and $D_2(z)$ using Eqs. (94) or (95)
4. scale $N_1(\omega)$ and $N_2(\omega)$ so that $|H_1(\omega)|^2 + |H_2(\omega)|^2 = 1$
5. calculate the all-pass phase α using Eqs. (96) or (97)
6. find the coupling or reflection strengths and phases for each all-pass filter

The optical parameters in step 6 depend on the architecture chosen. Table 6-7 gives the coupler and phase values for both all-pass functions of the multi-wavelength selector assuming a cascade of single stages. The all-pass filters have different orders, $A_1(z)$ is ninth-order while $A_2(z)$ is seventh-order.

6.4.1.3   *Filters with Arbitrary Magnitude Responses*   The all-pass decomposition filter architecture is not limited to bandpass designs. A simple modification is to replace the 3 dB couplers in the MZI with other values. Let the coupling ratios be given by $\kappa = 0.3$ (5.2 dB) and assume that each arm has a four-stage all-pass filter

TABLE 6-7   All-Pass Filter Design Values for the ARMA Multichannel Selector

κ_1	ϕ_1	κ_2	ϕ_2
0.926	0.346	0.246	0.518
0.718	−2.584	0.724	0.392
0.474	0.452	0.342	−2.548
0.077	0.556	0.048	2.162
0.466	2.483	0.294	2.281
0.092	−2.569	0.355	2.843
0.002	−2.496	0.995	−2.806
0.106	2.923		
0.155	2.197		

as designed for the previous eighth-order elliptic filter in Section 6.4.1.1. The square magnitude response depends on the coupling ratios and all-pass filter phase responses as follows:

$$|G_o(\omega)|^2 = 4\kappa(1 - \kappa) \cos^2\left(\frac{\Phi_1(\omega) - \Phi_2(\omega)}{2} \right) \tag{98}$$

The spectral response shape is the same as before except that $|G_o(\omega)|^2$ has a maximum equal to –0.8 dB while its power complementary response has a minimum of –8 dB. Thus, each spectral response has two levels that can be varied by choosing the splitting and combining coupling ratios.

Now, we consider the design of a filter with an arbitrary ARMA spectral response using all-pass decomposition. The problem is to determine $A_1(z)$ and $A_2(z)$ so that one of the filter responses closely approximates the desired square magnitude response, $|H_D(\omega)|^2$. The filter order is equal to the sum of the number of stages in $A_1(z)$ and $A_2(z)$. If the desired response fluctuates between two levels, a bandpass filter design approach can be used with the MZI coupling ratios chosen using Eq. (98) and power conservation for the other response. For more general filter func-

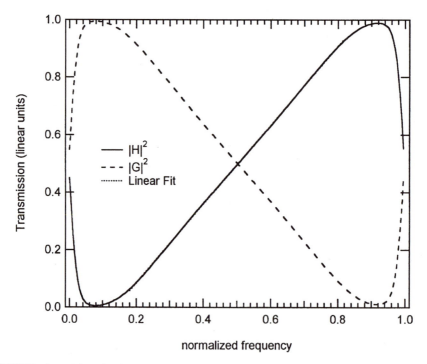

FIGURE 6-27 Linearized response using an interferometer with single-stage all-pass filters in the arms.

tions, the characteristic function approach is used to set the zeros for the numerator polynomials, $N_1(z)$ and $N_2(z)$. The numerator polynomials must be linear-phase so that Eqs. (94) and (95) are satisfied for the same value of ϕ.

As an example, a filter with a linear square-magnitude response over $0.15 \geq \nu \geq 0.85$ was designed. Only frequencies within this range were used in the numerical optimization. The square-magnitude response for a second-order filter is shown in Figure 6-27. The response is linear to within ± 0.005 over $0.18 \geq \nu \geq 0.82$. Each arm of the MZI contains a single-stage all-pass filter. The pole magnitudes and phases (rad) for the all-pass filters are $0.2925 \angle 0.0088$ and $0.7865 \angle -0.0099$. The remaining design parameters are $\kappa = 0.5$, $\phi = -0.7121$, and $2\alpha = -1.5896$ rad. Applications for such a response include frequency discriminators for laser wavelength stabilization and modulators which respond linearly to the input signal for analog transmission.

6.4.2   An $N \times N$ Router

Using all-pass filters, an $N \times N$ router with inherently low loss and flat passbands can be designed. By comparison, most approaches for flattening the passbands of WGRs trade off excess loss for passband flatness, as discussed in Chapter 4. The architecture using all-pass filters is discussed for devices with small N, for example $<$ 10. Demultiplexers realized with MA filters were discussed in Chapter 4. For small N, their spectral response lacks passband flatness and has large cross-talk between adjacent bands as shown for a 1×4 device in Figure 6-28a. This frequency response is obtained by connecting two MMIs, or similar $N \times N$ uniform power splitters, as shown in Figure 6-29a. An equivalent response is obtained by cascading 2×2 MZIs with FSRs that vary by powers of two from stage to stage, as discussed in Section 4.2.

For the MMI architecture shown in Figure 6-29a, the origin of the frequency response in Figure 6-28a is examined in detail to provide insight into the changes that need to be introduced by the all-pass architecture. Ideally, the MMI splitter excites the grating arms uniformly. The grating arm lengths increase by ΔL between consecutively numbered arms as shown in Figure 6-29a for $N = 4$ [19] instead of adjacent arms as in the typical slab coupler WGR. The combiner then images the input at one of the output ports depending on wavelength. This imaging may be described using a discrete Fourier transform. Let the grating arms be designated by k and the input and output port by i and n, respectively. Assuming that the splitter and combiner form a perfect imaging system, the frequency response is given by:

$$H_{ni}(\Omega) = \frac{1}{N} \sum_{k=0}^{N-1} e^{j[\Phi_k^G(\Omega) + \psi_{ik} + \theta_k + \psi_{kn}]} \tag{99}$$

where the phase introduced by each grating arm is represented by $\Phi_k^G(\Omega)$, the phase introduced by the splitter and combiner are given by ψ_{ik} and ψ_{kn}, respectively, and θ_k is a bias phase to route a signal from a particular input to output. For example, to

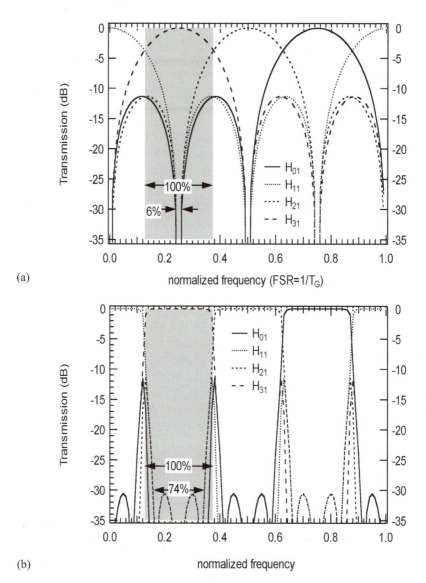

FIGURE 6-28   Spectral response for (a) a standard 4-stage MA filter and (b) the same filter with a single-stage all-pass filter in each arm.

route the center frequency from i' to n', $\theta_k = -(\psi_{i'k} + \psi_{kn'})$. For uniformly varying arm lengths, the phase is linear with frequency so that $\Phi_k^G(\Omega) = -k\Omega T_G$ where $T_G = n_g \Delta L/c$. The response is periodic with a FSR $= 1/T_G$. The square-magnitude response, $|H_{n1}(\Omega)|^2$, is shown in Figure 6-28a. Uniform splitting and combining is accounted for by the $1/N$ factor in Eq. (99). The linear-phase response $\Phi_k^G(\Omega)$ is shown in Figure 6-30a. The passband centers are marked by vertical lines. The phase dif-

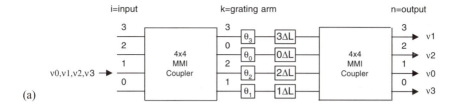

(a)

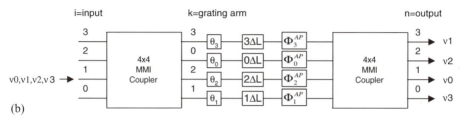

(b)

FIGURE 6-29   An $N \times N$ frequency selector: (a) standard design and (b) new architecture with all-pass filters.

ference between the arms determines the output port. To achieve a squarer passband response, we need to maintain the phase difference at each passband center over a larger bandwidth. As discussed in Section 6.1, a ring resonator in one arm of a MZI flattens the passband response for one choice of the coupling ratio. Extending this concept to a multiport interferometric device, one or more all-pass filters is added to each grating arm as shown in Figure 6-29b. A single-stage all-pass filter modifies the phase response by locally increasing the slope at frequencies near the resonance and flattening the slope away from resonance as discussed in Section 6.1. Using the all-pass filters off-resonance is the key to flattening the magnitude response of the router. Transmissive or reflective all-pass filters may be used. In the reflective case, the architecture in Figure 6-29b is folded about the all-pass filter. The phase values Φ_k^G and θ_k are reduced by a factor of 2 since a second pass is made through each one after reflecting from the all-pass filter.

In the conventional router, each grating arm has a group delay of

$$\tau_k^G(\Omega) = -\frac{d\Phi_k^G(\Omega)}{d\Omega} = kT_G$$

Let the router have a FSR equal to N times the channel spacing, so FSR $= 1/T_G = N\Delta f_{ch}$. To flatten the passband, we introduce an all-pass filter into each grating arm with a FSR equal to the channel spacing, $\Delta f_{ch} = 1/T_{AP}$, and set the filter's anti-resonance condition to the design center frequency, Ω_0. These two conditions give $T_{AP} = NT_G$. Next, we find values for each ρ_k to make the group delay of every grating arm identical at Ω_0. Since we can only add positive delays using the all-pass filters, we choose the new group delay for each arm to be equal to the longest arm of the con-

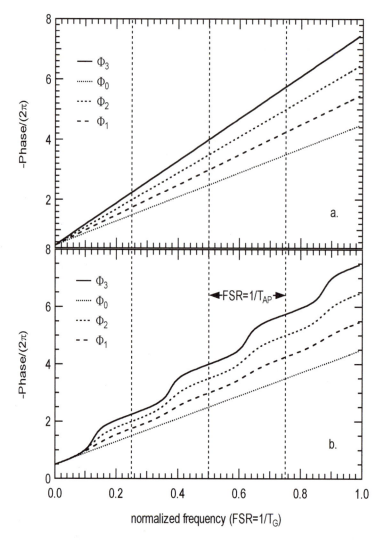

FIGURE 6-30   The phase response for a 4 × 4 MMI-WGR with an all-pass filter in each arm.

ventional design, which is NT_G. The group delay for the kth arm is now $kT_G + \tau_k^{AP}(\Omega_0) = NT_G$, which simplifies to

$$N = \left(\frac{1 - \rho_k}{1 + \rho_k}\right)N + k \qquad (100)$$

Solving for ρ_k gives

$$\rho_k = \frac{k}{2N - k} \tag{101}$$

for $k = 0, \ldots, N-1$. The range of reflection coefficients is $0 \leq \rho_k \leq (N-1)/(N+1)$. The all-pass filter phase is determined by the antiresonance condition, $\phi_k + \Omega_0 T_{AP} = \pi\{2m + 1\}$, which is independent of k. For constructive interference at the design center wavelength, $\phi_k = \pi$.

The phase response, $\Phi_k^G(\omega) + \Phi_k^{AP}(\omega)$ where $\omega = \Omega T_G$ modulo 2π, for the grating arms of a 4×4 router with a single-stage all-pass filter in each arm is shown in Figure 6-30b. The all-pass filter design values are: $\rho_k = [0, 0.1429, 0.3333, 0.6000]$. This 4×4 filter has a significantly flatter passband and the cross-talk is much lower compared to the response in Figure 6-28a. The usable width of each band, defined by the 30 dB cross-talk points, increased from 6% without the all-pass filters to 52%. For the same number of all-pass filters, a more optimal choice of pole magnitudes can be obtained by trading cross-talk for bandwidth in the Chebyshev sense of defining an acceptable stopband ripple. For a minimum 30 dB cross-talk loss, a stopband width of 74% is achieved as shown in Figure 6-28b. The reflectance values are: $\rho = [0.1484, 0.3828, 0.6112, 0.8512]$. The design is easily extended to larger N as long as uniform splitting and combining ratios are maintained. The passband width and cross-talk performance can be improved further by increasing the number of all-pass filter stages in each grating arm.

Next, we evaluate the impact of fabrication variations for the optimized 4×4

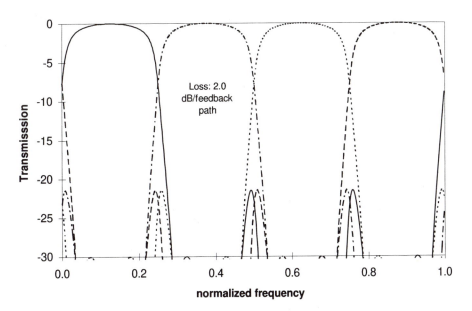

FIGURE 6-31 Spectral response of an optimized 4 ×4 MMI-WGR with 2 dB loss per feedback path in each all-pass filter.

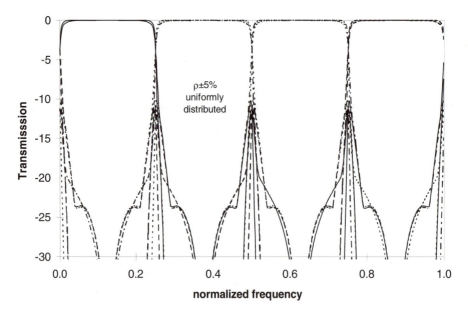

FIGURE 6-32 Simulation of the minimum and maximum spectral responses for each port given random variations on the partial reflection coefficients.

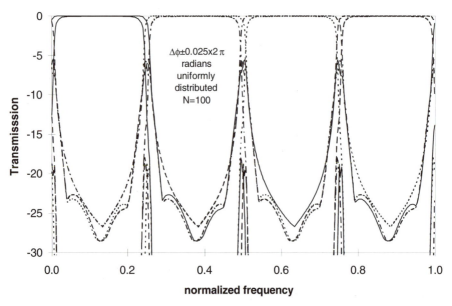

FIGURE 6-33 Simulation of the minimum and maximum spectral responses for each port given random variations on the phase of the all-pass filters.

design. First, a 2 dB loss per feedback path, which is quite large for current silica waveguide fabrication processes where losses are typically less than 0.1 dB/cm, is simulated. The results are shown in Figure 6-31. The general filter shape is maintained, although the passbands are rounded. Variations in the partial reflectors from their nominal values were simulated by assuming a uniformly distributed random variation of ±0.05 in each partial reflector. The maximum and minimum transmission at each frequency over 100 simulations is shown in Figure 6-32. Next, we simulated phase variations in the all-pass filters by allowing the phase to vary uniformly over the range ±0.025*2π radians from the nominal value of π. The maximum and minimum transmission for 100 runs are shown in Figure 6-33. For both the reflector and phase variations, >50% of the band maintains a crosstalk loss of >20 dB.

Next, we consider the impact of loss for a single-stage all-pass filter response. Off-resonance, the transmission is maximum and is given by the following expression:

$$T_{max} = \left| \frac{\rho + \gamma}{1 + \rho\gamma} \right|^2 \tag{102}$$

As an example, the maximum transmission is calculated for round-trip resonator

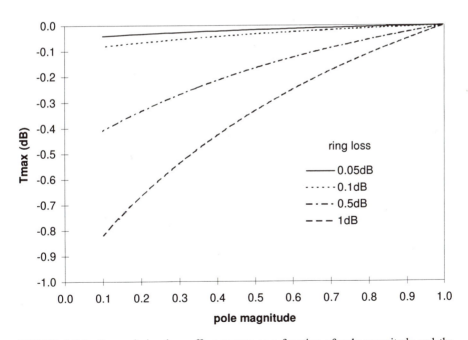

FIGURE 6-34   Transmission loss off-resonance as a function of pole magnitude and the ring loss.

losses of 0.05, 0.10, 0.50 and 1.0 dB. The results are shown as a function of the pole magnitude in Figure 6-34. Over the range of pole magnitudes of interest, the transmission loss is less than 1 dB in all cases and improves as the pole approaches the unit circle. *The use of all-pass filters off-resonance to create or modify bandpass responses is quite insensitive to resonator loss!*

The $N \times N$ router with all-pass filters functions as a router with low loss, flat passbands and low crosstalk. In the conventional architecture, WGRs with small N suffer from large cross-talk, and the passbands of MMI-based WGRs cannot be flattened with standard techniques. Both issues are addressed by the introduction of all-pass filters in the interferometer arms. The periodic frequency response of the device lends itself to demultiplexing very closely spaced channels for downstream filtering by courser devices. The routing capability is the same as the conventional WGR. One channel from the N channels in the FSR can be routed to any output by changing the bias phases, and the routing of the other channels in the FSR is determined from this choice.

REFERENCES

1. K. Oda, N. Takato, H. Toba, and K. Nosu, "A Wide-Band Guided-Wave Periodic Multi/Demultiplexer with a Ring Resonator for Optical FDM Transmission Systems," *J. Lightw. Technol.,* vol. 6, no. 6, pp. 1016–1022, 1988.

2. S. Suzuki, M. Yanagisawa, Y. Hibino, and K. Oda, "High-Density Integrated Planar Lightwave Circuits Using SiO_2–GeO_2 Waveguides with a High Refractive Index Difference," *J. Lightw. Technol.,* vol. 12, no. 5, pp. 790–796, 1994.

3. K. Jinguji, "Synthesis of Coherent Two-Port Optical Delay-Line Circuit with Ring Waveguides," *J. Lightw. Technol.,* vol. 14, no. 8, pp. 1882–1898, 1996.

4. B. Moslehi, "Cascaded fiber optic lattice filter," *U.S. Patent,* no. 4,768,850, 1988.

5. F. Gires and P. Tournois, "Interferometre utilisable pour la compression d'impulsions lumineuses modulees en frequence," *C. R. Acad. Sci,* vol. 258, no. 5, pp. 6112–6115, 1964.

6. C. Madsen and G. Lenz, "Optical All-pass Filters for Phase Response Design with Applications for Dispersion Compensation," *IEEE Photonics Technol. Lett.,* vol. 10, no. 7, pp. 994–996, 1998.

7. A. Gnauck, L. Cimini, J. Stone, and L. Stulz, "Optical equalization of fiber chromatic dispersion in a 5-Gb/s transmission system," *IEEE Photon. Technol. Lett.,* vol. 2, no. 8, pp. 585–587, 1990.

8. S. Dilwali and G. Pandian, "Pulse response of a fiber dispersion equalizing scheme based on an optical resonator," *IEEE Photonics Technol. Lett.,* vol. 4, no. 8, pp. 942–944, 1992.

9. B. Eggleton, G. Lenz, N. Litchinitser, D. Patterson, and R. Slusher, "Implications of fiber grating dispersion for WDM communication systems," *IEEE Photon. Technol. Lett.,* vol. 9, no. 10, pp. 1403–1405, 1997.

10. G. Lenz, B. Eggleton, C. Giles, C. Henry, R. Slusher, and C. Madsen, "Dispersive properties of optical filters for WDM systems," *J. Quantum Electron.,* vol. 34, no. 8, pp. 1390–1402, 1998.

11. M. Scalora, R. J. Flynn, S. Reinhardt, R. Fork, M. Bloemer, M. Tocci, C. Bowden, H.

Ledbetter, J. Bendickson, J. Dowling, and R. Leavitt, "Ultrashort pulse propagation at the photonic band edge: Large tunable group delay with minimal distortion and loss," *Phys. Rev. E54,* pp. 1078–1081, 1996.

12. K. Deng, K. Kang, I. Glask, and P. Prucnal, "A 1024-channel fast tunable delay line for ultrafast all-optical TDM networks," *IEEE Photon. Technol. Lett.,* vol. 9, no. 11, pp. 1496–1499, 1997.

13. L. Boivin, M. Nuss, W. Knox, and J. Stark, "206-Channel Chirped Pulse Wavelength Division Multiplexed Transmitter," *Electronics Lett.,* vol. 33, no. 10, pp. 827–829, 1997.

14. C. Madsen, "Efficient Architectures for Exactly Realizing Optical Filters with Optimum Bandpass Designs," *IEEE Photonics Technol. Lett.,* vol. 10, no. 8, pp. 1136–1138, 1998.

15. S. Mitra and J. Kaiser, *Handbook for Digital Signal Processing.* New York: John Wiley & Sons, 1993.

16. K. Sasayama, M. Okuno, and K. Habara, "Photonic FDM Multichannel Selector Using Coherent Optical Transversal Filter," *J. Lightw. Technol.,* vol. 12, no. 4, pp. 664–669, 1994.

17. A. Willson and H. Orchard, "Insights into digital filters made as the sum of two all-pass functions," *IEEE Trans. Circuits and Systems,* vol. 42, no. 3, pp. 129–137, 1995.

18. L. Rabiner and B. Gold, Theory and Application of Digital Signal Processing. Englewood Cliffs, N.J.: Prentice-Hall, Inc., 1974, p. 284.

19. P. Besse, M. Bachmann, C. Nadler, and H. Melchior, "The integrated prism interpretation of multileg Mach–Zehnder interferometers based on multimode interference couplers," *Optical and Quantum Electronics,* vol. 27, pp. 909–920, 1995.

PROBLEMS

1. Design a fourth-order GEF filter using all-pass filter decomposition. Assume the desired response is given by the third-order ARMA response listed in Table 6-2.

2. Design a 5th-order all-pass filter for a constant group delay of 10 times the unit delay over 10% of the FSR. Determine the minimum group delay ripple over the passband. Repeat for a constant group delay of 20 times the unit delay.

3. Using the filter designed in problem 2 for $\tau_g = 10T$, calculate the coupling ratios and phases for a lattice filter for (a) a lossless filter, (b) the same design with 0.5 dB loss/feedback path, and (c) redesign the filter for a constant magnitude response assuming the loss in part (b). In addition, design a cascade of MA and AR multi-stage filters assuming a lossless filter and calculate the magnitude response.

4. Design a filter with a 3-level staircase square magnitude response using a 2×2 MZI with a single-stage all-pass filter in one arm. Hint: Use the architecture discussed in Section 6.1 with different ratios of the delays in the ring to the through-path arm and a different ring coupling ratio.

5. Design a notch filter with $H(z) = 1/2[1 + A(z)]$ where $A(z)$ is a single-stage all-pass filter with a pole magnitude of 0.9. Compare the power complementary responses to a single-stage AR filter with the same pole magnitude.

CHAPTER SEVEN

Optical Measurements and Filter Analysis

In this chapter, an introduction to optical measurements is given which focuses on the characterization of filters. For a broader discussion on optical measurements, see [1]. We begin with the direct measurement of loss versus wavelength followed by techniques for measuring polarization dependent loss. For planar waveguide filters, the uncertainty in the fiber-to-waveguide coupling loss makes it difficult to characterize loss over small lengths. An indirect method for characterizing loss and coupling ratios is presented that avoids the coupling loss uncertainty. Complete characterization of optical filters requires information on both the magnitude and phase (or group delay) response. Direct and indirect methods for group delay measurement are presented.

In Section 7.2, the inverse problem of filter analysis is treated, whereby measurements are made to infer the filter's constituent components. For a waveguide filter, the constituent elements are the coupling ratios and phases for each stage. For a reflective filter, the components are the partial reflectance values and phases for each stage. As discussed in Chapters 4–6, multi-stage narrowband filters are sensitive to fabrication variations. Filter analysis techniques combined with post-fabrication tuning methods are critical to successfully demonstrating complex optical filters with tight performance specifications. Three analysis algorithms are discussed which have been demonstrated on fabricated devices. The first two are pertinent to MA filters and the latter addresses AR filter analysis.

7.1 OPTICAL MEASUREMENTS

7.1.1 Spectral loss

A filter's magnitude response is found by measuring its loss as a function of wavelength. The loss at a single wavelength is referred to as the insertion loss. A direct method for measuring insertion loss is addressed first, followed by a discussion of various choices for the source and detector.

7.1.1.1 *Insertion Loss*

Spectral loss requires the measurement of a device's insertion loss as a function of wavelength. Insertion loss (IL) is defined as the ratio of the power out of a device (P_{out}) to the power into the device (P_{in}) for one wavelength, input state of polarization, and typically for the fundamental mode of the waveguide. A basic insertion loss measurement involves a source, the device under test (DUT), and a detector as shown in Figure 7-1. Two measurements are required, one without the DUT for P_{in} and one with the DUT for P_{out}. The former measurement is referred to as the reference measurement. In the reference measurement, there may be some loss at the point of coupling between the source and detector (or input and output fiber). For the DUT measurement, there are two sources of excess loss: the coupling between the source (or input fiber) and DUT and between the DUT and detector (or output fiber). Variations in each of these excess losses contribute to the overall uncertainty in the insertion loss measurement for the DUT. As an example, consider the measurement of the coupling ratio for a planar waveguide directional coupler. The measured insertion loss includes two fiber-to-waveguide coupling losses (IL_{w-f}), the waveguide propagation loss (αL), and loss of the coupler, which is designated as the DUT (IL_{DUT}) in this example.

$$IL_{tot}(dB) = -10 \log_{10}\left(\frac{P_{out}}{P_{in}}\right) = 2IL_{w-f} + \alpha L + IL_{DUT} \tag{1}$$

The DUT may have multiple loss mechanisms such as transition loss at abrupt changes in curvature, bend loss, and coupling loss for a directional coupler. All of these are included in IL_{DUT}. Transmission through a straight waveguide of the same length is a convenient way to obtain a reference power level for calculating IL_{DUT}.

$$IL_{ref}(dB) = -10 \log_{10}\left(\frac{P_{ref}}{P_{in}}\right) = 2IL_{w-f} + \alpha L \tag{2}$$

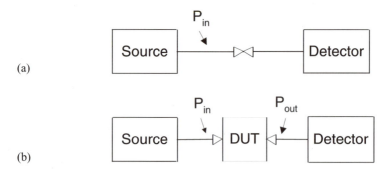

(a)

(b)

FIGURE 7-1 Basic insertion loss measurement: (a) reference and (b) with DUT.

The measurement of IL_{DUT} is independent of the absolute power into the waveguide for a linear device; so, only P_{out} and P_{ref} need to be measured.

$$IL_{DUT} = -10 \log_{10}\left(\frac{P_{out}}{P_{ref}}\right) \tag{3}$$

The repeatability of coupling losses to the DUT can dominate the uncertainty associated with insertion loss measurements. Slow power fluctuations of the source are tolerated by using an optical tap and an additional detector so that P_{in} and P_{out} are measured relative to the source power, P_{src}, at the time of the measurement. A directional coupler with a small coupling ratio is an example of an optical tap.

Other sources of measurement uncertainty include detector noise, polarization dependence, and coherence effects. When the insertion loss is large, causing the received power to be low, detector noise contributes to measurement uncertainty. At a specific wavelength, the difference in the maximum and minimum insertion loss with respect to polarization is defined as the polarization dependent loss (PDL) of the DUT. For a partially polarized source, any variation of the state of polarization over time into an element with PDL introduces power fluctuations. To minimize these effects, taps and other components with minimum PDL are used. Finally, highly coherent sources combined with discrete reflections, for example from connectors, contribute interference noise. A remedy is to minimize reflections by using fusion splices where possible, angled connectors, and isolators between reflections that cannot be eliminated or substantially reduced. Fusion splices can be optimized to have losses <0.1 dB and unmeasurably small PDL. Angle connectors are convenient for testing different components, but they introduce PDL. To reduce the impact of reflections, one can also broaden the source linewidth as discussed in Section 7.1.1.2.

For Bragg gratings, the reflection spectrum must be measured. Since the input and output are on the same port, a method is needed to separate them. One method, shown in Figure 7-2, is to use a circulator (or an isolator and a coupler). An isolator

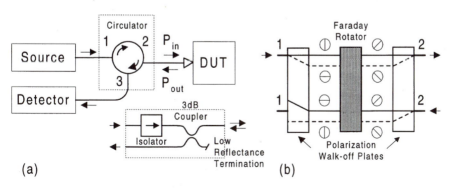

FIGURE 7-2   (a) Reflection (or return loss) measurement setup and (b) isolator principle of operation.

has negligible loss from the input to the output (Port 1→2), but substantially attenuates light traveling from Port 2→1 as illustrated in Figure 7-2b. A birefringent walk-off plate spatially separates the polarizations. The Faraday rotator is a non-reciprocal element that rotates the polarization by 45°. A second walk-off plate combines both polarizations at the output port. In the oposite direction, each polarization is rotated by –45 degrees and does not image onto the input port. A circulator uses similar principles. In a 3-port circulator, the low loss paths are Port 1→2 and 2→3 while the high attenuation paths are in the opposite direction.

7.1.1.2 Source and Detector Choices

There are many choices for sources and detectors. Two popular combinations are: (1) a broadband source and an optical spectrum analyzer (OSA), and (2) a tunable laser source and a power meter. For resolutions ≥ 0.1 nm, the broadband source and OSA measurement is fast. For higher resolutions or applications demanding more dynamic range, a tunable laser combined with either a power meter or OSA is desirable. For filter measurements, we are interested in the wavelength resolution and the source's output power for dynamic range considerations, spectral range, stability, and coherence.

Germanium or InGaAs p-i-n detectors are used for the 1550 nm range. Incident power levels between 0 dBm and –90 dBm can be accommodated. Higher incident powers require a well characterized, polarization independent attenuator in front of the detector. The detected signal-to-noise ratio (SNR) decreases as the incident power decreases; so, the lower limit on the incident power depends on the required SNR. For example, an incident power of –70 dBm may result in an SNR of 10 dB [1]. Germanium detectors exhibit wavelength and temperature dependence near 1550 nm, while InGaAs detectors have a nearly flat response over the 1550 nm band. Large area detectors allow repeatable coupling with low polarization dependence from a singlemode fiber; however, their response time is in the ms range. Faster detectors, which have smaller areas, are required for group delay measurements. To improve the sensitivity, the source can be intensity modulated. The detector output is then fed to a lock-in amplifier where it is mixed with the modulator drive signal and filtered to obtain a very narrow electrical detection bandwidth that reduces the receiver noise.

Broadband sources include lamps, light-emitting diodes (LEDs), and erbium-doped fiber amplified spontaneous emission (ASE) noise sources. For filter measurements, the bandwidth and spectral density that can be coupled into a DUT are critical. A tungsten lamp covers the infrared range (> 600 nm); however, its spectral density depends on the lamp temperature [2]. A spectral density of –63 dBm/nm at a lamp temperature of 2700 K can be coupled into a fiber. For higher temperatures, the lamp lifetime decreases. LEDs provide a larger spectral density and bandwidths of 70 to 100 nm. Edge emitting LEDs can couple –25 dBm/nm with bandwidths of 50 to 100 nanometers into a standard singlemode fiber [1]. Relative to the other broadband sources, edge emitting LEDs tend to have a higher degree of polarization. Erbium-doped fiber amplifiers pumped by semiconductor lasers at 980 nm or 1480 nm offer power spectral densities in the range of –10 dBm/nm [1] over a wavelength range of 1525 to 1565 nm.

Tunable lasers offer resolutions in the 1 to 20 pm range and large spectral densities. Tunable external cavity lasers (ECL) are commercially available for either the 1300 nm or 1550 nm bands with tuning ranges of 100 nm and output powers of 1 mW and higher. Wavelength discontinuities may result during tuning from two competing modes of the external cavity as shown in Figure 7-3, which compares the wavelength setting of the laser and the wavelength measured by a wavemeter. For picometer resolution measurements, this behavior is unacceptable since the mode hopping may occur during a measurement and lead to erroneous results.

An isolator on the source output prevents reflections in the test setup from affecting the source. Any pair of reflections in the measurement setup that are separated by a distance shorter than the coherence length of the source interfere and lead to fluctuations at the detector, resulting in measurement uncertainty. For a Lorentzian linewidth of $\Delta f_{\mathrm{FWHM}} = 100$ MHz, $L_c = 0.44\ c/n_g\Delta f_{\mathrm{FWHM}} \approx 0.9$ m in a fiber (ng ~ 1.47). Typical linewidths are <100 kHz for an ECL and 20 to 50 MHz for a DFB. To minimize coherence effects, the ECL or DFB may be modulated to broaden its linewidth. The relationship between coherence length and the spectral width is discussed in detail in Section 7.2.2.

As channel spacings in dense WDM systems become increasingly smaller, the need for precise wavelength calibration arises. Wavemeters with a resolution and accuracy on the order of a pm are commercially available. Many wavemeters are based on the Michelson interferometer, which is a folded MZI as shown in Figure 7-3. The number of fringes resulting when one path length is moved by a fixed distance are counted for the source wavelength and compared to the fringes of an inter-

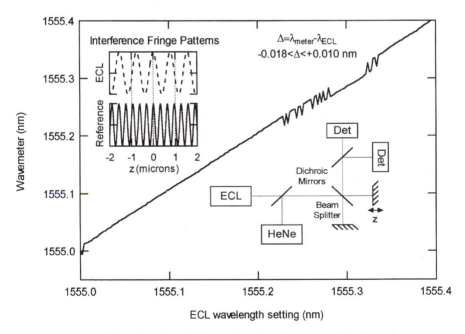

FIGURE 7-3   Mode hopping in an ECL wavelength tuning curve. Also shown is a wavemeter schematic employing a Michelson interferometer and the resulting interference fringes.

nal reference laser, such as a helium–neon laser. The ratio of fringes gives the ratio of the unknown wavelength to the HeNe wavelength ($\lambda_{vacuum} = 632.99076$ nm [1]) as follows: $\lambda_{meas} = (N_{ref}/N_{meas})(n_{meas}/n_{ref})\lambda_{ref}$ where N is the number of fringes, n is the refractive index, and the subscripts denote the measured and reference wavelengths. The refractive index of air at 15°C, sea level and no humidity is 1.000273 at 1550 nm [1]. The difference between the air and vacuum wavelength is 0.4 nm at 1550 nm, which is half a channel for a 100 GHz spaced WDM system. Reporting the wavelength in vacuum avoids questions about the air temperature, humidity and pressure. Also, the temperature at which the device is measured may be critical to making an accurate spectral loss measurement. Silica-based devices have a temperature dependence of 1.25 GHz/°C. For a ±1 pm measurement uncertainty due to temperature, the temperature must be specified to ±0.1°C!

In addition to dynamic range, an issue for measuring notch filters and stopbands in general is out-of-band light. For example, ECLs have a broad spectrum of ASE noise. The lasing wavelength and ASE noise are shown in Figure 7-4 over a 200 nm range with a 1 nm resolution. When the lasing wavelength is within the filter's notch, the integrated noise power detected by a power meter can be significant compared to the transmission of the lasing wavelength, so the null appears to be shallower than it actually is. A similar result may occur with a broadband source and OSA due to the finite stopband rejection of the OSA's diffraction grating. The combination of ECL and spectrum analyzer or ECL plus a tunable filter provides sufficient rejection of the out-of-band noise to measure deep transmission notches, typical of fiber gratings. A tunable filter which tracks the laser center wavelength can be used to significantly reduce the broadband noise, as shown in Figure 7-4.

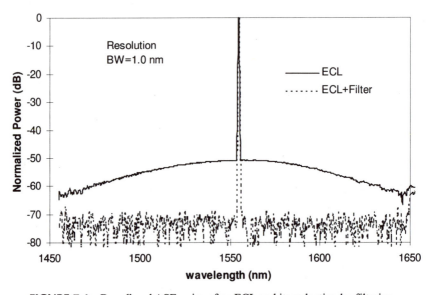

FIGURE 7-4 Broadband ASE noise of an ECL and its reduction by filtering.

7.1.2 Polarization Dependent Loss

An interferometric device realized in a birefringent material exhibits PDL. To characterize the filter, we need both the insertion loss as a function of polarization and wavelength. If the wavelength is fixed and we vary the input state of polarization to find the maximum and minimum loss, the measurement resembles a PDL measurement. If we fix the polarization state and sweep the wavelength, the measurement resembles a spectral loss measurement. The latter approach is appropriate for planar waveguide devices where there are two well defined principal states of polarization. The relative wavelength shift between the spectral responses for the TE and TM polarizations is called the polarization wavelength dependence (PWD).

Birefringence introduces another concern, polarization mode dispersion (PMD). PMD for transmission fiber increases as the square root of distance due to random coupling and orientation of the birefringence along the fiber length. High quality transmission fiber has a PMD of ≤ 0.1 ps/$\sqrt{\text{km}}$. A span of 100 km contributes 1 ps; consequently, components are typically specified to have minimal PMD by comparison, for example <0.1 ps. For a waveguide filter with PWD = 0.1 nm, its birefringence is $\Delta n = 1 \times 10^{-4}$. For a 10 cm device, the relative delay is $\Delta \tau = \Delta n / c$ which corresponds to 0.03 ps for glass. For short filter lengths, polarization dependent wavelength shifts of the spectral response are a bigger concern for system applications than the PMD.

Two methods are introduced to characterize loss versus input state of polarization. The difference between them is that a particular input state is well defined in the first case; whereas, knowledge of the orientation of the principal states is not required in the second. When a desired polarization must be maintained, for example from the source to the DUT, a polarization maintaining fiber (PMF) may be used in lieu of free space coupling optics. A PMF has very different propagation constants for two well-defined axis. The birefringence may be created by deforming the core shape or introducing an asymmetric stress across the core. PMF is characterized by an extinction ratio describing the coupling to the orthogonal mode when light is launched into one state. Extinction ratios of 30 dB are available commercially. Light must be launched along one of the principal axes in order for the polarization to be maintained. A linear polarizer may be used to define the polarization state launched into the PMF. Components for setting or modifying the state of polarization are also needed for PDL measurements. Waveplates provide this functionality. They are made of birefringent material and have an optical thickness measured as a fraction of the wavelength, $\Delta n L = \lambda / m$ where Δn is the birefringence, L is the plate thickness, $m = 2$ for a half-wave plate and $m = 4$ for a quarter-wave plate. A polarizer combined with a rotatable $\lambda/4$ and $\lambda/2$ waveplate, as shown in Figure 7-5, allows any state of polarization to be defined in a known manner. Another way to vary the polarization is to use fiber paddles, whereby loops of fiber are rotated relative to each other to change the output state of polarization [3]. The stress in the fiber bend provides the birefringence, and the number of bends is chosen to correspond to $\lambda/4$ or $\lambda/2$ waveplates. Bend diameters are on the order of 1″ for λ ~1550 nm. The relative tilt between the paddles provides the equivalent of rotating the waveplates. The exact output state is not known, but one is able to sample all of the polarization

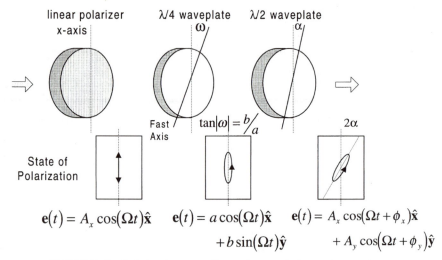

FIGURE 7-5 Polarization controllers using a polarizer and waveplates.

states to obtain a sufficiently accurate maximum and minimum transmission through a DUT.

For many waveguide devices, the spectral loss depends on the input polarization. When a device has a well defined axis for the polarization dependence, then a measurement solution is to launch power into one of the principal axis of a PMF and align it to the axis of the DUT, for example, vertically or horizontally with respect to the substrate plane of a planar waveguide. To determine the orientation of the PMF's axis to the plane of the substrate, a polarizer is placed between the PMF and a detector. The PMF is rotated to find the maximum and minimum transmittance positions. A measurement setup for polarization resolved spectral loss is shown in Figure 7-6, where the polarization controller may be realized in several different ways depending on the source degree of polarization and fiber type between the source

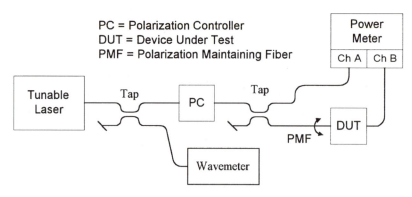

FIGURE 7-6 A TE/TM resolved spectral loss measurement setup.

and polarization controller. The power into the DUT is monitored in Channel A, and the power out of the DUT is detected on Channel B. The wavelength is scanned for the TE and TM polarizations, separately. An alternate measurement method, described next, allows both the maximum and minimum transmission to be measured at once.

Measurement of the PDL versus wavelength yields the maximum and minimum transmission responses. In devices where the input polarization state corresponding to each response may not be known a priori, this approach is necessary. Although PDL measurements can be made by sampling all polarization states to find the maximum and minimum transmittance, a faster approach relies on four input states of polarization to define the PDL [4,5]. A polarization state for partially polarized light is described by a Stokes vector. When a device operates on the state of polarization, the input and output states are related by a Mueller matrix. The Stokes vector is defined by the power measured for different states as follows [6]:

$$
\begin{bmatrix} S_0 \\ S_1 \\ S_2 \\ S_3 \end{bmatrix} = S_0 \begin{bmatrix} 1 \\ \cos(2\omega)\cos(2\alpha) \\ \cos(2\omega)\sin(2\alpha) \\ \sin(2\omega) \end{bmatrix} = \begin{bmatrix} I \\ I_H - I_V \\ I_{+45} - I_{-45} \\ I_{RCP} - I_{LCP} \end{bmatrix}
\tag{4}
$$

where S_0 is the total power in all states of polarization of the electromagnetic wave being described. The coefficients S_1, S_2, and S_3 represent differences in power between orthogonal states of polarization as defined in the last equality of Eq. (4). The subscripts denote the following polarizations: horizontal (H), vertical (V), linearly polarized at $+45°$ and $-45°$, right circularly polarized (RCP), and left circularly polarized (LCP). The angles α and ω are the polarization state azimuth and ellipticity which define a unique point on the Poincare sphere [6], a geometrical representation of the polarization state. For completely polarized light, $S_0 = \sqrt{S_1^2 + S_2^2 + S_3^2}$. For a partially polarized source, the degree of polarization (DOP) is defined as DOP $= \sqrt{S_1^2 + S_2^2 + S_3^2}/S_0$.

The Mueller matrix is a 4×4 matrix that transforms the input polarization denoted by the Stokes vector **S** into an output polarization denoted by the Stokes vector **T** as shown below.

$$
\begin{bmatrix} T_0 \\ T_1 \\ T_2 \\ T_3 \end{bmatrix} = \begin{bmatrix} m_{00} & m_{01} & m_{02} & m_{03} \\ - & - & - & - \\ - & - & - & - \\ - & - & - & - \end{bmatrix} \begin{bmatrix} S_0 \\ S_1 \\ S_2 \\ S_3 \end{bmatrix}
\tag{5}
$$

Each optical device that operates on the state of polarization is described by a Mueller matrix. Since the power is given by the first element of the Stokes vector, only the top row of the Mueller matrix is needed to calculate the transmission through the DUT. The Mueller matrix elements for a DUT can be determined if known input states of polarization with an input power S_0 for each state are launched into the DUT, and the transmitted power T_0 is measured for each state. For example, let the following states be input to the device one at a time:

$$\text{vertical: } S_V = [S_0, -S_0, 0, 0] \quad \text{horizontal: } S_H = [S_0, S_0, 0, 0]$$

$$45°: S_{+45} = [S_0, 0, S_0, 0] \quad \text{circular: } S_{\text{RHC}} = [S_0, 0, 0, S_0]$$

The measured output is $T_{0,H} = (m_{00} + m_{01})S_0$ when S_H is launched into the DUT and $T_{0,V} = (m_{00} - m_{01})S_0$ when S_V is launched. From these two measurements, the first element of the Mueller matrix is given by $m_{00} = (T_{0,H} + T_{0,V})/2S_0$. Similarly, the remaining elements of the first row are determined as follows:

$$m_{01} = \frac{1}{2S_0}(T_{0,H} - T_{0,V}) \tag{6}$$

$$m_{02} = \frac{T_{0,45}}{S_0} - m_{00} \tag{7}$$

$$m_{03} = \frac{T_{0,\text{RHC}}}{S_0} - m_{00} \tag{8}$$

The input state of polarization for the maximum and minimum transmission through the device are used to measure a device's PDL. Lagrange multipliers allow a concise derivation for these input states of polarization. To minimize (or maximize) T_0 given the constraint that $S_0 = \sqrt{S_1^2 + S_2^2 + S_3^2}$, the partial derivatives of $T_0^D \equiv T_0 - \lambda(S_1^2 + S_2^2 + S_3^2)$ with respect to S_i are set to zero for $i = 1, 2$, and 3, where λ is the Lagrange multiplier. The solutions for $\partial T_0^D / \partial S_i = 0$ are $S_i = m_{0i}/2\lambda$, and the constraint on S_0 defines two solutions, $\lambda = \pm(1/2S_0)\sqrt{m_{01}^2 + m_{02}^2 + m_{03}^2}$. The transmitted power for the polarization states defined by λ are given by

$$T_0 = \left[m_{00}S_0 \pm \frac{1}{2\lambda}(m_{01}^2 + m_{02}^2 + m_{03}^2)\right] = [m_{00} \pm a]S_0 \tag{9}$$

where $a \equiv \sqrt{m_{01}^2 + m_{02}^2 + m_{03}^2}$. The maximum transmission state corresponds to $+\lambda$ and the minimum to $-\lambda$. The PDL is defined as follows:

$$\text{PDL} = 10 \log_{10}\left(\frac{m_{00} + a}{m_{00} - a}\right) \tag{10}$$

The input states of polarization for the maximum and minimum transmission through the DUT are

$$S_{\text{max}} = S_0\left\{1, \frac{m_{01}}{a}, \frac{m_{02}}{a}, \frac{m_{03}}{a}\right\} \tag{11}$$

$$S_{\text{min}} = S_0\left\{1, \frac{-m_{01}}{a}, \frac{-m_{02}}{a}, \frac{-m_{03}}{a}\right\} \tag{12}$$

Test sets using this algorithm are commercially available. A polarization controller employing a polarizer and waveplates are typically used. A tap after the polarization controller allows the detected power from the DUT to be normalized to the relative input power, removing the effects of slow variations in the output power of the source over time. For highly polarized sources connected by non-PMF, a fiber paddle is placed between the source and polarization controller to maximize the power through the polarizer. Waveplates are used to obtain the four states of polarization for the PDL measurement. Because the different waveplate settings lead to different S_0's for each of the four input states of polarization, a reference measurement is made for each polarization state at each wavelength of interest before the DUT is measured. Any extraneous sources of PDL contribute to uncertainty on the measured PDL value. This uncertainty arises because the orientation of the DUT's PDL and other sources of PDL are not known, so we don't know whether they are adding, subtracting or at an intermediate orientation.

The steps in the spectral four-state PDL measurement are outlined as follows. A reference measurement is done to record the detected power for the four output states of polarization at each wavelength of interest without the DUT. Then, the measurement is repeated with the DUT. For each wavelength, the first row of the Mueller matrix is calculated, after which, the PDL and IL are calculated. A PDL measurement gives the maximum and minimum polarization for each wavelength. We may wish to obtain the spectral response for the principal polarization states. In this case, the Stokes vector for the maximum and minimum transmittance can be used to separate the polarization responses. Figure 7-7a shows the maximum and minimum IL from a planar waveguide ring measurement, and Figure 7-7b shows the individual polarization responses obtained by using the Stokes vector to define the wavelengths where the responses cross over.

7.1.3 Indirect Loss Measurements

As previously discussed, variations in fiber-to-waveguide coupling may introduce significant measurement uncertainty. To overcome this problem, methods that use the wavelength dependence of ring resonators [7] and Fabry–Perot cavities [8] have been devised to measure the propagation loss of planar waveguides. In this section, the use of ring resonators to measure both ring loss and coupling ratios is presented. The ring configuration is a natural candidate for studying constant radius bend loss. Expressions are derived in terms of quantities obtainable from the measured spectral insertion loss. By employing this ring resonator technique, the coupling ratios and ring loss are evaluated independently of the fiber-to-waveguide coupling loss and repeatability.

7.1.3.1 *The Ring Resonator Method*   A simple ring resonator with one coupler is shown in Figure 7-8 with its pole-zero diagram. As described in Chapter 6 for an all-pass filter, the following transfer functions for ports 1 to 2 and ports 1 to 3 are derived using the sum of all optical paths:

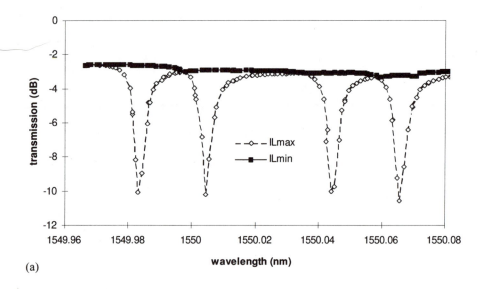

(a)

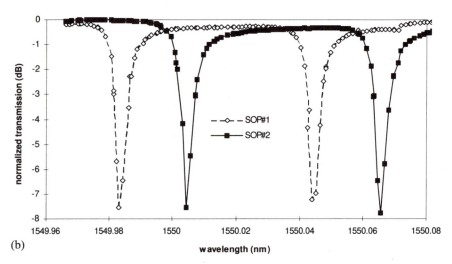

(b)

FIGURE 7-7 Spectral loss or a ring resonator: (a) maximum and minimum polarization, (b) individual polarization responses resolved using the change in Stokes vector.

$$H_{12}(z) = \frac{c - \gamma e^{-j\phi}z^{-1}}{1 - c\gamma e^{-j\phi}z^{-1}} \tag{13}$$

$$H_{13}(z) = \frac{-js}{1 - c\gamma e^{-j\phi}z^{-1}} \tag{14}$$

where $c = 1 \sqrt{-\kappa}$, $s = \sqrt{\kappa}$, the round trip loss of the ring is $\gamma = \exp(-\alpha L)$ for an av-

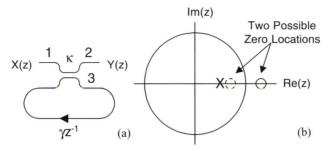

FIGURE 7-8 One coupler ring resonator (a) and its pole-zero diagram (b).

erage ring loss per unit length of α and ring perimeter L, and ϕ is the phase of the pole. The phase is set to $\phi = 0$ since we are only considering one stage. The ring loss contains all sources of loss including propagation loss, bend loss, and loss due to transitions in curvature. The pole magnitude is given by $p = \gamma c$. The transfer function H_{12} has zero transmittance if $c = \gamma$. If $c \neq \gamma$, then there are two possible locations for the zero that have the same magnitude response for H_{12}, namely $z_1 = \gamma/c$ and $z_2 = 1/z_1$. This ambiguity results in two possible solutions for c and γ given only the magnitude response for Port 1–2. The pole-zero diagram in Figure 7-8b shows the possible zero locations. The solid circle is the maximum-phase solution, and the dashed circle is the minimum-phase solution.

The ring's transmission is defined as the square magnitude of the transfer function $T(z) = |H_{12}(z)|^2$ evaluated on the unit circle ($z = e^{j2\pi\nu}$) where the normalized frequency is defined as $\nu = (f - f_0)/\text{FSR}$. The beginning of the FSR is defined by the frequency f_0. By choosing f_0 at the frequency where $T_{\min}$ occurs for a given FSR, $\phi = 0$. The contrast ratio is defined as the ratio of maximum-to-minimum transmission over one FSR. It does not depend on the fiber-to-waveguide coupling loss, which is assumed to be constant over the measured FSR. Using the definition of $H_{12}(z)$ from Eq. (13), the contrast ratio is expressed in terms of c and γ as follows:

$$C_{12} = \frac{T_{\max}}{T_{\min}} = \frac{[(1 + \gamma/c)(1 - c\gamma)]^2}{[(1 - \gamma/c)(1 + c\gamma)]^2} \qquad (15)$$

where $T_{\max} = |H_{12}(-1)|^2$ and $T_{\min} = |H_{12}(1)|^2$. Likewise, the contrast ratio of $H_{13}(z)$ is

$$C_{13} = \frac{|H_{13}|_{\max}^2}{|H_{13}|_{\min}^2} = \frac{|H_{13}(1)|^2}{|H_{13}(-1)|^2} = \frac{[1 + c\gamma]^2}{[1 - c\gamma]^2} \qquad (16)$$

It is convenient to define the full width resonance bandwidth, $\Delta\nu$, at the 3dB down points of $|H_{13}(z)|^2$ as follows:

$$|H_{13}(z_h)|^2 = \frac{1}{2}|H_{13}(1)|^2 \qquad (17)$$

where $z_h = \exp(j\pi\Delta\nu)$. Using this relationship, the following expression for $\Delta\nu$ results [9]:

$$0 = c\gamma + 2\sqrt{c\gamma}\,\sin\left(\frac{\pi}{2}\Delta\nu\right) - 1 \tag{18}$$

$$= P^2 + 2P\,\sin\left(\frac{\pi}{2}\Delta\nu\right) - 1$$

where $P = \sqrt{c\gamma}$. For a given $\Delta\nu$, Eq. (18) is used to solve for P. One solution is negative, and consequently neglected. Equation (18) is valid only if the contrast ratio for Port 1–3 is >3dB, i.e. $|H_{13}(1)|^2/2 > |H_{13}(-1)|^2$, which translates to $\gamma c \geq 0.1716$. For example, a coupling ratio of $\kappa = 0.10$ allows a ring loss up to 14.9 dB to be measured. The contrast ratio C_{12} for this coupling ratio and ring loss is 0.33 dB. The contrast is much higher for port 1–3 than for port 1–2. Substituting P into the definition for C_{12} yields the following expression for the ratio $R = \gamma/c$:

$$R_{1,2} = \frac{(1+P) \pm \sqrt{C_{12}^{-1}}(1-P)}{(1+P) \mp \sqrt{C_{12}^{-1}}(1-P)} \tag{19}$$

Because of the squares in the definition for C_{12}, two solutions result. The solutions are inverses of each other, i.e. $R_1 = 1/R_2$.

The procedure for determining the ring loss and coupling ratio from the measured spectral loss is now defined. First, the quantities Δf, FSR and C_{12} are obtained from the measured spectral insertion loss. These quantities do not depend on the fiber-to-waveguide coupling loss. Next, $\Delta\nu$ and L are calculated. Equation (18) is solved for $P = \gamma_1 c_1$. Then, $R = \gamma_1/c_1$ is found using Eq. (19). Finally, the product and quotient of P and R yield the solutions $\gamma_1 = \sqrt{PR} = c_2$ and $c_1 = \sqrt{P/R} = \gamma_2$. One catch in this procedure, as defined so far, is that $\Delta\nu$ must be found from measurements on H_{12} not H_{13}. To remedy this situation, we evaluate $T(z) = |H_{12}(z)|^2$ at z_h:

$$T(z_h) = \frac{c^2 - 2\gamma c\,\cos(\pi\Delta\nu) + \gamma^2}{1 - 2\gamma c\,\cos(\pi\Delta\nu) + (\gamma c)^2} \tag{20}$$

The denominator of Eq. (20) is the same as the denominator of $|H_{13}(z_h)|^2$. Using Eq. (17), the denominator simplifies to $2(1 - \gamma c)^2$. Using Eq. (18), the numerator is rewritten as $(c - \gamma)^2 + 4\gamma c\,\sin^2(\pi\Delta\nu/2)$, which simplifies to $(c - \gamma)^2 + (1 - \gamma c)^2$. Substituting these simplifications into Eq. (20) results in the following expression:

$$T(z_h) = \frac{1}{2}[1 + T_{\min}] \tag{21}$$

where $T_{\min} = (c - \gamma)^2/(1 - \gamma c)^2$. Note that $T(z)$ cannot be measured directly without the uncertainty associated with the fiber-to-waveguide coupling loss. To avoid this problem, the approximation $T_{max} \approx 1$ is made, and the normalized transmission is used to define z_h as follows:

$$T(z_h) \cong \frac{1}{2}\left[1 + \frac{1}{C_{12}}\right] \tag{22}$$

With this approximation, the ring loss and coupling ratio are determined from the normalized spectral loss of the ring resonator. Given that the ring loss has very little effect on the transmission maximum as seen in Chapter 6, Eq. (22) is typically a good approximation.

A small loss, small coupling approximation for the ring analysis was published by Adar et al. [7]. The coupling is approximated by $c \approx 1 - \kappa/2$ and the loss by $\gamma \approx 1 - \alpha L$. They reported results for the finesse, $F \equiv FSR/\Delta f$, and T_{min} of rings with various radii for phosphorous-doped silica waveguides made using a CVD process [10]. Table 7-1 compares the loss in dB/m for the small loss and small coupling approximation against the more general formulation. The assumption $T_{max} \approx 1$ is good for this data. The approximate method predicts a slightly larger loss than the general method.

7.1.3.2   Application to the Study of WGMs   Ring-based filters require small bend radii to achieve FSRs of practical interest. The bend radius depends dominantly on the index of refraction difference between the core and cladding. Very high index contrast waveguides, Δ's of several percent and larger, typically have large propagation and fiber-to-waveguide coupling losses and large polarization dependence. Thus, we are motivated to find ways to extend the FSR achievable with lower Δ waveguides. As discussed in Chapter 2, the ring waveguide width can be increased until whispering gallery modes (WGM) are supported. Although increasing the width allows for tighter bends, the drawback is working in such a highly multimoded regime. By using the ring method, WGM loss and coupling to a singlemode waveguide are studied in a practical configuration.

Experimental results were reported for a series of rings with varying waveguide widths and a bend radius of 2.75 mm [11]. The waveguides had a height of 6 μm and $\Delta \sim 0.7\%$, which limits the bend radius to ~4 mm for low loss singlemode propagation. Waveguide widths of 6 to 20 μm were explored. One of the key issues for filter design is the range of coupling ratios that can be achieved for a given ring ra-

TABLE 7-1   Comparison of Approximate and General Ring Analysis

R (mm)	F	T_{min} (dB)	T_{max} (dB)	Approximate α (dB/m)	General α (dB/m)
30	132	−6.78	−0.000	0.85	0.80
20	108	−13.47	−0.001	1.22	1.22
10	68.3	−11.14	−0.002	4.72	4.06
5	4.7	−1.25	−0.116	173	166

dius and fabrication process. As discussed in Chapter 2, maximum coupling is achieved only when the two waveguides have identical propagation constants. Since the effective index of a bent waveguide is lower than that of a straight waveguide [12], a larger coupling is expected if a narrower input/output waveguide is used to match the effective index of the ring. Two sets of rings were designed so that the coupling could be investigated using different input/output waveguide widths, w_{io}. Set #1 had $w_{io} = 6.0$ μm and Set #2 had $w_{io} = 4.0$ μm. The maximum to minimum transmission varied from 4 to 12 dB depending on the ring waveguide width and coupling. A best fit of $|H_{12}(\omega)|^2$ as a function of c and γ to the measured magnitude response was used to obtain values for the ring loss and coupling ratio. Two solutions, the maximum- and minimum-phase solutions, were obtained for each ring. To identify the particular solution, the group delay was measured. The measured group delay for the 12 μm wide rings from each set is shown in Figure 7-9. For Set #1, all of the rings had negative group delays near the transmission minimum while rings in Set #2 had positive group delays. These results indicate that the zeros are minimum-phase ($\gamma/c < 1$) for Set #1 and maximum-phase ($\gamma/c > 1$) for Set #2. The calculated ring loss and coupling ratio solutions are given in Figure 7-10 and Figure 7-11. The solutions identified by the group delay measurement are shown with solid lines and the other solutions with broken lines. As the waveguide width increased from 6 to 8 μms, the loss decreases by a factor of 2. A further factor of 2 decrease is obtained by increasing the ring guide width to 20 μm. The coupling ratios are significantly larger for Set #2 with the narrower input waveguides, as predicted. The coupling ratios do not vary significantly with waveguide width for the 12, 16 and 20 μm cases, indicating that the effective index of the wide waveguides does not vary

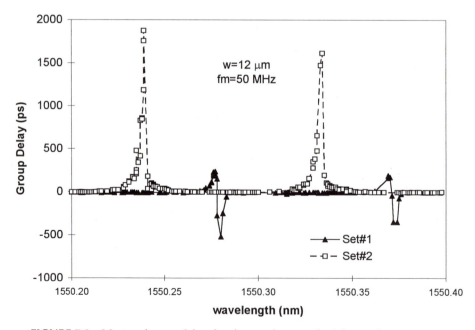

FIGURE 7-9 Measured group delay showing maximum and minimum phase responses.

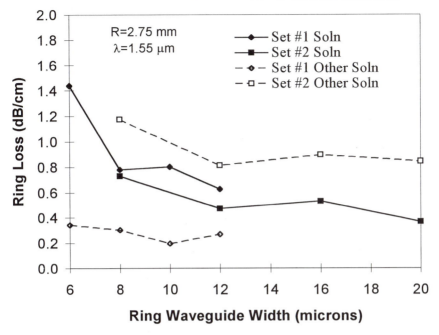

FIGURE 7-10   Bend loss as a function of the ring waveguide width.

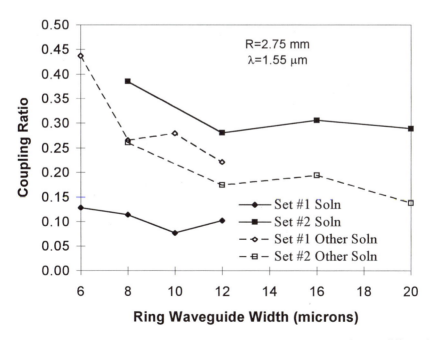

FIGURE 7-11   Coupling ratio as a function of the ring waveguide width for two different input waveguide widths: Set #1 ($w_{io} = 4 \ \mu m$) and Set #2 ($w_{io} = 6 \ \mu m$).

significantly. These results demonstrate the use of the ring method for simultane-ously evaluating coupling and loss while avoiding uncertainties associated with ab-solute loss measurements.

7.1.4 Group Delay

There are two approaches for determining the group delay: (1) it can be measured directly, or (2) it can be calculated for a minimum-phase filter from the filter's mag-nitude response. Since it is typically easier to measure a filter's magnitude response than the group delay, the method for calculating the group delay is discussed first, followed by direct methods for measuring the group delay.

7.1.4.1 Hilbert Transform Method   In Chapter 3, it was shown that for a causal, real impulse response, the real and imaginary parts of the transfer function are Hilbert transform pairs. In this section, the relationship between the magnitude and phase of the transfer function is explored. Let $H(\omega) = |H(\omega)|e^{j\Phi(\omega)} = e^{-\alpha(\omega)}e^{j\Phi(\omega)}$. Then, the natural logarithm of $H(\omega)$ is defined in terms of two new functions as fol-lows:

$$\hat{H}(\omega) = \log[H(\omega)] = -\alpha(\omega) + j\Phi(\omega) \tag{23}$$

For the real and imaginary parts of $\hat{H}(\omega)$ to be Hilbert transforms, the sequence $\hat{h}(n)$ must be causal. This translates in the Z-domain to the requirement on $H(z)$ that there be no zeros outside the unit circle. Stability requires that there are no poles outside the unit circle. Therefore, if $H(z)$ is minimum-phase, then its magnitude and phase are Hilbert transform pairs.

$$\Phi(\omega) = \frac{1}{2\pi}\int_{-\pi}^{\pi}\alpha(\omega)\cot\left(\frac{\omega - \theta}{2}\right)d\theta \tag{24}$$

$$\alpha(\omega) = -\frac{1}{2\pi}\int_{-\pi}^{\pi}\Phi(\omega)\cot\left(\frac{\omega - \theta}{2}\right)d\theta + \alpha(0) \tag{25}$$

Given the magnitude response, the phase response can be calculated. Given the phase response, the magnitude response can be calculated to within a constant which is equal to its dc component. To avoid calculating the integral, a time domain approach is employed. From Chapter 3, we recall that causality allows the impulse response to be expressed in terms of the unit step response, $u(n)$. For the magnitude and phase, these relationships are

$$\hat{h}(n) = \hat{h}_e(n) + \hat{h}_o(n) \tag{26}$$
$$= 2\hat{h}_e(n)u(n)$$
$$= 2\hat{h}_o(n)u(n) + \hat{h}(0)\delta(n)$$

where $\hat{H}_e(\omega) = -\alpha(\omega)$ and $\hat{H}_o(\omega) = \Phi(\omega)$. By substituting $u(n) = [1 + \text{sgn}(n)]/2$ into the above equation, we obtain the following result:

$$\hat{h}_o(n) = \hat{h}_e(n)\ \text{sgn}(n) \tag{27}$$

$$\hat{h}_e(n) = \hat{h}_o(n)\ \text{sgn}(n) + \hat{h}(0)\delta(n)$$

By inverse Fourier transforming $-\alpha(\omega)$, multiplying by $-j\text{sgn}(n)$ and Fourier transforming, $\Phi(\omega)$ is obtained without performing the integrals in Eqs. (24) and (25) directly. Since AR filter responses are minimum-phase, their magnitude and phase are Hilbert transform pairs. This approach applies to the transmission response of gratings and thin film stacks as well as the AR response of ring filters. The Weiner-Lee transform uses this approach along with frequency warping to find the imaginary part of an aperiodic function with a real, causal impulse response [13]. The Weiner-Lee transform has been applied to symmetric Bragg gratings [14]. In addition to the phase response, qualitative information about the coupling strength was obtained by calculating the impulse response of the grating from its complex transfer function. For continuous-time functions (aperiodic frequency response), the Kramers-Kronig relations replace Eqs. (24)–(25) and Eqs. (26)–(27) are written as a function of t [see 13, 15, 28].

Although the Hilbert transform relationship is limited to minimum-phase functions, it does not require that the magnitude response have even symmetry. A filter may be physically symmetric, but have an asymmetric magnitude response. An example is a Bragg grating with a Gaussian "ac" and "dc" index profile. As shown in Chapter 5, its reflection spectrum contains Fabry–Perot modes on the short wavelength side and a smooth transition on the long wavelength side. Such asymmetry arises when there is a nonzero average index change across the filter. Figure 7-12 shows the magnitude and phase responses for a 5 mm long, Gaussian grating with a maximum index change of $2\overline{\delta n}_e = 1 \times 10^{-3}$ calculated using coupled-mode theory. In transmission, the calculated phase using the Hilbert transform is identical to the phase response obtained using coupled-mode theory, as expected. In reflection, the calculated phase using the Hilbert transform differs from the actual phase response. This difference can be understood by noting that a linear-phase response may have zeros outside the unit circle, as long as each one is matched by a zero located at a reciprocal position within the unit circle. The Hilbert transform reflects all maximum-phase zeros inside the unit circle, thereby destroying the linear-phase nature of the numerator polynomial and returning the minimum-phase response instead.

Next, we explore the information that can be gained from the magnitude response and Hilbert transform for lattice ring resonators, thin film filters and Bragg gratings. Whether the filter has any physical symmetry or not, the group delay of the two ARMA responses (mixed-phase) is related to the group delay of the AR response (minimum-phase). By differentiating Eq. (60) of Chapter 5, we obtain

$$\tau_{AR}(\omega) = \frac{1}{2}[\tau_{ARMA\#1}(\omega) + \tau_{ARMA\#2}(\omega)] \tag{28}$$

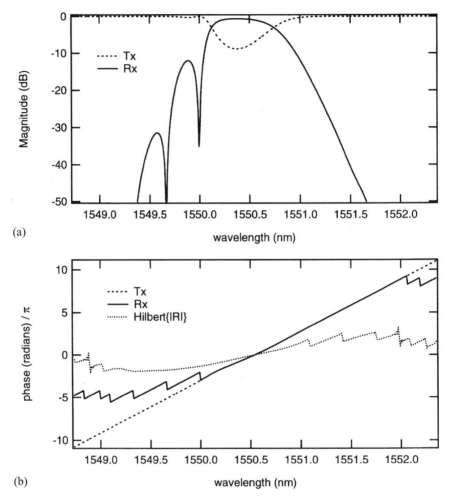

FIGURE 7-12   (a) Magnitude and (b) phase relative to the center wavelength of a grating with a Gaussian profile.

For a lossless filter, the ARMA and AR magnitude responses are related by power conservation. Once one response is measured, the other can easily be calculated. The AR magnitude response is Hilbert transformed to define the AR phase, group delay, and dispersion responses. The Hilbert transform can also be used to bound the group delay of the ARMA responses [15]. The lower bound is set by the minimum-phase ARMA response, calculated by Hilbert transforming the ARMA magnitude response. Then, the maximum-phase response is found from Eq. (28). The length of the filter (or number of stages) must be known to calculate the maximum-phase group delay. The evaluation of the underlying filter structure (index profile, coupling rates, etc.) from the magnitude and phase response is discussed in Section 7.2.

7.1.4.2   *Direct Measurement*

Group delay is typically measured using the phase shift method [16]. The optical signal is amplitude modulated at a radian frequency Ω_m before propagating through the DUT. At the receiver, the detected current of the modulated signal, $i_m(t) \propto \cos(\Omega_m t - \phi)$, has a phase $\phi(\lambda)$ that is a function of wavelength because of dispersion in the DUT. The group delay is expressed in terms of the phase as follows [17]:

$$\tau_g(\lambda) = \frac{\phi(\lambda)}{\Omega_m} \qquad (29)$$

The measurement setup is calibrated using a reference delay, such as a fiber with a known length. The minimum detectable phase change limits the resolution of the group delay measurement. Higher modulation frequencies allow smaller changes in delay to be measured. The group delay for a 2π phase change is $1/f_m$, where f_m is the modulation frequency, so higher modulation frequencies reduce the maximum measurable group delay.

An underlying assumption for this technique is that the magnitude and group delay response of the DUT is constant over the bandwidth of the modulated signal. For measuring broadband devices, such as optical fibers, this is not a limiting assumption. For extremely narrowband optical filters or very high resolution measurements, it can be restrictive. The modulated signal consists of a carrier and two sidebands. When the sidebands experience a different amplitude and phase response, the detected phase is not a true representation of the filter's group delay. The dependence of the measured group delay on the modulation frequency was simulated for a narrowband filter consisting of a ring resonator with $2\pi R = 2.5$ cm, a ring loss of 0.2 dB/cm, and a single coupler having $\kappa = 20\%$ [18]. Figure 7-13 shows the theoretically calculated group delay as the solid line. The magnitude of the zero is 1.056, so the response is very sharp. The peak group delay in this example is $T \times 24 = 2900$ ps where $T = n_g L/c = 120$ ps. Figure 7-13 also shows the calculated group delay for modulation frequencies of 100, 200 and 400 MHz. The FSR of the simulated ring is 8.3 GHz, so $\nu = \pm 0.1$ corresponds to ± 830 MHz. The peak measured response is smaller than the true group delay, and it decreases as the modulation frequency increases. For large modulation frequencies, there is also a distortion of the measured response from the actual group delay.

The group delay measurement setup used for the ring measurements discussed in Section 7.1.3.2 is shown in Figure 7-14. The optical circuit is denoted in bold and the electrical circuit in lighter lines. A setup capable of resolving the TE and TM responses was used. A modulator was placed after the laser, and fiber paddles were used to optimize the polarization into the modulator for optimum modulation depth. A heterodyne technique was employed to allow the phase change to be detected by a lock-in amplifier [19]. Two RF signal generators operating at 100 MHz and offset by 10 kHz were used. Splitting of the first RF signal allowed a portion to be used to

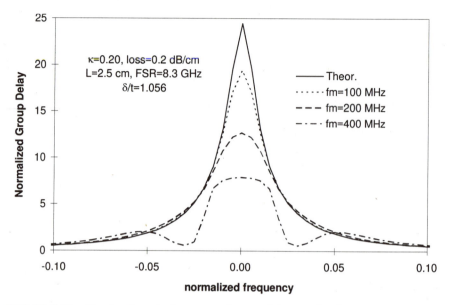

FIGURE 7-13 Simulated results for measured group delay as a function of modulation frequency.

drive the modulator and another portion to be mixed with the offset frequency to create the low frequency reference signal for the lock-in. A high speed detector, with a bandwidth from DC to over 1 GHz, was used. The detected signal was mixed with the offset RF to obtain a 10 kHz signal with the same phase as the RF signal after transmission through the DUT. The modulation frequency and the phase resolution of the lock-in, ~0.1°, determine the group delay resolution. Both the magni-

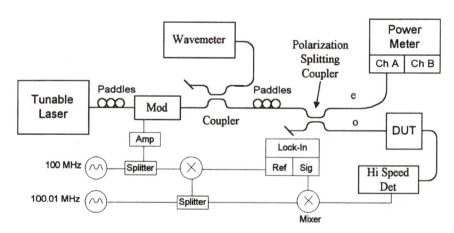

FIGURE 7-14 Group delay measurement setup.

tude and group delay can be measured simultaneously; however, the width of the modulated optical signal distorts the magnitude response for very narrowband filters in a manner similar to the group delay distortion.

A single sideband (SSB) technique overcomes the limitations of double sideband modulation for measuring the group delay of narrowband filters. Optical SSB modulation has been demonstrated using an optical filter to suppress one of the sidebands [20]; however, this approach requires a very narrow optical filter that must be centered on the sideband so as not to affect the carrier or opposite sideband. A technique that avoids this complexity achieves optical SSB modulation by employing a dual electrode MZI modulator [21]. The electrodes are driven in quadrature by an ac voltage, V_{ac}. The voltage required to completely switch the MZI is designated by V_π. For example, the output of the cross-port is given by

$$|H_{12}(V)|^2 = \cos^2\left(\frac{\pi}{2}\frac{V}{V_\pi}\right)$$

A bias voltage, V_{dc}, equal to half the switching voltage is applied to one of the electrodes. The output signal is given by [21]

$$E_0(t) = \frac{A}{2}\left[\cos\left\{\Omega_c t + \frac{\pi}{2} + \alpha\pi\cos(\Omega_m t)\right\} + \cos\{\Omega_c t + \alpha\pi\sin(\Omega_m t)\}\right] \quad (30)$$

where Ω_c is the optical center frequency, Ω_m is the modulation frequency in radians per second, and $\alpha = V_{ac}/V_\pi$. To illustrate the sideband cancellation, the following relationships are needed to expand Eq. (30) in terms of Bessel functions [22].

$$\cos(z\sin\theta) = J_0(z) + 2\sum_{k=1}^{\infty} J_{2k}(z)\cos(2k\theta) \quad (31)$$

$$\sin(z\sin\theta) = 2\sum_{k=0}^{\infty} J_{2k+1}(z)\sin\{(2k+1)\theta\}$$

$$\cos(z\cos\theta) = J_0(z) + 2\sum_{k=1}^{\infty}(-1)^k J_{2k}(z)\cos(2k\theta)$$

$$\sin(z\cos\theta) = 2\sum_{k=0}^{\infty}(-1)^k J_{2k+1}(z)\cos\{(2k+1)\theta\}$$

For $\alpha < 1/\pi$, the Bessel functions of order 2 and greater can be neglected, resulting in the following approximate expression:

$$E_0(t) \cong \frac{A}{2}[J_0(\alpha\pi)\cos(\Omega_c t) - J_0(\alpha\pi)\sin(\Omega_c t) - 2J_1(\alpha\pi)\cos\{(\Omega_c - \Omega_m)t\}] \quad (32)$$

The power spectral density is the square magnitude of the Fourier transform of $E_0(t)$.

$$S(\Omega) \cong \frac{A^2}{2} J_0^2(\alpha\pi)\pi^2\delta(\Omega - \Omega_c) + A^2 J_1^2(\alpha\pi)\pi^2\delta\{\Omega - (\Omega_c - \Omega_m)\} \qquad (33)$$

The upper sideband disappears leaving the carrier and the lower sideband. The SSB modulation technique offers a fast, high resolution method to measure a device's phase response by sweeping the modulation frequency and detecting the amplitude and phase on a network analyzer. A SSB measurement has been applied to measure the phase of a chirped grating [23]. Ripples in the group delay response arise from an effective Fabry–Perot cavity whose reflectors occur at the ends of the grating where the average index change is smaller than the center of the grating. For a meter-long grating, the ripple period is very small, on the order of 100 MHz. With the SSB technique, resolution in the kHz range is possible. A 6–18 GHz frequency sweep was used in combination with stepping the laser center wavelength to characterize the grating over a broad frequency range.

7.2 FILTER ANALYSIS

Because of variations introduced during fabrication, multi-stage filters require a method to determine where the errors have occurred. This information may then be used to change the fabrication process or to tune the device to the desired response. Examples of postfabrication tuning are the use of phase shifters to correct for phase variations between stages, patch processes where material is removed by laser ablation to change the phase, and UV-induced index changes. A complete description of the filter is obtained if the amplitude and phase of its frequency response are known over the full FSR. Likewise, if the magnitude and phase of its impulse response in the time domain are known, then the filter is fully characterized. Three analysis methods are described in this section that have been demonstrated on fabricated devices. The first two lend themselves in a straightforward manner to MA filter analysis and the latter to AR filter analysis.

7.2.1 Time Domain Measurement

For MA filters, a time domain approach allows the coefficients of each stage to be measured. A pulse that is shorter than the unit delay is input to the filter. For an N-stage MA filter, there are N output pulses, nominally separated by the unit delay. Since the input pulse width is shorter than the unit delay, the pulse bandwidth is larger than the filter's FSR. Sasayama [24] demonstrated the time domain approach on a 4-stage transversal filter with a unit delay of 200 ps. The homodyne detection measurement setup is shown in Figure 7-15. The source was a 1.3 μm YAG laser, and a 100 ps FWHM pulse was created using a lithium niobate modulator. The device was temperature stabilized to within 0.1° C, and the reference path was stabilized using a PZT phase modulator. Each splitter was realized using a symmetric MZI with a phase shifter so that by changing its phase, the coupling coefficient could be adjusted. The tap coefficients were set to $-1, 2, -1, 0$ as shown in the streak

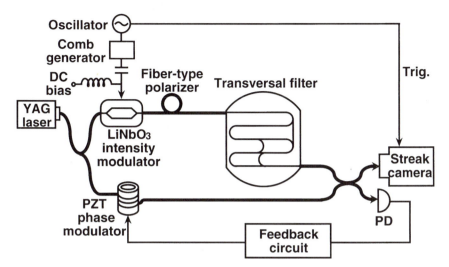

FIGURE 7-15 Homodyne delay measurement [24]. Reprinted by permission. Copyright ©
1991 IEEE.

camera response in Figure 7-16. Since the sum of the coefficients is zero, the device
functions as a high pass filter. Frequencies near $\Omega T = 2\pi m$ are attenuated and those
near $\Omega T = \pi(2m + 1)/2$ are passed, where m is the filter order. The filter's FSR was
5 GHz, and it was used to remove a 150 MHz tone that had been frequency multiplexed with a 2.4 GHz signal. This measurement technique gives the complex impulse response since both the magnitude and phase are measured. One drawback is
that shorter pulses are needed for larger FSRs. An 8 tap filter with a unit delay of 50
ps (FSR = 20 GHz) was subsequently demonstrated [24]. A pulse width of 20 ps
was acquired from a gain-switched 1.3 μm DFB laser. A square-law detector was

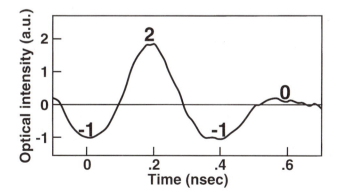

FIGURE 7-16 Measured coefficient values for a 4-tap high pass filter [24]. Reprinted by
permission. Copyright © 1991 IEEE.

used instead of the homodyne setup, so phase information was not directly measured. The measured square magnitudes of the coefficients were used to set the taps, and information from previously calibrated heaters was used to set the phase.

7.2.2   Optical Low-Coherence Interferometry and Fourier Spectroscopy

Optical low coherence interferometry (OLCI) uses a continuous-wave broadband source and a variable-delay interferometer to obtain a spatially-localized interference or finge pattern. When the fringe pattern is Fourier transformed, the spectral density of the source is obtained. The frequency response (magnitude and phase) of a DUT can also be measured by placing it in one arm of an interferometer. This method is called Fourier spectroscopy. A schematic of the basic setup is shown in Figure 7-17. The source has a wide spectral width with a spectral density of $|B(f)|^2$, and is split equally between two paths, a reference path and a path containing the DUT. An interference pattern results when the reference path length is varied. First, we consider the interference of a monochromatic wave given by $e_v(t) = \mathbb{R}e\{E_v e^{j2\pi ft}\}$ where

$$E_v(x) = \frac{B(f)}{2}\{e^{-j2\pi f\frac{xn_e}{c}} + A(f)e^{-j\phi(f)}\} \tag{34}$$

An ideal 50% splitter and combiner are assumed, and the frequency response of the DUT is denoted by $A(f)e^{-j\phi(f)}$. It is also assumed that the polarizations are identical. No interference occurs between orthogonal polarizations. The detected power contains two "dc" terms and one "ac" term.

$$I_v(\tau) = \frac{|B(f)|^2}{4}[1 + |A(f)|^2 + 2A(f)\cos\{\phi(f) - 2\pi f\tau\}] \tag{35}$$

where $\tau = xn_e/c$. Next, we integrate over the frequency content of the source. Neglecting the "dc" terms, the interference pattern is given by

$$I(\tau) = \frac{1}{2}\int_0^\infty |B(f)|^2 A(f)\cos\{\phi(f) - 2\pi f\tau\}df \tag{36}$$

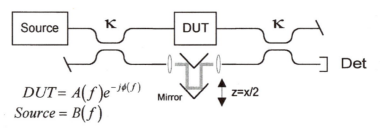

$$DUT = A(f)e^{-j\phi(f)}$$
$$Source = B(f)$$

FIGURE 7-17   Basic schematic of a low coherence interferometer for Fourier spectroscopy.

The integral is nonzero only when the reference path and DUT path are matched to within a coherence length of the source. The coherence time of the source is inversely proportional to its bandwidth. For a Gaussian spectral density, the product of the bandwidth and coherence time is $\Delta\tau\Delta f = 0.88$, where the widths are FWHM in each domain. For a rough estimate, the product is often approximated as one. An LED centered at 1550 nm with a FWHM of 80 nm has a coherence length $L_c = 0.88\lambda^2c/n_g\Delta\lambda = 0.88c/n_g\Delta f$ of 26 μm in air. Its interference pattern is shown in Figure 7-18. Now, we introduce a DUT with a Gaussian magnitude response having a FWHM = 40 nm and a path length difference of 0.5 μm from the reference path. The resulting interference pattern is shown in Figure 7-19. Note that the interference pattern is asymmetric compared to the pattern of the source by itself, and it is much wider due to the narrowing of the spectrum by the DUT.

Next, we calculate the frequency response of the DUT from the interference pattern. For computational ease, Eq. (36) is rewritten as an inverse Fourier transform by assuming an even magnitude response, $|B(f)|^2A(f) = |B(-f)|^2A(-f)$, and an odd phase response, $\phi(f) = -\phi(-f)$.

$$I(\tau) = \int_{-\infty}^{\infty} \frac{|B(f)|^2}{4} A(f)e^{+j\{2\pi ft - \phi(f)\}}df \qquad (37)$$

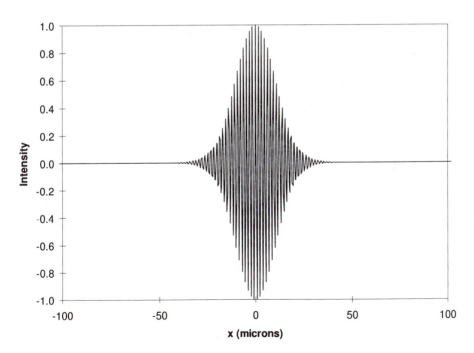

FIGURE 7-18   Interference pattern for a Gaussian source centered at $\lambda = 1550$ nm with FWHM = 80 nm.

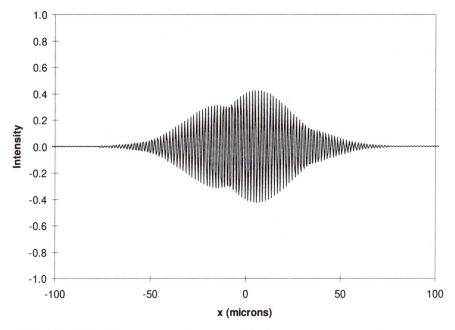

FIGURE 7-19   Interference pattern for a DUT with a FWHM = 40 nm Gaussian magnitude response and a path length difference of 0.5 μm relative to the reference path.

The Fourier transform of $I(\tau)$ yields the DUT's magnitude and phase response, assuming that the source spectral density is known.

$$S(f) = \frac{|B(f)|^2}{4} A(f)e^{-j\phi(f)} = \int_{-\infty}^{\infty} I(\tau)e^{-j2\pi f\tau}d\tau \qquad (38)$$

The spatial sampling frequency must be high enough to accommodate the largest source frequency. When $I(\tau)$ is sampled at $\tau = n\Delta\tau$, $S(f)$ is periodic with a Nyquist frequency corresponding to $f_{max} = 1/2\Delta\tau = c/2n_e\Delta x$, which implies that the largest spatial sampling interval for the interference pattern is related to the shortest source wavelength by $\Delta x_{max} = \lambda_{min}/2n_e$. The frequency response was calculated from the interferometric pattern of the source and DUT using $N = 1024$ and $\Delta x = 391$ nm. The results are compared to the input values in Figures 7-20 and 7-21 for the magnitude and phase responses, respectively. The phase response is zero for the source and linear for the DUT as expected. The spectral resolution is 6 nm near $\lambda = 1550$ nm. The frequency resolution depends on the total distance sampled, $\Delta f = c/n_g N\Delta x$; so, a longer interference pattern provides a higher spectral resolution. Wavemeters that measure the wavelength of several channels on a single input use a long interferogram to obtain resolutions of 20 GHz [1]. Since truncating the interferogram is equivalent to multiplying it by $\Pi(\tau)$, other window functions can be applied to $I(\tau)$ to decrease the sidelobes created by the window function on the calculated $S(f)$.

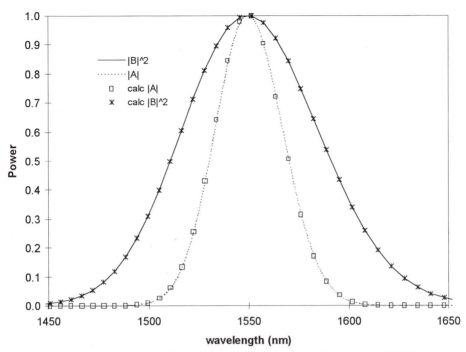

FIGURE 7-20 Square magnitude and magnitude response for the source and DUT, respectively, compared to the calculated response.

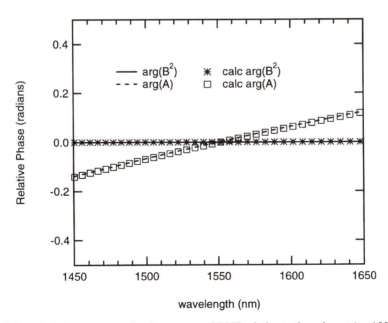

FIGURE 7-21 Phase response for the source and DUT relative to the pahse at $\lambda = 1550$ nm.

Fourier spectroscopy has been applied to measuring the phase errors introduced during fabrication in the grating arms of a WGR. As the channel spacing decreases, the order of the WGR increases and the relative path length difference between each arm increases. It is difficult to maintain the relative phase differences when the path length is large. Fabrication variations that introduce path length errors are the dominant source of cross-talk degradation in WGR devices. OLCI and Fourier spectroscopy were applied to determine the phase and amplitude of each grating arm using the experimental setup shown in Figure 7-22 [25]. The source was an edge emitting LED with a FWHM of 80 nm centered at $\lambda = 1.55$ μm. A stabilized 1.3 μm laser was also launched into the interferometer to provide a reference for nulling out phase variations between the interferometer arms due to temperature fluctuations and vibration. A single input and output of a WGR are connected in the DUT arm. The results are discussed for a WGR with 100 GHz channel spacing and a crosstalk of −31 dB. The WGR has 101 grating arms with a path length difference between adjacent arms of 126 μm. The FSR is 1600 GHz (12.8 nm). An interferogram results when the reference arm length is approximately equal to one of the grating arm lengths. The interferogram was sampled at $\Delta x = 652.89$ nm, and a waveform recorder with 4×2^{13} memory locations was used. The square amplitude response versus wavelength for the central 80 grating arms is shown in Figure 7-23. The quasi-periodic variation arises from coupling between the waveguides in the arrays exciting the slab regions. The optical path length difference between the grating arms is much longer than the source coherence length, so the interference pattern for each arm is isolated and Fourier transformed independently of its neighbors. The measured phase errors vary by ±30 degrees as shown in Figure 7-24. The measurement error on the phase is ±1° and the measurement error on the square of the am-

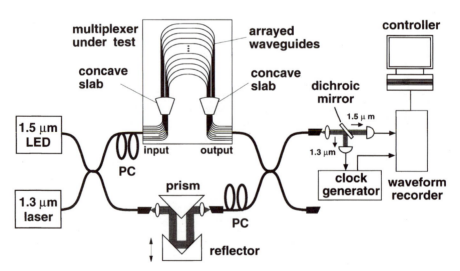

FIGURE 7-22 Experimental setup for measuring the WGR grating arms [25]. Reprinted by permission. Copyright © 1996 IEEE.

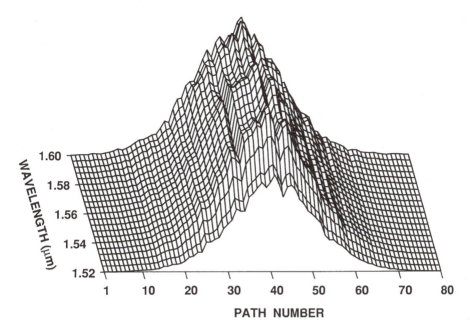

FIGURE 7-23 Amplitude coefficients for each grating arm of a 100-GHz channel spacing WGR [25]. Reprinted by permission. Copyright © 1996 IEEE.

plitude coefficients is ±5%. The contribution of the magnitude and phase errors can now be analyzed separately. First, the magnitude errors are neglected by assuming that the amplitude coefficients have a Gaussian distribution. Using the measured phase error, the calculated spectral response shown in Figure 7-25 is in excellent agreement with the measured spectral response. The worst case crosstalk is –30 dB. Figure 7-26 shows the impact of amplitude errors assuming no phase errors. The worst case crosstalk is –39 dB. By controlling the phase errors, the crosstalk can be significantly reduced.

Phase errors become increasingly worse for smaller channel spacings which require longer path length differences between the grating arms. The phase errors were measured for a 5 × 5, 10 GHz channel spacing WGR designed with a length difference between adjacent arms of 3312 μm. The phases were then corrected by applying power to thin film heaters deposited on each of the 21 grating arms [26]. Without compensation, the phase errors ranged over 270° and the crosstalk was –5 dB. With the compensation, the phase errors were reduced to ±3° and the crosstalk was reduced to –25 dB. Using thin film heaters is a good approach for demonstrating that phase errors can be compensated; however, a permanent means of compensating the errors is needed. To address this need, a patching process using amorphous Si was demonstrated [27]. The a-Si induces a stress that changes the local index through the photoelastic effect. The phase errors were minimized in a 16 channel, 10 GHz spacing WGR with $\Delta L = 1271$ μm. The phase was compensated in

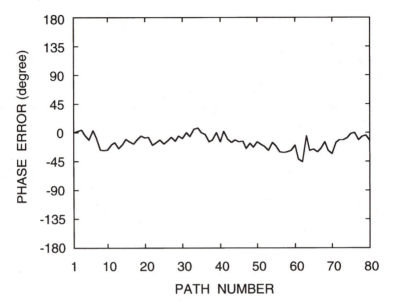

FIGURE 7-24 Phase errors over the grating waveguide arms for a 100 GHz WGR [25]. Reprinted by permission. Copyright © 1996 IEEE.

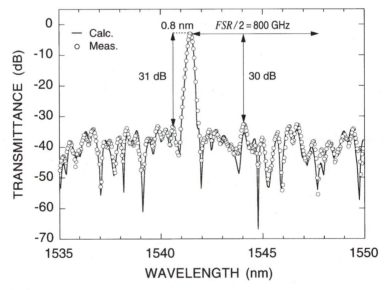

FIGURE 7-25 Comparison of measured spectral response to calculated response using the measured phase errors and assuming Gaussian distributed amplitude coefficients for a 100 GHz WGR [25]. Reprinted by permission. Copyright © 1996 IEEE.

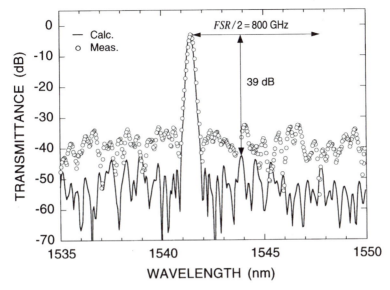

FIGURE 7-26 Comparison of measured spectral response to calculated response using the measured amplitude errors and assuming no phase errors for a 100 GHz WGR [25]. Reprinted by permission. Copyright © 1996 IEEE.

56 of the 64 grating arms to less than 2° RMS by trimming a portion of the film away using an Ar ion laser. The remaining arms carry very little power, so they do not have a major effect on the device performance. Before compensation, the phase errors were distributed over 360° and the spectral response was unusable as a bandpass filter. After compensation, the crosstalk was reduced to less than −30 dB.

OLCI is also used for reflectance measurements, where it is called optical low coherence reflectometry (OLCR). High spatial resolution (<2 μm) and high sensitivity (−160 dB) can be obtained for point reflectors [1]. For such high spatial resolutions, the dispersion of the optical paths comprising the interferometer arms must be balanced over the broadband source spectrum. OLCR can be used to determine the complex reflectivity of a filter using an interative algorithm [28].

7.2.3   The AR Analysis Algorithm

Filter analysis requires two things: (1) the measured complex frequency or time domain response, and (2) an algorithm that relates the filter coefficients and phases to the frequency or time domain response. These requirements also characterize the general design algorithm when the measured response is replaced by the desired response. AR filter analysis is complicated by the feedback that produces an infinite impulse response. The previous analysis methods that measure each stage independently are not applicable. The AR analysis method discussed next focuses on the lattice ring architecture. It clearly identifies the problem of uniqueness when trying to determine the underlying filter structure given only magnitude response information. As noted in Section 7.1.4.1, the ARMA group delay responses can be bound for ring, thin film, and Bragg grating filters from knowledge of the AR magnitude response, but they cannot be uniquely determined. The group delay for one of the ARMA responses must be measured, then the other ARMA response is fully specified (for a lossless filter) and filter analysis can be accomplished given an algorithm to translate between the complex frequency response and the physical structure.

For a ring resonator lattice filter, we shall find Γ and the coefficients of $A_N(z)$ from measurements of $|H_{AR}(\omega)|^2$. By solving the set of normal equations, described in Chapter 3, the best fit AR model is determined. If the magnitude response is an even function with respect to the start of the FSR, then the autocorrelation function is real; otherwise, it is complex with Hermitian symmetry. There are $N + 1$ unknowns, $[a_{1N}, \ldots, a_{NN}, \Gamma_N]$ and $N + 1$ equations. First, we estimate the autocorrelation function by taking the inverse transform of the measured square magnitude re-

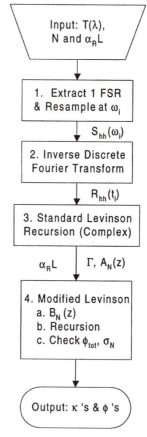

FIGURE 7-27   AR analysis algorithm.

sponse over one period to obtain $R(k)$ for $k = 0, \ldots, N$. Then, the standard Levinson recursion algorithm is used to calculate the solution for the unknowns $[a_{1N}, \ldots, a_{NN}, \Gamma_N]$ with $[R(0), \ldots, R(N)]$ as input. Once $A_N(z)$ is found, the coupling ratios are calculated using the modified Levinson algorithm discussed in Chapter 5.

The steps of this analysis algorithm are summarized in Figure 7-27. The inputs are the measured AR magnitude response, $T(\lambda) = |H_{AR}(\lambda)|^2$, the number of stages in the filter, and the nominal feedback path loss αL. The transmission response includes only the filter's response and not the excess loss arising from fiber-to-waveguide coupling losses or propagation losses associated with the input and output waveguides. In Step 1, one FSR is extracted from the measured data. The unit length L_U can then be calculated from the FSR. The data is resampled at evenly spaced frequency intervals for input to the inverse DFT in Step 2. The resulting autocorrelation function is complex in general, due to any asymmetry in the spectral loss. For Step 3, a standard Levinson algorithm which handles complex inputs is used to obtain the AR model from the autocorrelation function. A detailed description of the standard Levinson algorithm is given in Chapter 3. The outputs of Step 3 are Γ and $A_N(z)$ which are input along with the loss to the modified Levinson algorithm in Step 4. The value for Γ depends on the filter's measured peak transmission. If only $A_N(z)$ is of interest, the measured response can be normalized to unity peak transmission. The modified Levinson algorithm then calculates σ_{max} and uses it to find $A_N(z)$.

For filter synthesis, the minimum-phase roots of $B_N B_N^R$ were chosen to uniquely define B_N; however, for filter analysis there are a total of 2^N possible solutions for the amplitude coefficients and phase terms that lead to physically realizable filters. In the analysis algorithm, all solutions are determined in the modified Levinson algorithm. Additional information is required to identify the particular solution. A special case where this degeneracy is avoided is for a symmetrical filter. Then, there is only one solution for $B_N(z)$, which is linear-phase. A symmetric, folded architecture was discussed in Chapter 5. Only the magnitude response would be required to completely determine its coupling ratios and phases.

For an asymmetric filter, the various solutions for $B_N(z)$ cannot be distinguished by the ARMA magnitude response for the lossless case as discussed in Chapter 5, and it is difficult to distinguish some solutions even with loss. If a prior knowledge is available on the coupling ratios, the difference in coupling ratio solutions may allow the particular solution to be identified. The definitive approach is to measure the ARMA group delay, which is unique since maximum- and minimum-phase zeros produce group delays of opposite signs independent of the loss. The group delay characteristics are needed in the regions of low transmission and where the transmission is changing abruptly for the ARMA response, which places demanding requirements on the dynamic range and resolution of the group delay measurement technique.

The AR analysis algorithm has been applied to the fabricated lattice ring filters of order $N = 2$ and 3, which were discussed in Chapter 5. Heaters were deposited on each ring for phase control. Directional couplers with the same coupling ratios used in the filters were included on the wafer, adjacent to the filters, to obtain an inde-

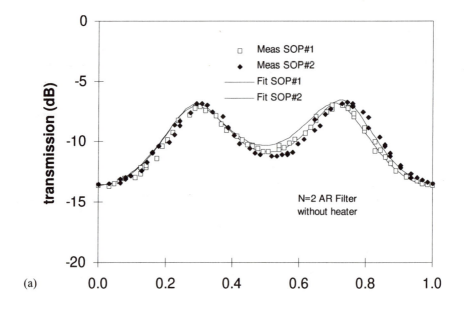

(a)

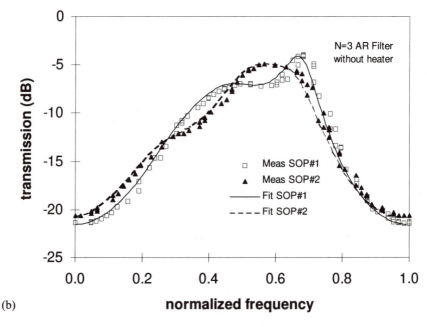

(b)

FIGURE 7-28   Measured AR response for the (a) $N = 2$ lattice filter and (b) $N = 3$ lattice filter without the waveguide heaters, showing the impact of fabrication errors.

TABLE 7-2   Analysis Results for As-Fabricated N = 2 Lattice Filter

Solution x: $\|z_k\|<1$ o: $\|z_k\|>1$	Coupling Ratios			Phase	
	$\kappa1$	$\kappa2$	$\kappa3$	$\phi1$	$\phi2$
1 (xx)	0.14	0.96	0.88	−3.05	3.04
⇒ 2 (xo)	0.68	0.26	0.67	1.83	−1.84
Test Coupler	0.59	0.33	0.60	—	—

pendent measurement of the coupling ratios. The waveguides were Ge-doped silica on a silicon substrates with $\Delta = 0.75\%$ and a core size of $6 \times 6.5 \ \mu m^2$. The minimum bend radius was 4 mm, which limited the maximum FSR to 8.2 GHz, or 0.066 nm. Measurements were made over multiple FSRs near 1550 nm for each of the two polarization responses. Figure 7-28 shows the measured magnitude responses for the $N = 2$ and $N = 3$ filters as-fabricated along with their AR fit, $\Gamma_2/A_2(z)$ and $\Gamma_3/A_3(z)$. The frequency axis is normalized to one FSR, ~63 pm. The minimum transmission wavelengths were chosen to define one FSR from each polarization response, so the true wavelength offset between the different polarization passbands is not represented on the graph. The phase errors introduced during fabrication are evident in the transfer function shape.

The possible coupling ratios and phase error solutions from the analysis algorithm are shown in Table 7-2 and Table 7-3 along with measured values for the test couplers. The minimum- and maximum-phase solutions are indicated by x and o, respectively, in the tables. Using the test coupler measurements, the solution with the closest coupling ratios, indicated by the arrow in the tables, was chosen.

To compensate the relative phase errors, the phase change induced for a given applied heater power must be known. For a cascade AR filter where the stages are independent, determining the phase errors is straightforward; however, for a lattice filter, a change in the phase on one stage changes all of the filter's pole locations. Figure 7-29 shows a simulation of the trajectories of the poles and zeros for an $N = 3$ lattice filter with coupling ratios equal to the test coupler values. The phase on the first stage, ϕ_1, was varied from 0 to 0.4π while the phase on the remaining stages

TABLE 7-3.   Analysis Results for As-Fabricated N = 3 Lattice Filter

Solution x: $\|z_k\|<1$ o: $\|z_k\|>1$	Coupling Ratios				Phase		
	$\kappa1$	$\kappa2$	$\kappa3$	$\kappa4$	$\phi1$	$\phi2$	$\phi3$
1 (xxx)	0.96	0.68	0.52	0.17	2.48	2.95	−2.61
⇒ 2 (xox)	0.81	0.36	0.25	0.84	3.11	−2.36	2.07
3 (xoo)	0.35	0.31	0.57	0.95	−2.25	2.68	2.40
4 (xxo)	0.88	0.53	0.17	0.75	2.19	−2.32	−3.33
Test Coupler	0.74	0.33	0.33	0.74	—	—	—

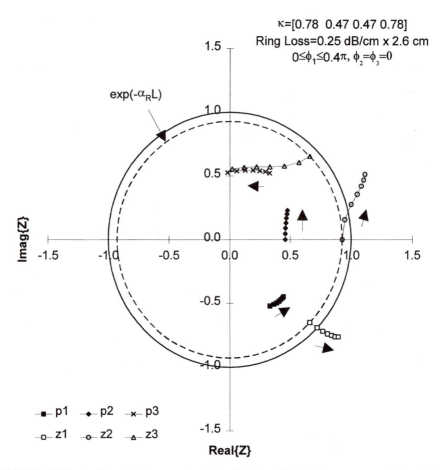

FIGURE 7-29 An $N = 3$ pole-zero simulation for the ARMA#1 response as ϕ_1 is varied from 0 to 0.4π while $\phi_2 = \phi_3 = 0$.

was set to zero, $\phi_2 = \phi_3 = 0$. All of the poles and zeros move. Without the analysis algorithm, it would not be straightforward to compensate the phase errors. The zeros may change from minimum to maximum phase or vice versa as the phase on a given stage is varied; therefore, multiple solutions must be considered at each heater setting. The waveguide heaters were calibrated by applying power to a single heater, measuring the AR spectral loss, and calculating the phase using the analysis algorithm. A slope of –0.94±0.06 FSR/Watt was obtained, where one FSR represents a 2π radian change in phase.

Using the analysis algorithm and the heater calibration results, the compensated responses shown in Figures 5-22 and 5-23 were obtained. The filter's peak transmission for the optimized $N = 3$ lattice was –2.0 dB compared to –4.9 dB for the as-fabricated case, and the measured maximum to minimum transmission contrast was 21.4 dB compared to 15.9 dB for the as-fabricated case. For the $N = 2$ case, the peak transmission improved from -6.8 dB to -2.5 dB, and the contrast ratio increased from 6.8 dB to 14.1 dB. The maximum difference in the phase errors between stages was reduced from 3.7 to 0.3 radians for the $N = 2$ case and from 1.85 to 0.1 radians for the $N = 3$ case.

REFERENCES

1. D. Derickson, *Fiber Optic Test and Measurement.* Upper Saddle River, NJ: Prentice Hall, 1998.

2. L. Stokes, "Coupling Light from Incoherent Sources to Optical Waveguides," *IEEE Circuits and Devices,* vol. 10, no. 1, pp. 46–47, 1994.

3. H. Lefevre, "Singlemode fibre fractional wave devices and polarization controllers," *Electronics Letters,* vol. 16, no. 20, pp. 778–780, 1980.

4. B. Nyman, "Automated System for Measuring Polarization-Dependent Loss," in *OSA Technical Digest.* Optical Fiber Communication Conference. San Jose, CA, 1994, p. 230.

5. D.L. Favin, B.M. Nyman, and G.M. Wolter, "System and Method for Measuring Polarization Dependent Loss," *U.S. Patent No.* 5 371 597 (6 December 1994).

6. D.S. Kliger, J.W. Lewis and C.E. Randall, *Polarized Light in Optics and Spectroscopy,* San Diego: Academic Press, 1990, p. 128.

7. R. Adar, Y. Shani, C.H. Henry, R.C. Kistler, G.E. Blonder, and N.A. Olsson, "Measurement of very low-loss silica on silicon waveguides with a ring resonator," *Appl. Physics Lett.,* vol. 58, no. 5, pp. 444–445, 1991.

8. T. Feuchter and C. Thirstup, "High Precision Planar Waveguide Propagation Loss Measurement Technique Using a Fabry–Perot Cavity," *IEEE Photonics Technol. Lett.,* vol. 6, no. 10, pp. 1244–1247, 1994.

9. L.F. Stokes, M. Chodorow and H.J. Shaw, "All-single-mode fiber resonator," *Optics Lett.,* vol. 7, no. 6, pp. 288–290, 1982.

10. R. Adar, M.R. Serbin, and V. Mizrahi, "Less than 1 dB per meter propagation loss of silica waveguides measured using a ring resonator," *J. Lightw. Technol.,* vol. 12, no. 8, pp. 1369–1372, 1994.

11. C. Madsen and J. Zhao, "Increasing The Free Spectral Range Of Silica Wave-Guide Rings For Filter Applications," *Optics Lett.,* vol. 23, no. 3, pp. 186–188, 1998.

12. D. Rowland and J. Love, "Evanescent Wave Coupling of Whispering Gallery Modes of a Dielectric Cylinder," *IEE Proc.,* vol. 140, no. 3, pp. 177–188, 1993.

13. A. Papoulis, *The Fourier Integral and Its Applications.* New York: McGraw-Hill, 1962.

14. M. Muriel and A. Carballar, "Phase Reconstruction from Reflectivity in Uniform Fiber Bragg Gratings," *Optics Lett.,* vol. 22, no. 2, pp. 93–95, 1997.

15. L. Poladian, "Group-Delay Reconstruction for Fiber Bragg Gratings in Reflection and Transmission," *Optics Lett.,* vol. 22, no. 20, pp. 1571–1573, 1997.

16. K. Daikoku and A. Sugimura, "Direct measurement of wavelength dispersion in optical fibres - difference method," *Electron. Lett.,* vol. 14, p. 367, 1978.

17. D. Marcuse, *Principles of Optical Fiber Measurements,* New York: Academic Press, 1981, p. 279.

18. C. Madsen, "Demonstration of the First General Planar Waveguide Autoregressive Optical Filter," PhD dissertation. New Brunswick, NJ: Rutgers University, 1996.

19. G.T. Harvey, measurement development (AT&T Bell Laboratories, Holmdel, NJ, 1995).

20. K. Yonenaga and N. Takachio, "A Fiber Chromatic Dispersion Compensation Technique with an Optical SSB Transmission in Optical Homodyne Detection System," *IEEE Photonics Technol. Lett.,* vol. 5, no. 18, pp. 949–951, 1993.

21. G. Smith, D. Novak, and Z. Ahmed, "Technique for optical SSB generation to overcome dispersion penalties in fibre-radio systems," *Electron. Lett.,* vol. 33, no. 1, pp. 74–75, 1997.

22. M. Abramowitz and A. Stegun, *Handbook of Mathematical Functions with Formulas, Graphs, and Mathematical Tables.* New York: Dover, 1972.

23. J. Roman, M. Frankel, and R. Esman, "High resolution technique for characterizing chirped fiber gratings," in *Vol. 2 1998 OSA Technical Digest Series* (Optical Society of America, Washington, D.C., 1998), Optical Fiber Communication Conference 1998, pp. 6–7.

24. K. Sasayama, M. Okuno, and K. Habara, "Coherent Optical Transversal Filter Using Silica-Based Waveguides for High-Speed Signal Processing," *J. Lightw. Technol.,* vol. 9, no. 10, pp. 1225–1230, 1991.

25. K. Takada, H. Yamada, and Y. Inoue, "Optical Low Coherence Method for Characterizing Silica-Based Arrayed-Waveguide Grating Multiplexers," *J. Lightw. Technol.,* vol. 14, no. 7, pp. 1677–1689, 1996.

26. H. Yamada, K. Takada, Y. Inoue, Y. Hibino, and M. Horiguchi, "10 GHz-spaced arrayed-waveguide grating multiplexer with phase-error-compensating thin-film heaters," *Electronics Lett.,* vol. 31, pp. 360–361, 1995.

27. H. Yamada, K. Takada, Y. Inoue, Y. Ohmori, and S. Mitachi, "Statically-phase-compensated 10 GHz-spaced arrayed-waveguide grating," *Electronics Lett.,* vol. 32, no. 17, pp. 1580–1581, 1996.

28. E. Brinkmeyer, "Simple algorithm for reconstructing fiber gratings from reflectometric data," *Optics Lett.,* vol. 20, no. 8, pp. 810–912, 1995.

PROBLEMS

1. Calculate the noise resulting from the interference of two point reflectors as a function of their separation, their reflectance (assuming equal reflectance values), and the source linewidth. Assume that the source has a Lorentzian line shape.

2. Derive Eqs. (32) and (33) for optical SSB modulation using an MZI driven in quadrature. Derive an expression for the power in the carrier and lower sideband as a function of the modulation depth.

3. Calculate the interferogram for a 4 channel input with 100 GHz frequency spac-

ing and a center wavelength of 1550 nm. Determine the sampling length required for a resolution of 10 GHz.

4. Calculate the interference pattern for a third-order AR filter with poles located at $0.9 \angle 0$ and $0.8 \angle \pm\pi/20$ rad for two cases: (a) the unit delay is 10 times the source coherence time and (b) the unit delay is 0.1 times the source coherence time.

CHAPTER EIGHT

Future Directions

Optical filters are evolving at a rapid pace. This transformation is arising from both a pull by the development needs of optical communication systems and the technology push coming from research in materials and fabrication processes. These two areas are discussed with respect to their impact on future directions for optical filters.

8.1 COMMUNICATION SYSTEM APPLICATIONS

8.1.1 Ultra-Dense WDM Systems and Optical Networks

Dense WDM systems are widely deployed and provide a wealth of applications for optical filters. The filter requirements for the bandpass, gain equalization, and dispersion compensating filters become more demanding as the number of channels, the bandwidth utilization, and the system wavelength range increase.

Different transmission formats are used in WDM systems, for example, non-return-to-zero (NRZ) and solitons. The transmission format impacts channel spacing, dispersion compensation and gain equalization; thus, impacting filter requirements. The best spectral efficiency of 0.6 bits/s/Hz was demonstrated using duo-binary coding [1]. A 2.6 Tb/s WDM system was achieved using 132 channels at 20 Gb/s each over 120 km of standard fiber. The channels were separated by 33.3 GHz and demultiplexing was accomplished by using three WGRs with FWHM = 0.32 nm, temperature control for center wavelength alignment, and manually tuned double cavity thin-film filters with 0.3 nm bandwidths.

To date, point-to-point and fixed add/drop commercial WDM systems have been deployed. Optical networks are on the horizon and bring with them a host of challenges. The major advantage of networks is to achieve routing flexibility [2]. Routing capability is accompanied by added complexity to the overall optical design [3]. For example, gain equalization and dispersion compensation are complicated by the fact that channels originate from different locations (thus experiencing a different number of amplifiers and dispersion maps) and their routing can change. Optical monitoring is critical, and active gain equalization and dispersion compensation may be necessary. Lower bit rates allow more channels to be carried on a network; however, this smaller granularity implies filters with narrower, stable passbands and very precise center wavelengths. Other transmission effects also surface in a WDM network. In systems with many amplifiers operating in saturation, changing the power into an amplifier (for example, by adding or dropping channels) can induce power transients on the other channels that are extremely fast [4]. Dynamic gain control at high speeds may be necessary to suppress this effect.

A crucial new component for networks is an optical cross-connect where channels on one fiber can be routed onto another fiber and vice-versa. Consider a general optical cross-connect for a WDM system with M fibers, each with a capacity of N WDM channels for a total of $M \times N$ channels. A channel on any fiber must be allowed to switch onto any other fiber. For four fibers with 80 channels each, there are a total of 320 channels to route. Wavelength conversion is required so that an add channel will not interfere with a channel already on the output fiber. Various combinations of demultiplexers, switches, and multiplexers may be used. For example, an optical cross-connect consisting of 2M WGRs with an intervening switch is shown in Figure 8-1. Optical micro-electromechanical (MEMs) devices are one option for realizing large switch arrays [5,6]. The functionality and cost of an optical cross-connect must be equivalent or better than what can be achieved with electrical cross-connects for it to be commercially viable.

8.1.2 Ultra-Fast TDM and Optical Codes

Dense WDM systems already employ time division multiplexing (TDM) to combine many lower bit rate signals into a single high bit rate channel of several Gb/s.

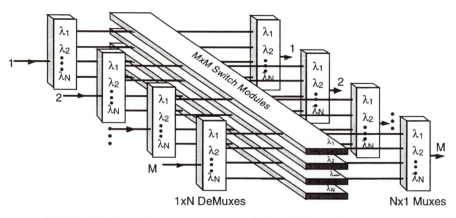

FIGURE 8-1 An optical cross-connect employing WGRs and space switches.

The ongoing race is to demonstrate the highest bit rate over the longest propagation distance. For the propagation of ultra-short pulses, broadband dispersion compensation is required. For example, an MA lattice filter has been used to compensate for third-order dispersion and reduce the power penalty by 4 dB for a 200 Gb/s signal transmitted over 100 km [7]. The 9-stage filter had a FSR = 300GHz, an insertion loss of 11.4 dB, and a dispersion slope of -7.9ps/nm^2.

Variable delay structures are critical for TDM applications. A delay buffer is needed to multiplex the lower bit rate stream up to the higher one so that it is placed in its correct time slot. For demultiplexing, the delay must be chosen to select the desired channel. To multiplex a large number of channels, a wide tuning range for the delay is needed. The delay must be tuned to within a fraction of the time slot, i.e. have high precision. A 1024-channel fast tunable delay line has been demonstrated using Mach–Zehnder delay lines with 2×2 couplers and fiber delay lines [8]. An aggregate rate of 50 Gb/s was experimentally demonstrated using a five stage delay structure with two electro-optic modulators, a 20 ns reconfiguration time, and a loss of 22 dB. Transmission on the band-edge of gratings where the dispersion is very high is being explored for delay lines as well. A short grating with high index contrast can provide a large group delay. For example, a delay of 5 ns was demonstrated using a 10 mm grating in a semiconductor chip [9]. Simulation results for a 30 period grating in GaAs/AlAs showed a group index of 13.5, three times the value for the bulk material, with a resonance bandwidth of 2.3 nm and negligible degradation in amplitude and shape of a 2 ps pulse [10].

Given the success of code division multiple access (CDMA) modulation format in wireless communication systems, optical systems employing CDMA have been investigated. Although users share the same wavelength band, each user's information is encoded so that an autocorrelation with the matched code is required at the receiver to decode it. Various optical filters have been used for code generation including MA transversal [11] and lattice [12] structures, thin-film coupled cavities [13,14], and gratings plus phase masks [15]. Two filters are needed for each user, one for multiplexing and demultiplexing. Orthogonal codes are required to minimize cross-talk between users. The feasibility of optical CDMA systems to handle a large number of users is a topic of research [16]. A capacity of 4 Gb/s with four 1 Gb/s channels has been demonstrated [17]. The main limiting factor for incoherent systems is optical noise at the square law detector, which is exacerbated over WDM and TDM systems because a large wavelength range is used with multiple users transmitting over the same wavelength range.

Besides CDMA, codes on optical pulses may find other applications such as routing of channels in the optical level. For example, a matched filter for generating an optical code was demonstrated using the fiber-optic delay line shown schematically in Figure 8-2 [18]. The experimental versus theoretical autocorrelation output is shown in Figure 8-3 for an 8 bit optical 10 Gb/s packet. In another experiment, an 8-bit 100 Gb/s self-synchronizing pattern generator was demonstrated having two 15 ps marker pulses and six 10 ps data pulses using a 1×8 silica waveguide splitter, an array of semiconductor laser amplifiers and a silica waveguide circuit of delay lines and a combiner [19].

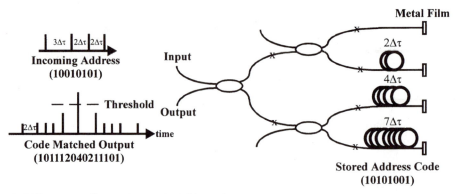

FIGURE 8-2 Fiber-optic matched filter schematic [18]. Reprinted by permission. Copyright © 1996 IEEE.

8.2 MATERIALS AND PROCESSING

While silica has set the standard for low loss optical waveguides, additional functionality is desired for active device control. For example, faster and lower power consumption effects for tuning, switching, and modulation are needed than afforded by the thermo-optic effect. Nonlinear optical polymers [20], lithium niobate, and semiconductor materials such as InGaAsP and silicon [21,22] offer rapid tuning ef-

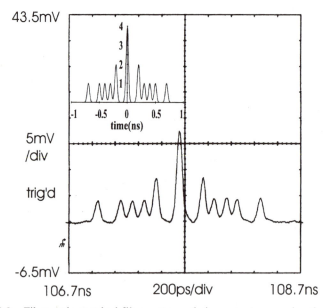

FIGURE 8-3 Fiber-optic matched filter autocorrelation output comparing the experimental results to the theoretical results shown in the inset [18]. Reprinted by permission. Copyright © 1996 IEEE.

fects. Hybridization, or integrating multiple materials, is attractive for improving device functionality and performance. An area for research is the design of filters with gain.

On the device side, microresonators that combine whispering gallery mode operation [23] and asymmetric cavity resonator designs [24] have been demonstrated. The focus has been on lasers so far, but these geometries offer a new coupling mechanism for optical filters as well.

Another new area is photonic bandgap (PBG) engineering, which employs periodic structures with sub-micron element sizes to make devices. The PBGs are created by fabricating a periodic structure with a high index contrast so that it reflects over a large wavelength range, the photonic bandgap [25]. One-dimensional PBGs are high index contrast gratings [26], as discussed in Chapter 5. By creating a line defect in a two-dimensional periodic array, waveguiding is achieved. Figure 8-4 illustrates a

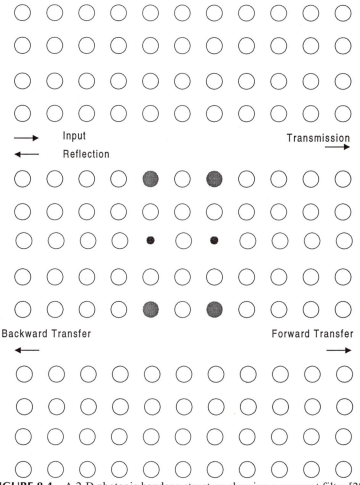

FIGURE 8-4   A 2-D photonic bandgap structure showing a resonant filter [28].

periodic lattice of dielectric rods where two rows have been removed on the top and bottom to act as waveguides. Simulations show that 90° bends can be achieved with practically no loss in 2-D PBG waveguides [27]. To achieve very small radii bends in conventional waveguides requires very large Δ's; however, propagation and scattering losses increase with Δ. The PBG structures are expected to be robust to random fabrication variations since the field can be localized in air regions away from the high dielectric contrast interfaces. A new resonator structure is illustrated by the two defects between the waveguides in Figure 8-4 [28]. The resonator has a symmetric and an antisymmetric mode that are degenerate (i.e., resonate at the same frequency) and identical coupling strengths to the input/output waveguides [29]. The size or index of the defects between the input/output waveguides and the resonator defects (shown as gray circles) are designed to produce these conditions so that the structure functions as an add/drop filter. Many details must be explored for PBG filter applications including coupling from/to a standard waveguide into the PBG, minimizing the polarization dependence, developing and analyzing defect architectures for controlled coupling between waveguides, and methods to change the optical phase or amplitude for tuning and switching applications.

8.3   SUMMARY

While many of the filters discussed in this book and the technologies for realizing them are quite mature, there are also many new designs and concepts that need experimental investigation. With the demands for bandwidth being driven by applications such as the World Wide Web, it is a given that optical communication systems will continue to push the state-of-the-art in optical filter technology. We can expect the number of channels and the routing and control functions done at the optical layer to increase. When combined with advances in materials, increases in hybridization and integration, and advances in fabrication processes, the field of optical filters has a strong outlook for the future with a large potential for growth.

REFERENCES

1. Y. Yano, T. Ono, K. Fukuchi, T. Ito, H. Yamazaki, M. Yamaguchi, and K. Emura, "2.6 Terabit/s WDM transmission experiment using optical duobinary coding," 22nd European Conference on Optical Communication. Oslo, Norway, 1996, p. ThB.3.1.

2. R. Wagner, R. Alferness, A. Saleh, and M. Goodman, "MONET: Multiwavelength optical networking," *J. Lightw. Technol.*, vol. 14, no. 6, p. 1349, 1996.

3. J. Zyskind, J. Nagel, and H. Kidorf, "Erbium-doped fiber amplifiers for optical communications," in *Optical Fiber Telecommunications IIIB,* I. Kaminow and T. Koch, Eds. San Diego: Academic Press, 1997, pp. 13–68.

4. J. Zyskind, Y. Sun, A. Srivastava, J. Sulhoff, A. Lucero, C. Wolf, and R. Tkach, "Fast power transients in optically amplified multiwavelength optical networks," in *OSA Technical Digest Series,* Optical Fiber Communications Conference 1996, p. PD31.

5. H. Ukita, "Micromechanical Photonics," *Optical Review,* vol. 4, no. 6, p. 623–633, 1997.

6. L. Lin, "Micromachined free-space matrix switches with submillisecond switching time for large-scale optical crossconnect," in *Vol. 2 1998 OSA Technical Digest Series* (Optical Society of America, Washington, DC, 1998), Optical Fiber Communication Conference1998, pp. 147–148.

7. K. Takiguchi, K. Jinguji, K. Okamoto, and Y. Ohmori, "Variable Group-Delay Dispersion Equalizer Using Lattice-Form Programmable Optical Filter on Planar Lightwave Circuit," *IEEE J. Selected Areas Commun.,* vol. 2, no. 2, pp. 270–276, 1996.

8. K. Deng, K. Kang, I. Glask, and P. Prucnal, "A 1024-channel fast tunable delay line for ultrafast all-optical TDM networks," *IEEE Photon. Technol. Lett.,* vol. 9, no. 11, pp. 1496–1499, 1997.

9. W. Stewart, "Optical devices incorporating slow wave structures," *U.S. Patent,* no. 5,311,605, 1994.

10. M. Scalora, R. J. Flynn, S. Reinhardt, R. Fork, M. Bloemer, M. Tocci, C. Bowden, H. Ledbetter, J. Bendickson, J. Dowling, and R. Leavitt, "Ultrashort pulse propagation at the photonic band edge: Large tunable group delay with minimal distortion and loss," *Phys. Rev. E54,* pp. 1078–1081, 1996.

11. P. Prucnal, M. Santoro, and T. Fan, "Spread spectrum fiber-optic local area network using optical processing," *J. Lightw. Technol.,* vol. 4, no. 5, pp. 547–554, 1986.

12. A. Holmes and R. Syms, "All-optical CDMA using "quasi-prime" codes," *J. Lightw. Technol.,* vol. 10, no. 2, pp. 279–286, 1992.

13. V. Narayan, E. Dowling, and D. MacFarlane, "Design of Multi-Mirror Structures for High Frequency Bursts and Codes of Ultrashort Pulses," *IEEE J. Quantum Electron.,* vol. 30, no. 7, pp. 1671–1680, 1994.

14. V. Narayan, D. MacFarlane, and E. Dowling, "High-Speed Discrete-Time Optical Filtering," *IEEE Photonics Technol. Lett.,* vol. 7, no. 9, pp. 1042–1044, 1995.

15. R. Griffin and D. Sampson, "Coherence coding of optical pulses for code-division multiple access," OFC/IOOC. OSA Technical Digest Series r, 1993, pp. 200–201.

16. D. Sampson, G. Pendock, and R. Griffin, "Photonic code-division multiple-access communications," *Fiber and Integrated Optics,* vol. 16, pp. 129–157, 1997.

17. G. Pendock and D. Sampson, "Increasing the transmission capacity of coherence multipled communication systems by using differential detection," *IEEE Photon. Technol. Lett.,* vol. 7, pp. 1504–1507, 1995.

18. J. Shin, M. Jeon, and C. Kang, "Fiber-Optic Matched Filters with Metal Films Deposited on Fiber Delay-Line Ends for Optical Packet Address Detection," *IEEE Photonics Technol. Lett.,* vol. 8, no. 7, pp. 941–943, 1996.

19. D. Rogers, J. Collins, C. Ford, J. Lucek, M. Shabeer, G. Sherlock, D. Cotter, K. Smith, C. Peed, A. Kelly, P. Gunning, D. Nesset, and I. Lealman, "Optical pulse pattern generation for self-synchronizing 100 Gbit/s networks," Optical Fiber Communications Conference, San Jose, CA, 1996, pp. 98–99.

20. Wang, D. Chen, H. Fetterman, Y. Shi, W. Steier, and L. Dalton, "40-GHz polymer electrooptic phase modulators," *IEEE Photon. Technol. Lett.,* vol. 7, no. 6, pp. 638–640, 1995.

21. P. Trinh, S. Yegnanarayanan, and B. Jalali, "5 × 9 Integrated Optical Star Coupler in Silicon-on-Insulator Technology," *IEEE Photonics Technol. Lett.,* vol. 8, no. 6, pp. 794–796, 1996.

22. P. D. Trinh, S. Yegnanarayanan, F. Coppinger, and B. Jalali, "Silicon-on-insulator (soi)

Phased-array Wavelength Multi/demultiplexer With Extremely Low-polarization Sensitivity," *IEEE Photon. Technol. Lett.,* p. 940, 1997.

23. Y. Yamamoto and R. Slusher, "Optical processes in microcavities," *Physics Today,* June, pp. 66–73, 1993.

24. J. Nockel, A. Stone, G. Chen, H. Grossman, and R. Chang, "Directional emission from asymmetric resonant cavities," *Optics Lett.,* vol. 21, no. 19, pp. 1609–1611, 1996.

25. J. Joannopoulos, R. Meade, and J. Winn, *Photonic Crystals: Molding the Flow of Light.* Princeton, NJ: Princeton University Press, 1995.

26. J. Foresi, P. Villeneuve, F. Ferrara, E. Thoen, G. Steinmeyer, S. Fan, J. Joannopoulos, L. Kimerling, H. Smith, and E. Ippen, "Photonic-bandgap microcavities in optical waveguides," *Nature,* vol. 390, 13 November, pp. 143–145, 1997.

27. J. Joannopoulos, P. Villeneuve, and S. Fan, "Photonic crystals: putting a new twist on light," *Nature,* vol. 386, 13 March, pp. 143–149, 1997.

28. S. Fan, P. Villeneuve, and J. Joannopoulos, "Channel Drop Tunneling Through Localized States," *Phys. Rev. Lett.,* vol. 80, no. 5, pp. 960–963, 1998.

29. H. Haus, S. Fan, J. Foresi, P. Villeneuve, J. Joannopoulos, and B. Little, "Optical-wavelength-scale filters," in *Vol. 2,* LEOS 1997, pp. 96–97.

Index